Introduction to Statistics
for the
Social Sciences

Introduction to Statistics for the Social Sciences

Frederick D. Herzon
Temple University

Michael Hooper
Temple University

Thomas Y. Crowell Company
New York   Established 1834

Library of Congress Cataloging in Publication Data

Herzon, Frederick D
  Introduction to statistics for the social sciences.

  Includes index.
  1. Statistics. 2. Social sciences – Statistical methods. I. Hooper, Michael, joint author.
II. Title.
HA29.H48   1976   519.5   75-44483
ISBN 0-690-00854-6

Thomas Y. Crowell Company, Inc.
666 Fifth Avenue
New York, New York 10019

Design by A Good Thing, Inc.

Manufactured in the United States of America

Contents

Chapter 6

Chapter 7

Chapter 8

Chapter 9

Contents

Preface

Our intent is to present the fundamental concepts and the reasoning that underlie the use of some of the most common statistical techniques employed in the social sciences. We have deliberately not attempted to cover all the techniques that are prominent in social research, nor even all the ramifications of those that are covered. For example, we do not present formulas for computations with grouped data because they involve too many details which interfere with students gaining a firm understanding of the basic statistic.

We did not intend, however, to write a cookbook. We try to communicate the reasoning behind the use of a statistical technique, not just how a particular statistic is calculated. While mathematical derivations and proofs are rarely used, much time is devoted to providing a conceptual understanding of the basis for all techniques. Thus, we do not present the algebra of expected values, but we explain carefully and at length the construction and use of sampling distributions.

This book is written for students with relatively little background in mathematics. All that is required is arithmetic, a little basic algebra, and the ability to follow formulas. For several years, we used books which made similar claims in our own courses. Yet we found many students had difficulty and complained about the amount of math required. While this was in part attributable to the "culture shock" social science students are likely to experience in taking any problem-oriented course, we felt something more was involved.

Although no great level of mathematical sophistication was required, students with more math generally did better in our statistics courses. Yet since advanced mathematics are not used in such courses, these students could not have done better because they knew more. Rather we came to believe their advantage arose from their being accustomed to and therefore skilled at manipulating numbers and symbols. The organization of this book is based in part upon this analysis.

We provide many opportunities for students easily to manipulate the numbers and symbols of statistics by placing questions within the body of each chapter, immediately after a section of new material. These questions, which begin in Chapter 2, are problem oriented and usually require the student to do calculations. The questions break the calculations into steps and provide instructions for how to proceed. Thus the students get immediate practice on the material they have just learned. In addition, these questions allow students to practice and master the parts of a problem before working out a complete one. Since statistical problems are often complicated, we feel this approach should be helpful. The answers to these questions are given at the end of each chapter.

We also include a chapter that reviews basic arithmetic and algebraic operations. Our purpose is twofold. First, the chapter provides a review of all the arithmetic and algebra that will be needed to learn the statistics presented in this book. Many students have forgotten various things about arithmetic or algebra and consequently have difficulty in following the development of certain statistical techniques. The second reason for reviewing this material is to give students an opportunity to practice manipulating numbers and symbols on material with which they should be somewhat familiar. This should help increase the students' confidence in their ability to "do" math. The review can be completed in about two class meetings, and we have found it to be most worthwhile.

A word should be said about the examples in this book. Unless other-wise indicated, all the data used in the examples and illustrations is hypothetical. But we tried to make the content of the examples realistic in terms of the types of the problems to be encountered in social science. While most of these examples are taken from political science and sociology, all social science students should find their content familiar and easy to understand. The data we use approximates the real world, adjusted, however, so as to simplify calculations.

The authors are indebted to numerous people who in various ways contributed to the writing of this book. Not the least among these are our students upon whom we practiced various approaches and who educated us to the problems involved in an adequate presentation of statistics. We hope that we have learned our lesson well. A number of anonymous readers seemed only too happy to suggest changes and point out difficulties. Their comments, however, strengthened the book and we are grateful for the detailed attention they gave to the manuscript. We also wish to thank Daniel Reich for his comments on Chapter 3. Mrs. Zinna Schnee cheerfully typed, and retyped, a difficult manuscript. We wish to thank her for her efforts. Mr. Herbert J. Addison originally acted as editor for this book. His support for our approach to a social science statistics course was very important to the development of this book. Mrs. Joan Greene, who took over as editor, continued to provide us with a high level of support and encouragement. Finally, we wish to acknowledge our spouses, Alice and Nance, for their three year toleration of two very preoccupied husbands.

To The Student

It is likely that this course will be quite different from your other social science courses, for many of which three, four, five, or even more books may be assigned. It is possible that this book will be the only one used in your entire statistics course. The fact that there are fewer pages to be read does not mean that there is less material to be covered. As a branch of mathematics, statistics presents its information in a highly condensed form. Even though an effort has been made in this book to explain all techniques carefully, in many places you will need to go over the material more than once before you can really understand it.

Another major difference between a statistics course and many other social science courses is the necessity to work problems. It is impossible to overemphasize the importance of your working problems in order to learn statistics. To a great extent, you will be learning a skill. A person can be told, for example, how to drive a car: press the gas pedal to go faster, press the brake to slow down or stop, turn the steering wheel to turn. The person now intellectually understands how to drive a car. But we all know that without actual practice he will not be able to drive very well. The same is true of statistics. You can read the material and intellectually understand statistics. But without doing problems you will not be able to *do* statistics. And since the point of learning statistics is not just understanding, but doing something (research, for example), it is necessary to practice by doing problems.

It might also be mentioned that many students, after bitter experience, have told us that it is impossible to cram for statistics tests. The reason for this seems to be that everyone develops his or her own individual "hang-ups" about statistics. It is only by trying to work problems that these hang-ups can be identified. A student can then come to class and ask questions about the things that are holding him up. Obviously this cannot be done if you try to learn a lot of new material the night before a test. The moral then is to keep up with the material throughout the course and to work problems *for yourself.*

Introduction to Statistics
for the
Social Sciences

1
The Use of Statistics in Social Science

Disraeli once said that there were three kinds of lies: lies, damned lies, and statistics. In his way, Disraeli was expressing his awareness that people often abuse statistics for their own, often deceitful, purposes. Today many people agree with him and look upon statistics with great suspicion. There is both good and bad in this view. In a society that is perhaps overly impressed with the aura and trappings of science, it is good to be reminded that numbers in themselves do not make for truth. Thus, we should examine an argument that uses numbers and statistics just as critically as we would an argument expressed only in words. But a critical attitude does not require a dismissal of statistics. To dismiss all statistical information would be foolish and impractical. An extreme example of such a view is the saying: "Anything can be proved with statistics." This statement is not true. What is true is that some people lie and mislead others. And statistics, or numbers, do not prevent people from lying, just as elegant language does not.

Although it is useful to be reminded that we should be critical, more than a critical attitude is necessary to use the information that statistics can provide. To do this we must know the meaning of various statistics and be able to distinguish their proper and improper uses. To catch a lie, the correct information is needed. To protect ourselves from those who would mislead us through the improper use of statistics, we need to know the correct way to use statistics. In an age in which society relies heavily on statistical information, people need to be familiar with statistics so that they can make effective decisions based upon valid information.

Statistical information impinges upon us from all sides. Such things as the weather report; baseball hitting averages; public opinion polls on the political issues of the moment; what 7 of 10 doctors are or are not doing, eating, or drinking—all are instances of the use of statistics. One reason to gain a knowledge of statistics, then, is to deal more effectively with the statistical information that we encounter in our everyday lives.

While a familiarity with statistics is useful to us as private citizens,

our interest here is more academic. In the social sciences, statistics is perhaps the single most important tool of research. In political science, sociology, psychology, and economics, much of the current research cannot
be understood without a knowledge of statistics. In fact, statistics has become almost a prerequisite to the conduct of social science research.

Why is there such a heavy emphasis on statistics in social science?
To answer this question requires a brief discussion of the nature of both
statistics and knowledge in social science.

Statistics had its origins in two distinct sources. One was the interest of nations, especially England, in collecting facts about their populations. The word "statistics" originally came into the language as meaning the collection, compilation, and analysis of facts about the state.
Rulers wished to know such things as the potential soldier population
and the wealth of agricultural and manufacturing communities, and they
sent out people to collect these facts. Soon, information on other characteristics of the population, such as the number of suicides, was being collected. From these "facts about the state," political leaders were able to
obtain a *description* of their communities on matters of importance to
them. This function of description is still one of the primary uses of statistics, although its methods have become more sophisticated.

The second major source of statistics lay in the desire of certain
people who gambled to predict the outcome of games of chance. This
interest (supplemented by an interest in actuarial matters, which developed in Britain) led to the development of the field of probability. The
development of probability theory allowed statisticians to make inferences about the outcomes of situations in which a random mechanism
was present. This led to the function of statistics called *statistical inference,* which today is most commonly introduced as a way to make an
inference about the characteristic of a larger group from a sample of that
group when the sample is chosen in a random way.

In accord with its development, the discipline of statistics today can
be divided into two areas, description and inference, both of which are
important to the research process in social science. In its *descriptive*
function, statistics summarizes information in a body of data so that it is
comprehensible to the human mind. Suppose that we wished to get an
idea of the cumulative grade-point averages of students at a particular university. One way to do this would be to have a computer print the names
of the students, together with their cumulative grade-point averages.
Then we could read through the list of grade-point averages. But if the
university is even moderately large, there would be thousands of names,
and a headache is the only thing we would probably get out of such a
process. Even with as few as 100 observations, the amount of data will
probably be too large to make sense of if it is considered one observa-

tion at a time. The purpose of descriptive statistics is to *summarize* the information in a body of data so as to make it comprehensible. In the case of the students' cumulative grade-point averages, one descriptive statistic that would be appropriate to calculate is the arithmetic average. The arithmetic average is a descriptive staistic and would tell us in an easily understood way the central point around which all the students' grade-point averages are clustered. This would be a useful piece of information, one that could not be obtained by considering the data one student at a time. The arithmetic average is one of many descriptive statistics that can be calculated to summarize information in bodies of data.

Summary statistics are important to social science research because controlled investigations often require the study of a group of objects rather than a single object. If we wished to know whether increasing amounts of education lead to greater degrees of political participation, it is impossible to investigate the question by studying just one person. In fact, we would probably be dissatisfied if only 2, 3, 4, or even 10 persons were studied. We know that many other things probably affect why a person participates in politics, such as personality, amount of free time, and wealth; and we know that if we study only a few people we might get, say, some less-educated people who participate a great deal because they happen to be of a personality type that derives great satisfaction from participating, and some highly educated people who are so caught up in their jobs that they have little time to participate. What needs to be done is to select a representative group of people and see if education is related to participation within the group. And since many objects are involved, some way of summarizing the information present in the group on education and participation is necessary. The summarizing statistic then becomes the "fact" that is used as evidence to support or deny whether education is related to participation.

We are quite accustomed to using summary statistics as facts about the social, political, and economic life of communities. These statistics are often the basis of social thinking and research. Unemployment rates are descriptive statistics calculated in such a way as to allow comparisons to be made across both time (different years, say, in the same nation) and place (among different nations in the same year). Similar uses are made of inflation rates, divorce rates, suicide rates, crime rates, and participation rates in elections. By calculating such rates, say, suicide rates in rural and urban areas, we can see if there are differences. If different rates do exist, these "facts" might indicate that we are on the track of a "cause" of suicide. And by calculating rates with respect to other variables, say, religion, we may be led to an explanation of suicide. The point is that descriptive statistics often compose the basic facts about social life. It is therefore essential to be familiar with descriptive statistics and to know what information it is that they present.

The second function of statistics is *inference*. Statistical inference is used as a method against which to judge the outcomes of research. This statement requires elaboration. Empirical knowledge, knowledge obtained by observing the world, is formalized into statements of two types. Specific knowledge is encompassed in statements about a single event or object. Examples of such statements are: "Richard Nixon was the thirty-seventh president of the United States"; "Bill is a Catholic"; "Bill voted Democratic in 1972"; "Tom is 6 feet tall." General knowledge involves statements about classes of objects rather than a statement about a single object. Examples of such statements are: "Catholics (as a class) tended to vote Democratic in 1972"; "Lions (as a class) eat meat"; "Uranium atoms (as a class) decay into lead." General statements are often "built up" through the accumulation of many specific statements (for example, observing that many individuals who are Catholic voted Democratic in 1972).

If the evidence seems to warrant it, we might be willing to risk forming a generalization. Examples of generalizations might be: "Catholics tend to vote Democratic in presidential elections"; "All lions eat meat"; "All uranium atoms will (eventually) decay into lead." The important thing about generalizations is that they assert things about objects which we have not yet observed. The generalization about how Catholics vote says that in future presidential elections they will (as a class) vote Democratic. The generalization about lions says that lions we have not yet observed (including those not yet born) will eat meat. And the generalization about uranium says that all atoms of uranium will decay into lead, including uranium that has not yet been discovered. Generalizations, then, require a leap beyond the observable evidence. But if past, observed evidence seems to warrant it, we might be willing to make the leap.

Generalizations are among the most powerful types of knowledge. They are also necessary to explanation and prediction and to the solving of practical problems. Some generalizations are so useful that they are called *laws*. Newton's law of gravity is a generalization that has been found to be so useful it is called a law. But it is still a generalization, because it asserts things about events or objects which have not yet been observed, such as what will happen to a piece of chalk if tomorrow someone were to pick it up and let it go. In contrast, the generalization about lions eating meat is commonly accepted as true, but it is not important enough to be given the status of a law. It is by formulating and testing generalizations that useful knowledge about the world is accumulated.

The Use of Statistics in Social Science

One of the most important functions of research is to formulate and test generalizations. When a statement is originally offered as a generalization, its truth is usually not yet established. At this stage, the statement is called a *hypothesis*. A hypothesis is simply a formalized way of stating a guess that certain things are related in some way. If it is asserted that lower-class people are more prejudiced than upper-class people and little evidence has been adduced to support this assertion, we have a hypothesis relating social class to prejudice. Ultimately, all hypotheses are guesses, and evidence must be gathered to test their truth or falsity. Of course, we want to make guesses that turn out to be true. This is why researchers try to inform themselves as much as possible about the nature of the problem they are studying. By doing so they increase their chances of making guesses (hypotheses) that turn out to be true. If the hypothesis is supported by the evidence, we have a true generalization, and if it is found to be very useful, it may eventually be granted the status of a law.

Generalizations can be of two types: universal and statistical. A *universal generalization* asserts something about a class of objects without exception. Newton's law of gravity is a universal generalization. In one of its forms it says that any unsupported body near the surface of the earth will (without exception) fall toward the earth's center (until it becomes supported). *Statistical generalizations* differ from universal generalizations in that they imply that there will be exceptions to what the generalization itself asserts. The statement that if a person is exposed to the measles, then it is likely that he will catch the disease is a statistical generalization. It implies that there will be some people who when exposed to the measles will not get sick.

Statistical generalizations, although not as neat as universal ones, still embody useful knowledge. Statements concerning the half-life of uranium are really statistical generalizations, since there is no one who can predict which atoms of uranium will have decayed and which will not by the end of a half-life period. In social science, most, if not all, of the hypotheses and generalizations are statistical in form. Thus the statements "Catholics tend to vote Democratic," "more education usually leads to increased political participation," and "lower-class people tend to be more prejudiced than upper-class people" are all statistical generalizations (or hypotheses). One reason for the importance of statistics in social science research is that the generalizations and hypotheses in this area are usually statistical in nature.

The importance of statistics comes from the role that inferential statistics plays in the verification of statistical generalizations. With a universal generalization the standard of disconfirmation is quite clear: Find one instance that violates the generalization, and it is proved false (or at least must be amended in some way). But with a statistical generalization, 1, 2, 10, or 100 such instances may not disconfirm the generalization. Thus, if we know 1, 2, 10, or 100 Catholics who voted Republican, it still can be true that Catholics as a class tend to vote Democratic.

Inferential statistics helps to provide a standard of disconfirmation for statistical generalizations (or hypotheses). It does this by setting up a competing generalization to the one the researcher is asserting. This competing generalization says that the results which the researcher wishes to use as support for his generalization are only due to chance. Through various procedures, the researcher is able to compute the probability that the results could have occurred by chance. If this probability is fairly large, the results are, in fact, likely to be due to chance, and a result that is likely to have occurred by chance certainly cannot be interpreted as support for the researcher's generalization. But if the probability is very small, it is very unlikely that the results are due to chance, and some nonchance phenomenon must be at work. In this case the researcher has established that the results are worth using as evidence in support of his generalization because he has eliminated the possibility that they just occurred by chance. Of course, eliminating the chance hypothesis does not mean that the results prove the researcher's generalization. It only establishes that the results were not due to chance and are worth consideration. It could be, though, that a different generalization (which is consistent with the results) is really true, not the one being put forth by the researcher.

The usefulness of this procedure is that it can deal with generalizations to which there are exceptions (statistical generalizations). Suppose, for instance, that we are betting on the outcome of flipping a coin. We always bet heads, and in the first 10 flips the coin comes up tails 9 times. Now if the coin had come up tails 10 times in 10 flips, there would have been no exceptions to the statement that the coin is rigged in some way to always come up tails. But since the coin did come up heads once, our opponent can point to this exception as showing that nothing is wrong. If, however, we treat the statement as a statistical one—that the coin is rigged in some way so that it *tends* to come up tails more often than heads—then the exception would not necessarily disprove it. What we need now is some way by which we can decide whether the statistical statement is true or false. If we assume for the moment that the coin is perfectly honest, we must recognize, as our opponent is asserting, that by chance a perfectly honest coin could come up tails 9 times in 10 flips just by chance. But what is the probability of this happening to an honest coin just by chance? As will be shown later, this probability can be calculated easily; it is about 1 time in 100. Now, this probability is so small that it is not likely that the result of 9 tails in 10 flips would happen to us the one time we happen to be betting. Consequently, we can conclude that the result (9 tails in 10 flips) did not occur by chance and we had better get a new coin (or a new flipper).

In summary, whether one is interested in establishing facts, solving practical problems, discovering new knowledge, or explaining and predicting social phenomena, the nature of social life is such that the use of statistics is necessary. And it is because of its utility in all these areas that statistics plays such a prominent role in social science.

It should be pointed out that statistics is one of many methodological techniques used in social science. And different areas in the social sciences do tend to emphasize different techniques. But statistics is central or basic to all areas of social science in that most methodological techniques require the use of statistics in some way. A knowledge of statistics, then, will provide a good foundation for whatever area of the social sciences is chosen for study.

Finally, it might be noted that statistics is a methodology and not a substitute for creative thinking. Statistics cannot create knowledge or crank out interesting ideas or hypotheses through a mathematical process. The development of interesting hypotheses is dependent upon the creativity and ingenuity of the researcher. The major value of statistics is in the establishment of facts in a body of data and as a tool in the process of verification of hypotheses. But statistics cannot tell us what facts are worth looking at or what hypotheses are useful.

2
Basic Concepts in Statistics

This chapter presents a brief introduction to some of the major statistical concepts. A number of terms are introduced and given their technical statistical definition. It is important to master these terms so as to provide a common terminology through which the discussion of various statistical techniques can take place.

SAMPLING

One of the most important uses of statistics is to provide a means to infer characteristics of a large group of objects from the examination of a subgroup of those objects. The results of the examination of one object is called an *observation*. The subgroup that is actually examined is called the *sample*. And the total collection of all possible observations of the same type as obtained from the sample is called the *population*. The population is the larger group about which something is to be inferred. The process of making an inference from a sample to a population is called *statistical inference*.

The term "population" is a technical one in statistics. It refers to the larger group of objects in which a researcher is interested. A population *might* be the population of all adults in the United States, but it also might be all redheaded people in New York, or all milk-producing cows, or all registered Democratic voters in Kansas, or all automobiles in California. In other words, the population is determined by the problem the researcher is investigating.

The "sample" is any subgroup of the population which the researcher actually examines and on which he collects information. The researcher collects information on some property of the objects in his sample. He then makes an inference about the characteristics of this property in the larger population.

All of us use this kind of procedure throughout our lives. We taste a few servings of spinach (a sample) and perhaps decide that we dislike all spinach (the population), even though we will never be able to taste more than a very small proportion of the spinach in the world. A teacher scores a test (the sample) and infers the nature of the students' knowledge of the course (the population).

In the quality control of products, sampling is essential. A wine taster takes a sip of wine (a sample) and infers the results to the taste of all the wine in a large batch (the population). The only way to know how all the wine really tastes is to taste all of it, which of course would leave none for us. A manufacturer may want to know that his light bulbs will last for at least a certain period of use. The only way to know for sure that the light bulbs will last the required period is to test each one. After the testing procedure is over, there would be no light bulbs for us to put into our homes. The only practical method is for the manufacturer to test a subgroup of bulbs (the sample) and infer the quality of all his bulbs (the population) from the results of testing that subgroup.

All sampling is done for practical reasons of this kind. If a researcher wished to know whether adults in the United States were for or against more government welfare programs, he could interview every adult in the country and ascertain his or her opinion on this topic. At the end of the survey the researcher would know exactly what proportion of adults in the United States were in favor of more government welfare programs and what proportion were opposed. However, such a procedure would be fantastically expensive and time consuming. To cope with the question realistically, the researcher will actually interview a subgroup of adults and on the basis of the results of this sample attempt to generalize about the nature of the opinions of all adults in the United States.

In statistics, the result that a researcher obtains from observing a characteristic of a sample is called a *statistic*. The corresponding characteristic in the population is called a *parameter*. Assume that a professor at a university is interested in knowing what proportion of students at the university smoke. To investigate this question, he asks the 50 students in his class whether or not they smoke and finds that 20 of them do. Thus, the percentage of students who smoke in his class is 40%. On the basis of this information, the professor constructs the following chart to represent his problem:

Group	Characteristic of Concern (= Percentage Who Smoke)
Population = all students at the university	*Parameter* = ? percentage of students at the university who smoke
Sample = students in the professor's class	*Statistic* = 40% of class smoke

Notice that no question exists about the value of the sample statistic (assuming that no one lied). This is an important point about sample statistics; they are never in doubt, their values are always known precisely. What is in doubt is the value of the population parameter, the percentage of all students at the university who smoke. What the professor wants to do, of course, is to use the sample statistic to infer the value of the population parameter. Making this type of inference, from sample statistic to population parameter, as accurately as possible is the main point of inferential statistics. Although statistics cannot guarantee the accuracy of this sort of inference, it provides the only method that exists whereby an estimate of the accuracy of the inference can be calculated.

At this point, the question of whether some of the students might have lied about their smoking might be raised. Fortunately or unfortunately, statistics is not concerned with this problem. This is a problem for the field of research design called "interviewing and questionnaire construction." Statistics deals only with numbers, not with whether the numbers were obtained accurately. Of course, if the numbers were obtained inaccurately, any statistic calculated on them is not going to be very meaningful. A saying that computer programmers apply to their field may be cited here: garbage in–garbage out. There is, however, a lot known about the problem of obtaining accurate information from people about their attitudes and behavior. But we should recognize the need to deal with these problems before data are collected, since statistics can do nothing to correct poor information.

Returning to the professor and his question about smoking, can the professor accurately infer, on the basis of his sample statistic, what percentage of all students at the university smoke? Two factors bear on this point. The first is the size of the sample, and the second is whether the sample is a representative cross section of all students at the university. Intuitively, we feel that the larger a sample is, the more accurate it is likely to be. But this is only true if the sample is an accurate cross section of the population. In 1936, a magazine called the *Literary Digest* attempted to forecast the outcome of the presidential election between Franklin D. Roosevelt and Alfred E. Landon. Selecting people randomly through telephone directories, the *Literary Digest* sent out millions of postcards asking people how they would vote. Several million cards were returned to the magazine. On the basis of this huge sample, the magazine predicted that Landon would win the election. As we know, Roosevelt won by one of the biggest landslides in history, and soon afterward the *Literary Digest* went defunct.

Sheer sample size, then, is no guarantee of accuracy. The problem the *Literary Digest* had was that it did not sample an accurate cross section of the voting public. The year 1936 was in the middle of the Great Depression and only people who were relatively well off could afford telephones. Wealthy people in the United States tend to vote Republican. Thus, the telephone directories did not contain the same proportion of Democratic voters as there were in the voting population — hence the bad prediction. A large sample is not enough. Even more important than sample size is a representative or *unbiased* sample. For the last 20 years, reputable public opinion pollsters have accurately forecast election outcomes using samples of from 1500 to 2000 people. They owe their accuracy to selecting a good sample.

But how can we tell what is a good sample or an accurate cross section? One way would be to know all the factors that affected the thing we are interested in. In the professor's case, this would mean knowing all the things that affected whether or not students smoke. We then could select our sample in such a way as to include all the factors in the right proportion. For example, age and sex both might affect smoking: Older students might smoke more than younger students and male students might smoke more than female students. If we knew this, we could represent age and sex in our sample in the same proportion as they exist in the population. We would then know that our sample is an accurate cross section of the population on these two factors.

The problem should now be obvious. We usually do not know what factors affect the thing that is being investigated and thus cannot specify what is an accurate cross section. Indeed, it is often the purpose of a study to discover what these factors are. Moreover, in social science there are usually a great number of factors that affect something and the calculations would be very complex, even if they were possible. Thus, although a procedure of this sort might give us some guarantee of the accuracy of the cross section of the sample, it is usually impractical to use.

Statistics solves this problem by employing what is called random sampling. *A simple random sample is a sample that is selected in such a way that every member of the population has an equal and independent chance of being selected and where the selection of all possible samples that could be chosen is equally likely.*

The requirements of a simple random sample are very stringent. The "equal and independent chance" requirement means that each member of the population has the same chance of being chosen as the first element that is selected into the sample, regardless of which member of the population is chosen as the first element in the sample, or the second, or the third, and so on. To meet this requirement it is usually necessary to be able to identify or list every member in the population. Frequently, this is difficult to do. If we wanted to interview a sample of voters to predict the outcome of an election, how can we identify members of the population, the people who will vote on election day? Who will actually vote will not be known until the voting is completed. In this situation there exists a real problem in trying to identify members of the population.

Once the members of a population have been identified, members must be randomly selected in such a way so as to give each member an equal and independent chance of being selected into the sample. This requires the use of some mechanism to generate random selections. Newspapers sometimes send a reporter to sample the opinion of the "man in the street" on some issue. The reporter stations himself on a street corner, say in front of city hall, and asks passers-by for their opinions. Often the newspaper reports the results as being obtained from a random sample of people. But this is *not* a random-sampling procedure; it is a *haphazard* one. People who walk by city hall on their way to work or to lunch have a greater chance of being selected into the sample than do people who customarily do not walk by city hall. The idea that such a procedure yields a random sample is incorrect.

Another problem in random selections can be illustrated by an urn filled with 1000 marbles. Assume that we wish to draw a sample of 10 marbles from the urn. Shaking the marbles thoroughly should yield a "random" mixture. We select our sample by drawing out one marble at a time and perhaps shaking the urn after each draw. The chances of any marble being selected on the first draw are 1 out of 1000 since there are 1000 marbles in the urn. But the chances of any remaining marble being selected on the second draw are 1 out of 999, since only 999 marbles are in the urn at the time of the second draw. Similarly, the chances of any remaining marble being selected on the third draw are one out of 998; the fourth, one out of 997; and so on. The sampling procedure may seem to be random, but a sample selected by this procedure is not a simple random sample since the results of earlier draws change the probability of selecting marbles on later draws and the equal-and-independent-chance requirement is violated.

Examples of this problem happen in real surveys whenever respondents are allowed to decide for themselves if they are going to participate in the survey. This is particularly a problem in mail surveys or "postcard" polls. In such surveys, respondents who are more willing to participate, because of their outgoing personality or because they have more free time or whatever, have a better chance of being selected than people who for some reason are reluctant to participate. Consequently, the equal-chance requirement of a random sample is violated. It is essential for a researcher to do everything possible to obtain a response from every person that he selects through his random-selection mechanism.

Another bad practice is to substitute additional respondents for those who refuse to participate in the survey. This procedure again gives a greater chance of selection to people who are disposed to participate than to people who are reluctant.

Technically, problems of nonresponse are not part of statistics. Statistics starts at the point where a good random sample has been obtained. If a researcher wants to use statistics, he must ensure that the data meet the requirements.

The most reliable random-sampling mechanism is to number each member of the population sequentially. A table of random numbers such as Table A–1 in the Appendix is employed to select the sample. Random-number tables are usually generated by a computer and are checked for randomness in all possible directions. This means that we can flip open the table to any point, close our eyes, and place a finger on a number in the table. (This is done to safeguard against looking for any particular number to be the starting point.) We then can proceed in any direction, across rows or up and down columns, and select those members in the population whose numbers appear in the table until we reach the desired sample size. Care must be exercised to use the right number of digits. If there are 500 members in the population, three-digit numbers must be examined. If there are 5000 members in the population, four-digit numbers must be examined. If a number appears in the table which is larger than the number of members in the population (for example, number 501 when there are 500 members), it is ignored and we proceed to the next number. Similarly, if a number repeats itself, the repetition is ignored. Finally, we must always move on to a new row or column and not repeat ones that have already been examined. Technically, this process does not yield a theoretically perfect random sample but an approximation. The approximation is very good, however, so this is usually the preferred way of drawing a random sample in practice.

When a sample is chosen in this way, a good simple random sample will be obtained. Random sampling cannot guarantee us that any sample statistic will accurately estimate its corresponding population parameter. But various statistical procedures allow us to calculate an estimate of how accurate a sample statistic is likely to be when the sample is chosen in a random manner. This is the reason that good random sampling is so important. Without it, there is no way to estimate the accuracy of our sample statistics.

Simple random sampling is a special case of a more general type of random sampling called *probability sampling*. In simple random sampling, each member of the population has an equal chance of being selected into the sample. In other types of probability sampling, the members of the population each have a known chance of being selected but these chances may be unequal for different members. Through various techniques it is possible to select a good sample in these conditions, although the sampling procedure still must be a random one.

Two frequently used techniques of sampling in social science are stratified sampling and cluster sampling. *Stratified sampling* allows the researcher to ensure that certain groups in the population are reflected proportionally in the sample. A researcher, for example, may wish to ensure that the sample contains the same proportion of blacks as there are in the population. Assume that the researcher wants to draw a sample of 1000 people and he knows that 20% of the particular population in which he is interested is black. He can draw two random samples, one of 800 from a list of whites and the other of 200 from a list of blacks. These two samples can then be combined into one which guarantees that the representation of blacks in it accurately reflects the proportion of blacks in the population. Such a sample is called a stratified sample and can be treated as a simple random sample.

In *cluster sampling,* more than one sampling operation is performed. Each sampling operation is called a *stage.* A researcher may be interested in the opinion of adults in a particular city on increasing property taxes to fund local schools. A list of adults in a city is hard to get or compile (it is often not even possible). A list of housing units, though, is usually available. The researcher could first take a random sample of housing units, the first stage. The researcher, of course, is not interested in interviewing housing units. At each unit he ascertains the number of adults in it and then randomly selects one adult in the unit to be interviewed, the second sampling stage. Cluster sampling is a complicated process and special correction procedures must be employed before various statistical techniques can be used. But it does yield a random sample.

A procedure that seems similar to cluster sampling, *area sampling,* does not provide a valid random sample. In this procedure, a random sample of city blocks may be drawn and the interviewer instructed to interview someone in the seventh house on the block. This procedure does not meet the requirement that every possible sample have an equal chance of occurring and thus does not provide a good sample for statistical purposes.

To summarize our discussion of sampling, the purpose of sampling is to infer a characteristic of a population from knowledge of a characteristic in a sample. A characteristic of a population is a parameter, and the corresponding characteristic in a sample is a statistic. The process of estimating population parameters from sample statistics is statistical inference. Statistics can provide an estimate of the accuracy of such inferences only when the sample is a random sample.

Question 2–1. Suppose that a researcher is able to obtain a simple random sample of voters in a particular city during a mayoralty election. The final election results showed that 60% of the votes were cast for the Republican candidate, 35% for the Democratic candidate, and 5% of the votes for other candidates. In the researcher's sample, 55% said that they voted for the Democratic candidate, 37% for the Republican candidate, and 8% for other candidates.

a. What is the characteristic under study?
b. What is the population?
c. What is the value(s) of the population parameter?
d. What is the value(s) of the sample statistic?

PROBABILITY

The notion of probability lies at the heart of many statistical procedures. In Chapter 1 it was pointed out that statistical statements require a different standard of confirmation than do universal ones. Probability plays the central role in providing a tool by which to evaluate evidence that is being brought to bear upon a statistical (as opposed to universal) statement.

It will be recalled that statistics evaluates the evidence bearing on a researcher's statement by setting up the competing hypothesis that the result (the evidence) came about only by chance. Now if the researcher's hypothesis is true, it implies that in an appropriate situation a result should occur which is not likely to have occurred just by chance. (If, for some reason, the researcher's hypothesis implies a result that is also likely to occur just by chance, we could just as well accept the chance hypothesis as being true rather than the researcher's.) And if, in a situation appropriate to the researcher's hypothesis, a result occurred that was likely to have occurred by chance, we might well interpret this result as evidence disconfirming the researcher's hypothesis. Thus, to use the result as evidence in support for the researcher's hypothesis, the minimum requirement is that the result be one which is not likely to have occurred by chance. This process enables us to test statistical statements because it allows for the occurrence of exceptions to the researcher's hypothesis. It does not require that the researcher's hypothesis hold in every instance but only to produce a result that is not likely to have occurred by chance.

This process obviously requires the calculation of the probability that the result under consideration could have occurred by chance. This requires the construction of a *model* of chance which is appropriate to

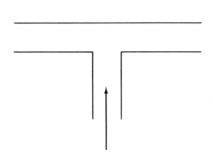

the situation. The following situations in a very rough way illustrate the use of probability calculations from different models of chance.

Suppose that an investigator thinks he has discovered a drug which affects people's choice of direction so that they will turn to the left when they have to make a change in their direction of walking. He tests his drug by injecting 10 people and having them each walk through a T maze. The T maze looks as shown. He finds that 7 of the 10 people do turn to the left and concludes that since a majority of the people turned left, his drug works. A critic points out, however, that by *chance alone* 5 of the 10 people could have been expected to turn left; and that 7 turning left is also fairly likely to have occurred by chance. The situation is analogous to flipping a coin 10 times and getting 7 heads. It is fairly likely that by chance alone 7 (or even more) heads would occur even with a perfectly honest coin. Consequently, the experimenter must conclude that the data cannot be used as support for the hypothesis.

The same experimenter now thinks that he has found a new type of music which will put people to sleep. He plays the music to 10 healthy and well-rested people and finds that 5 fall asleep immediately. Remembering his critic, though, the experimenter concludes that 5 out of 10 could occur by chance and rejects the idea that his music puts people to sleep. Again he has made a wrong decision! We know that the probability is very low that any well-rested person should just happen to fall asleep. And that 5 out of 10 people should all fall asleep at once would be an extremely unlikely thing to have happened by chance.

These experiments show how statistical generalizations are tested by *comparing the actual result or outcome of the data against a model of what could have happened by chance*. This requires setting up a model of chance that is appropriate to the situation, such as coin flipping in the first experiment. The model is then used to calculate the probability of the outcome occurring by chance. If the model says that there is a fairly high probability that the outcome could have occurred by chance, then we reject the generalization and say that the outcome was due to chance. If, however, the model says that there is only a very small probability that the outcome could have occurred by chance, we reject the idea that the outcome was due to chance and say the results can be used to support the generalization. To test statistical statements, then, we must be able to calculate probabilities of various results occurring by chance under a variety of models.

Most of us have an idea of what is meant by probability. We say that the probability of a head in flipping a coin is .5, or that the probability of rolling a 3 on a die (the singular of dice) is $1/6$. In statistics, probabilities are calculated by conducting experiments. An *experiment* can consist of anything, such as flipping a coin once or many times or rolling dice. A single flip of a coin is called a *trial* and many flips would be a series of trials. Each trial has an *outcome*. If the occurrence of a head in flipping a coin were of interest, then when a head appeared in a particular trial we say that the *event* occurred. It is conventional to say that the trial was a *success* if the event occurs on that trial (heads) and a *failure* if it does not (tails).

The probability of an event is equal to the ratio of the number of times the event occurs to the total number of outcomes or trials in the experiment in the long run. If A stands for the number of times a head appeared in a series of trials and T for the total number of trials, the probability of heads, P_H, flipping a coin is

$$P_H = \frac{A}{T}$$

If a coin was flipped 100 times, we might observe 58 heads. A would be equal to 58 and T equal to 100. The estimate of the probability of a head according to this experiment would be

$$\text{estimated } P_H = \frac{58}{100} = .58$$

This is not quite the .50 we expect. We would expect, however, that if we continued to flip the coin for a very long time, the ratio would approach .50 very closely. This is why we speak of probability as that which happens in the long run. Technically, it is inappropriate to speak in terms of the probability of a single event, such as: "The probability of obtaining a head on a single flip of a coin is .50." Such statements must be interpreted as a statement about the ratio or frequency of the number of events (successes) to the total number of trials in the long run. It is, however, convenient to talk in terms of the probabilities of single events and understand that we are really referring to what would happen over many repeated trials in the long run.

It is rare that the probabilities of events are actually determined empirically by observing the occurrence of actual events in repeated trials. An experiment may be carried out for a great many trials, but no experiment can be carried out indefinitely, which is what "long run" means. It is possible, however, to estimate probabilities by logically inferring what the outcomes of an idealized experiment would be in the long run. This is what we do intuitively when we say that the probability of getting a head in flipping an honest coin is .50, without actually having flipped the coin at all.

It is easy to logically infer probabilities in situations where all the outcomes to an "experiment" are equally likely. What is done is to list all the possible ways in which an experiment can turn out and all the possible ways in which the event of interest can occur. The probability of the event of interest is then calculated as follows, where E stands for the event of interest:

$$P_E = \frac{\text{number of ways E can occur}}{\text{total number of outcomes}}$$

For the experiment of flipping an honest coin, there is one way a head can occur and two ways the experiment can turn out (discounting the possibility of the coin landing on edge). Therefore, the probability of a head is 1 divided by 2, or .50. The probability of a 4 in rolling a die can be obtained in the same way. In rolling a die there is only one way a 4 can occur and six ways the experiment can turn out. Therefore, the probability of a 4 is 1 divided by 6, or .167.

Statistics uses models of chance to calculate probabilities, which in turn are used to accept or reject hypotheses. Sometimes the flipping of a coin or a number of coins can be used as the model of chance. These are relatively simple models, however, and are not appropriate for more complex and more interesting types of problems. Later we shall construct more complicated models of chance which can be applied to more complex problems. The calculation of probabilities under these models is also more complex and will be taken up in Chapter 6.

Question 2–2. Calculate the probabilities of the following events. (It should be kept in mind that the probability of all possible outcomes to an experiment must be 1.0. For example, in flipping a coin the sum of the probabilities of the separate outcomes, $P_H = .5$ and $P_T = .5$, adds up to 1.0.)

a. Write down all possible outcomes to the experiment of flipping two coins.
b. How many ways can the experiment turn out? How many ways are there for the experiment to give the result of two heads?

c. What is the probability of getting two heads in flipping two coins? Two tails?
d. What is the probability of *not* getting two heads or two tails in flipping two coins? _____

THE NORMAL CURVE

One of the most important models in statistics is the normal curve. Although we have not yet developed all the concepts necessary to a complete discussion of the normal curve, a brief indication of its properties is useful here. Assume that we survey the height of 1000 male students at a high school. Then we plot the height of each student on a graph which indicates how frequently any particular value of height occurs among the students. Figure 2–1 shows the values of height for 50 students plotted on a graph of this type. The figure shows that of the 50 students, 10 are 68 inches tall, 7 are 67 inches tall, and so on.

If we went on to plot the height of all 1000 students on the graph and then connect the top of each column of x's, we might get a curve that looks like that in Figure 2–2. (In Figure 2–2, the units on the vertical axis have been adjusted so that the graph can conveniently accommodate 1000 instead of 50 observations.) A *bell-shaped curve* of this sort is

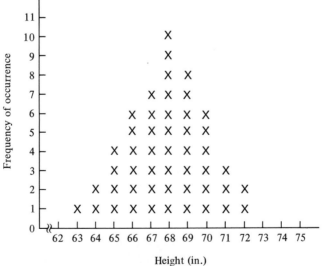

Height (in.)
Figure 2–1
Height of 50 High School Students

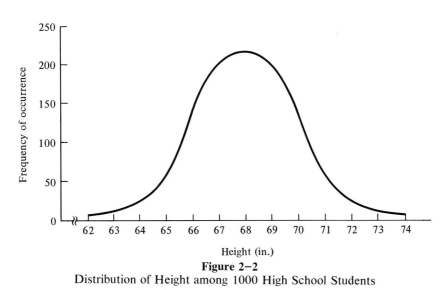

Figure 2–2
Distribution of Height among 1000 High School Students

called a *normal curve*. The curve is a representation of how frequently the various values of a variable, in this case height, occur among a particular set of observations. Notice that the normal curve is symmetric in shape and has one peak exactly in its middle.

The normal curve is a mathematical conception and is described by a precise mathematical formula. It is often used as a model in statistics, not because of any inherent quality but because many real-world variables, such as IQ, height, and weight, have distributions that closely approximate it.

The normal curve serves as a mathematical model for the frequency with which observations can be expected to occur. Since a probability is defined in terms of the frequency of an event in the total number of outcomes, the normal curve can be used to estimate the probability of occurrence of certain events. The curve in Figure 2–2, for example, tells us that if we were to select a high school boy at random, the probability of his height being between 66 and 70 inches is much greater than the probability of his height being between 70 and 74 inches. Similarly, it can be seen that if a random sample was selected from the 1000 boys, the average height of the sample would tend to be close to 68 inches. Later it will be possible for us to compute a probability concerning how close a sample average would be to a population average using the normal curve. Eventually, we shall use the normal curve to make very precise statements about probabilities concerning more interesting types of problems.

Question 2–3. In Figure 2–2 assume that the average height of the 1000 students is 68 inches and that this is the midpoint of the distribution: that half of the boys are taller than 68 inches and half are shorter.

a. What is the probability that if a "sample" of one boy were picked at random, he would be taller than 68 inches?
b. If a "sample" of two boys are picked at random from the distribution, write down all possible combinations of each being taller or shorter than 68 inches. (See Question 2–2a.)
c. What is the probability that both boys would be taller than 68 inches? Both shorter than 68 inches? One shorter and one taller?
d. If a "sample" of three boys are picked at random from the distribution, write down all possible combinations of each being taller or shorter than 68 inches.
e. What is the probability that all three boys would be taller than 68 inches? Shorter than 68 inches?
f. What is the probability that the boys would *not* all be taller or all shorter than 68 inches?
g. If the size of the sample were increased, what would happen to the probability that the boys in the sample would all be shorter or all taller than 68 inches? What do you think this indicates about the probability of obtaining a representative sample when the sample is selected in a random way? _____

VARIABLES

Statistics, as well as any of the sciences, approaches its problems by examining variables. The concept of a variable is so basic that it is usually employed without bothering to supply a definition for it. But it might be useful to examine this term in some detail.

Fundamentally, a *variable* is any property of an object that is singled out for study. For a certain political scientist the power of nations may be relevant to his problem, so he treats it as a variable. Another political scientist, concerned with a different problem, may treat the rate of political participation in nations as a variable and ignore their power entirely. The point is that every object, or animal, or person, or nation possesses a great many properties. But the researcher chooses to work only with those that he feels (for whatever reason) are relevant to the problem he has undertaken to study. The selected properties become the researcher's variables.

The term "variable" usually conjures up in the minds of most people the notion of numbers; that is, a variable is a property to which numbers have been attached. Technically, this need not be so. Any property may be treated as a variable even if quantification (the attaching of numbers) of the property is not possible. However, as the researcher is led to be more precise about his problem, he will find it useful to try to quantify his variables. Since statistics attempts to treat variables in a precise way, we will always attach numbers to variables.

Frequently, research is directed to investigating characteristics of a single variable. At other times, interest focuses on the relationship between two or more variables. Such research is usually concerned with ascertaining whether one variable "affects" or "causes" another variable and to what degree. In the process of analysis it is useful to distinguish two types of variables. Variables that are thought to do the causing or affecting in the situation are called *independent variables*. Variables which are thought to be the ones that are caused or affected are called *dependent variables*. Examples of such situations are the statements: "Education causes increased voting rates," and "Population density affects crime rates." Generally, an independent variable occurs *prior* in time to the dependent variable, and it *never* occurs after it. If it is hypothesized that education affects how people vote, the education must take place before the people vote, not after they cast their ballots.

There are lots of variables which occur prior in time to other variables but which are not referred to as independent and dependent variables. It has been found that sunspot activity increases prior to increases in stockmarket activity. But this does not mean that sunspot activity is an independent variable and stockmarket activity a dependent variable. Something else is needed before variables would be treated in such terms. This something else is an idea or hypothesis that someone has that one variable affects or causes another. And no one has seriously hypothesized that increasing sunspot activity causes people to buy or sell stocks. It is the guess of the researcher that selects variables and generates the use of the terminology independent and dependent within a particular problem.

The importance of the hypothesis in determining independent and dependent variables is illustrated by the fact that a variable may be an independent variable in one problem and a dependent variable in another. A strong sense of patriotism may make people more willing to fight for their country during time of war. In this hypothesis, patriotism is the independent variable and willingness to fight the dependent variable. Another hypothesis might be concerned with why people are more or less patriotic. An example of such an hypothesis might be that people who have experienced greater occupational and financial success (within their country) have stronger feelings of patriotism than those who have not been so successful. In the second hypothesis, patriotism is a dependent variable. Which variables are to be independent ones and which dependent ones, then, is determined by the guess or hypothesis that a researcher has about the problem he is studying.

Question 2–4. Specify the independent and dependent variables in each of the following situations.

a. The more education a person has, the more likely he is to participate in politics.
b. Older people vote more frequently than younger people.
c. The greater the population density of an area, the higher its crime rate.
d. Suicide increases in times of high unemployment, high international tension, and declining confidence in social institutions.

THE PROBLEMS
OF MEASUREMENT

In our everyday lives, we are quite accustomed to manipulating numbers. We add deposits and subtract expenditures from our checking accounts, we multiply a monthly salary by 12 to find a yearly income or divide it by 4 to find a weekly income, we calculate a percentage of our income to find our income tax, and students multiply their grades by course credits and divide by the total number of course credits they took in a semester to find their grade-point averages. The arithmetic operations involved in these calculations — adding, subtracting, multiplying, and dividing — are carried out on mathematical symbols: the digits 0, 1, 2, 3, 4, 5, 6, 7, 8, and 9. These symbols, of course, represent different variables in different problems. A 4 might represent $4 of money, 4 pounds of weight, or an A in a course. Rarely do we stop to consider how a number, which is a mathematical symbol, comes to stand for or represent a certain quantity or amount of a particular variable. Nor do we consider how a mathematical operation such as division gets a "meaning" in terms of the variable in a problem. Considerations such as these are termed *measurement problems.*

For most variables with which we are accustomed to dealing, such as money, height, and weight, numbers and the arithmetic operations have meaning in terms of the variable. Thus, we never have to worry about measurement problems. Unfortunately, many variables in social science are not as neat. These variables often do pose measurement problems, and this restricts us in our ability to manipulate them mathematically.

The most basic or lowest level of measurement is *nominal measure-*

ment. Nominal measurement is often called *classification,* since this is what is done at this level of measurement. The numbers on the backs of football players' jerseys are nominal measurement. In football it is customary to number quarterbacks from 1 to 19, to number ends in the 80s, centers and linebackers in the 50s, tackles and guards in the 60s and 70s, and backs in the 20s through 40s. The important thing about nominal measurement is that the numbers do *not* represent amounts of any underlying variable, so a higher number does not represent more of some variable than a lower number. Thus, the number on the back of an end (in the 80s) does not mean that he is bigger, or stronger, or heavier, or faster than a center (50s) or a back (20s through 40s). The numbers do not represent more or less of anything. What the numbers do is to put each player into a particular category, to classify them, and a different number is used to designate each of the categories that are being used.

A common question on political surveys is: What do you think is the most important problem facing the country today? Assume that a survey is done using this question. Various answers, such as taxes, war, pollution, corruption, welfare, and racial problems, will be given. To tabulate the number of respondents in the survey who mention a particular problem, "coders" are set to work reading the interviews. They are told to punch a number in an IBM card according to the following scheme:

1. Taxes
2. War
3. Pollution
4. Corruption
5. Welfare
6. Racial problems

These numbers are nominal measurements: They serve to identify to us (and the computer through which the data will be run) the particular category in which each respondent answered. The fact that "racial problems" is designated "6" and "taxes" "1" does not mean that racial problems are more or less important (or more or less of anything) than taxes.

The problem with nominal measurement is that the arithmetic operations of addition, subtraction, multiplication, and division do not make sense if they are applied to variables at this level of measurement. It is true mathematically that $1 + 6 = 7$, but what does this mean in terms of taxes and racial problems? Thus, we cannot arithmetically manipulate numbers which represent a variable at the nominal level of measurement.

One implication of this is that it is inappropriate to use statistical techniques which require these arithmetic operations (such as an average which requires addition and division) on nominal variables.

The second level of measurement is called *ordinal measurement*. In ordinal measurement numbers do represent more or less of some underlying dimension. A larger number, then, represents more of the variable than a smaller number. The problem with ordinal measurement is that we do not know how much more.

Questions that attempt to measure people's attitudes usually yield ordinal level measurements. Suppose that people were asked whether the federal government should see to it that there is a job for every person who wants one in this country. And their attitude on this question is ascertained by asking the respondents to locate themselves on the following scale:

Disagree strongly	Neutral	Agree strongly

| 1 | 2 | 3 | 4 | 5 | 6 | 7 |

As a result of this measuring procedure we would have a number for each respondent from 1 to 7. We could say that a person who was assigned a 6 was more in favor of the question than a person who was assigned a 5. Similarly, we could say that a 4 indicated a more favorable response than a 3. But can we say that going from 3 to 4 represents an increase in favorableness of attitude that is equal in size to the change in attitude in going the distance from 5 to 6? We cannot because we have no evidence which shows that the intervals between the numbers correspond to equal size changes on the underlying dimension.

The problem may be likened to the marking out of the numbers 1 through 7 on a piece of rubber band. Unstretched, the rubber band may look like this:

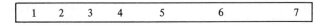

Now we could grasp the right half of the rubber band and stretch it so that it looked like this:

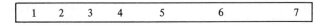

Or we could stretch the ends of the band and pinch the middle:

1		2	3	4	5		6		7

In every case, note that a larger number always retains its position as being farther to the right than any smaller number. This is what we mean when we say that in ordinal measurement a larger number represents more of the variable than a smaller number. The problem, though, is that we have no evidence to tell us whether our measurements are unstretched or stretched and, if stretched, exactly how they are stretched. There is no way to tell whether the interval between any two numbers is less than, equal to, or greater than the interval between any other two numbers in terms of the underlying dimension. Thus we cannot say that going from a 3 to a 4 on the scale represents the same amount of increase in favorableness to the question as going from 4 to 5 on the scale.

In ordinal level measurement, then, increases in the size of numbers represent increases in the amount of the underlying variable that is being measured. But it cannot be determined how much change an increase in the size of the numbers represents in terms of the underlying variable. This means that it is also nonsensical to perform arithmetic operations on numbers that measure ordinal level variables.

The third level of measurement is called *equal interval level measurement* (usually just called *interval measurement*). In interval measurement, increasing numbers represent increasing amounts of the underlying dimension as in ordinal measurement, but now the fact that the intervals between the numbers are equal in size is somehow established. An example of interval measurement is the Fahrenheit scale of temperature. In this scale the increase in temperature in going from 10°F to 20°F is equal to the increase in temperature in going from 75°F to 85°F. Because the property of equal intervals is present, it now makes sense to add, subtract, multiply, or divide numbers. A statement such as "If a room was 70°F and its temperature declined 10°F, it would then be 60°F" makes sense. Such statements do not make sense at the nominal or ordinal level.

Interval scales still do not allow the complete use of all possible arithmetic manipulations. The number 100 is twice as big as the number 50. But 100°F is *not* twice as hot as 50°F. In mathematical terms, the calculation of ratios cannot be done on interval level data. The problem is that the Fahrenheit scale lacks what is called a "fixed zero point." If 0°F actually represented a temperature of *no* heat, ratios could be calculated. In terms of temperature this point is known as *absolute zero* and in Fahrenheit degrees has the value of −459.69°. Only when the number 0 corresponds to the point of "nothing" of the variable can ratios be calculated.

This leads to the fourth and highest level of measurement, the *ratio level*. In ratio measurement, there are equal intervals between the numbers and the number 0 corresponds to a point where nothing of the variable is present. In ratio measurement the full array of mathematical manipulations can be used. In the measurement of temperature, the Kelvin scale is a ratio scale. Income in dollars, weight in pounds, and height in feet and inches are all ratio level scales. Usually, the most difficult problem in measurement is not the determination of a fixed zero point, but of equal intervals. Thus, if measurements are at the interval level, they are also likely to meet the ratio level requirements or can easily be made to do so (such as turning Fahrenheit into Kelvin). Consequently, some people only distinguish among nominal, ordinal, and ratio levels of measurement.

It can be seen that most of the things we are accustomed to measuring are at the ratio level. This is why the problems of nominal and ordinal measurements seem so foreign to us. But in social science it is necessary to pay strict attention to the level of measurement of the data.

How can the level of measurement of a particular set of numbers be determined? This can be done by examining the physical process by which the numbers were assigned. If numbers were assigned to objects in an arbitrary way only to distinguish certain categories among the objects and any other assignment of numbers would serve just as well, then the numbers are nominal measurement. Thus, if a researcher labels males 1 and females 2, this is nominal measurement because the researcher is only interested in distinguishing the two categories of being male or being female and the opposite assignment of the numbers would have served this purpose just as well.

It is usually fairly easy to see if numbers are nominal measurements. It is somewhat more difficult to distinguish between ordinal and interval measurements. Both of these levels "rank" objects along some underlying dimension. These levels can be distinguished by determining if there is a physical interpretation for the arithmetic operations of addition, subtraction, multiplication, and division. Suppose that we are interested in people's weights and employ a big balance scale as the physical mechanism by which we assign numbers to people to represent their weight. If two people each weigh 100 pounds and one person weighs 200, then putting the first two people on one side of the scale and the third person on the other side results in the scale balancing. What we have just done is to provide a physical interpretation in terms of the variable of weight for the arithmetic statement: $100 + 100 = 200$. When it is possible to think of physical interpretations for the arithmetic operations, the numbers must represent interval or ratio level measurements.

Now suppose that a teacher called his class to the front of the room and had the students arrange themselves in a line according to height. He then assigns numbers to students to represent their heights by having them count off. When this is done, it will be true that every student assigned a larger number will be taller than any student with a lower number. But can we think of any physical interpretations in terms of the variable whereby the arithmetic statement of, say, $4 + 5 = 9$ makes sense? No. Therefore, these numbers cannot be interval or ratio measurements and must only be ordinal level measurements.

Question 2–5. Determine if each of the following situations yields nominal, ordinal, interval, or ratio level measurements.

a. A teacher assigns students to class A, B, or C basketball leagues on the basis of their skill.
b. A farm worker divides eggs into two groups according to whether or not they are fertilized.
c. A worker takes a basket of apples and compares them two at a time for weight on a balance scale. Through this process he orders all the apples from lightest to heaviest.
d. A worker determines the weight of each apple in ounces.
e. A mechanic counts the number of each make of car he services every month.
f. IQ measurements.
g. A researcher assigns the numbers 1, 2, 3, or 4 to people according to whether their religion is Catholic, Protestant, Jewish, or other.
h. The times that people make in running the mile.
i. The "place" number each person finishes in, in running the mile.

CLASSIFICATION
OF STATISTICAL METHODS

It was seen in Chapter 1 that statistics can be divided into two areas: description and inference. Because there are so many statistical techniques it is sometimes difficult to decide which is appropriate in a particular situation. It is, therefore, useful to think of statistical techniques in terms of a classification. The first consideration in classifying statistics is whether their purpose is descriptive or inferential. Some statistics are designed to describe various characteristics of a set of data. Other statistics are concerned with making inferences from a sample to a population. (There is some overlap here, as descriptive statistics are often involved in the inference process.) But before any statistic can be selected, a researcher must determine what type of characteristic he is concerned about and whether he wants a description of that characteristic in a particular set of data or whether he wants to make an inference about that characteristic to some population from the set of data.

Statistical techniques can also be classified according to whether they are designed to deal with one variable, two variables, or more than two variables. The average is a statistic that describes a characteristic of one variable. Statistics that deal with one variable are called *univariate statistics*. Other statistics describe characteristics of the relationship between two or more variables. Statistics that deal with two variables are called *bivariate statistics*. Statistics that deal with more than two variables are called *multivariate statistics*. Table 2–1 shows how statistical techniques can be classified according to their purpose and the number of variables involved. This classification is somewhat arbitrary but it serves to orient us to an important problem: What is it that we wish to know when we go to apply "statistics" to a set of data? Only by explicitly formulating the question he wants to ask of the data can a researcher begin the process of selecting an appropriate statistical technique.

An additional consideration in selecting among statistical methods is the level of measurement of the data. Generally, a statistical technique is designed for use on data at a particular level of measurement. If we were interested in determining whether or not there is an association between two variables, the method that should be used to evaluate the data will depend upon whether the variables are nominal, ordinal, or interval (ratio data will be considered interval level here since they have interval properties) measurements or a combination.

Generally, a technique that is designed for a certain level of mea-

Table 2–1
Classification of Statistical Methods

	Univariate Statistics (One Variable)	Bivariate Statistics (Two Variables)	Multivariate Statistics (More than Two Variables)
Descriptive methods	I. Describing the characteristics of a single variable in a set of data	III. Describing the relationship between two variables in a set of data	V. Describing the relationships among many variables in a set of data
Inferential methods	II. Inferring the characteristics of a single variable in a population	IV. Inferring the nature of the relationship between two variables in a population	VI. Inferring the nature of the relationships among many variables in a population

surement implicitly assumes that the numbers do in fact represent that level of measurement and cannot be used on data that are at a lower level of measurement. It is, however, appropriate to use a statistic designed for lower level measurement on data that can meet higher level requirements. If this is the case, it might be wondered why all statistics are not just designed for the lowest level, nominal measurement. The answer is that higher level measurements contain much more information than do lower level measurements and it would be wasteful not to take advantage of the fact. Thus statistics designed for interval data can usually tell us much more about the data than statistics designed for use on nominal or ordinal data. Also, because numbers at the interval level contain more information than numbers at the nominal or ordinal level, the statistics designed for use on interval data are usually more powerful than statistics at the lower levels of measurement. "More powerful" here means, for example, that smaller sample sizes are needed to arrive at certain conclusions.

Quite often a distinction is made between *parametric* and *nonparametric statistical methods*. Technically, the distinction between these two types of statistics is rather complex. For most purposes, however, it is adequate to say that nonparametric statistics are designed for data that can only meet the requirements of nominal or ordinal measurement, whereas parametric statistics are designed for use on data that can meet the requirements of interval or ratio level measurement.

EXERCISES FOR CHAPTER 2

1. Define the following terms: observation, sample, population, parameter, statistic.
2. What is the reason for studying samples instead of populations? When would the study of populations be appropriate?
3. Define random sampling. Why are random samples used in statistics?
4. Define probability. What is wrong with the following statement? "In a single roll of a die, the probability of getting a 3 is $\frac{1}{6}$."
5. In an ordinary deck of 52 cards, let a king be 13 points, a queen 12 points, a jack 11 points, an ace 1 point, and the point value of the number cards be their number. What is the average point value of the cards in the deck? Shuffle a deck of cards thoroughly. Deal a "random sample" of 10 cards and compute the average point value of the 10 cards. What is the characteristic under study? What is the value of the population parameter? What is the value of the sample statistic?
6. Define a variable. Why are numbers usually associated with variables. Distinguish between independent and dependent variables. How do variables come to be designated as independent or dependent? What is a hypothesis?
7. Distinguish among nominal, ordinal, interval, and ratio level measurements.
8. Determine the level of measurement in the following situations. For each, determine if there is an underlying dimension being measured and, if there is, specify what it is.

   a. Sorting cards by suit.
   b. "Win," "place," and "show" in a horse race.
   c. Determining the age of trees by counting "rings" in tree trunks.
   d. Classifying books by the numbers of the Dewey decimal system.
   e. Rating meat as "prime," "choice," "good," or "commercial."
9. Determine whether the following variables would yield nominal, ordinal, interval, or ratio level measurements. Explain your choice.

   a. Weighing apples on a scale.
   b. Ranking apples by having a person weigh the apples by holding two at a time and judging which is heavier.
   c. Separating apples into two piles, spotted and unspotted.
   d. Having a class of students line up according to height with the tallest student at one end and the shortest at the other.
   e. Identifying people by their religion, such as Catholic, Protestant, Jew, and other.
   f. Identifying people by their social class on the scale lower, lower middle, middle, upper middle, upper.

g. Identifying people by their political affiliation, such as Republican, Democrat, and Socialist.

h. Identifying people as to their strength of party identification on the scale independent, weak party identifier, moderate party identifier, strong party identifier.

ANSWERS TO
QUESTIONS IN CHAPTER 2

2–1: a. The proportion of the vote cast for the various candidates.

    b. All people who voted in the election.

    c. 60%, 35%, 5%.

    d. 55%, 37%, 8%.

2–2: a. HH, HT, TH, TT.

    b. 4, 1.

    c. $P_{2H} = 1/4 = .25, P_{2T} = 1/4 = .25$.

    d. $P_{\text{(not getting 2H or 2T)}} = 1 - (P_{2H} + P_{2T}) = 1 - (.25 + .25) = 1 - .5 = .5$.

2–3: a. .5.

    b. Let T stand for taller and S for shorter; then TT, TS, ST, SS.

    c. $P_{\text{(both taller)}} = 1/4, P_{\text{(both shorter)}} = 1/4, P_{\text{(1 shorter, 1 taller)}} = 2/4$.

    d. $TTT, TTS, TST, TSS, STT, STS, SST, SSS$.

    e. $P_{\text{(all taller)}} = 1/8 = .125, P_{\text{(all shorter)}} = 1/8 = .125$.

    f. $P_{\text{(not all taller or all shorter)}} = 1 - (P_{\text{(all taller)}} + P_{\text{(all shorter)}})$
                              $= 1 - (.125 + .125) = 1 - .25 = .75$.

    g. As the sample size increases, the probability of the boys in the sample being all taller or all shorter decreases. The probability of the sample being unrepresentative decreases greatly as the sample size increases.

2–4: a. Independent, education; dependent, participation.

    b. Independent, age; dependent, frequency of voting.

    c. Independent, population density; dependent, crime rate.

    d. Independent, unemployment, international tension, confidence in social institutions; dependent, suicide.

2–5: a. Ordinal.   b. Nominal.

    c. Ordinal.   d. Ratio.

    e. Nominal.   f. Ordinal.

    g. Nominal.   h. Ratio.

    i. Ordinal.

3
Review of Basic Arithmetic and Algebraic Operations

In taking a course in statistics, some students feel that they are taking a mathematics course. This would be true of a mathematical statistics course. In such courses, statistics is presented as a branch of mathematics. But this is not true of an applied statistics course of the level presented in this book. In such a course the emphasis is on understanding statistical reasoning and on how to use statistics to answer questions by analyzing data. The application of statistics, however, does require a lot of arithmetic—adding, subtracting, multiplying, dividing, and taking squares and square roots. Although tedious, these operations are not "higher" mathematics or even intellectually difficult. But because these operations are performed repeatedly, it is useful to briefly review them.

The presentation of some statistical techniques, explaining how and why they work, requires a little algebra. Thus basic algebraic operations will also be reviewed here so that later we can concentrate on understanding the statistical technique and not have to worry about the algebra. Usually, once a technique is explained, algebra is not needed for its use on actual data. Finally, there are some types of elementary operations that often cause confusion: for example, the difference among proportions, percentages, and ratios, and the number of places to the right of a decimal point that should be carried in calculations. These points will also be covered in this chapter.

As will be recalled, algebra frequently uses letters to symbolize quantities. Variables are usually represented by letters. A *variable* is a property that can take on different values. Weight of persons, for example, may be treated as a variable. One person may weigh 125 pounds, another person 180 pounds, and another 150 pounds. The numerical value of a person's weight may be symbolized by a letter, say X. X then stands for the value of weight. It is because X can take on different numerical values (for different people) that it is called a variable. Any letter may be used to symbolize a variable. It is customary, however, to use letters toward the end of the alphabet, especially $x, y,$ and $z,$ to represent variables.

In many problems there are certain properties that have only one value, and this value does not change. In Einstein's famous equation $E = mc^2$, c is a constant and stands for the speed of light (186,000 miles per second). It is often convenient to use a letter to represent a constant instead of writing out its actual value. In some problems, it may be known that something is a constant but its actual numerical value may not be known. In such cases it is necessary to represent this constant by a letter. Usually the letters c or k or letters at the beginning of the alphabet are used to represent constants.

SYMBOLS
OF RELATION

Quite often it is necessary to indicate the type of relation that exists between two quantities. If we represent the quantities as a and b, some of the possible relations that could exist are: a equals b, a is not equal to b, a is greater than b, or a is less than b. Instead of verbally describing the type of relation, it is usually more convenient to use symbols to represent the various types of relations. Thus,

$$a = b \quad \text{means} \quad a \text{ equals } b$$
$$a \neq b \quad \text{means} \quad a \text{ is not equal to } b$$
$$a > b \quad \text{means} \quad a \text{ is greater than } b$$
$$a < b \quad \text{means} \quad a \text{ is less than } b$$

To remember the distinction between the symbols for "greater than" and "less than" it is helpful to note that the open part of the symbol is always placed next to the larger quantity:

$$3 < 5 \quad \text{means} \quad 3 \text{ is less than } 5$$
$$6 > 2 \quad \text{means} \quad 6 \text{ is greater than } 2$$

Sometimes it is necessary to express the relations "greater than or equal to" and "less than or equal to." This is done through the symbols $\geq$ and $\leq$. Thus, $X \geq 0$ means that X is greater than or equal to zero and $Y \leq 0$ means that Y is less than or equal to zero.

Sometimes we wish to question whether the relation between two quantities is one of equality. This can be represented by the symbol $\overset{?}{=}$. The statement "Is a equal to b?" can be symbolized as

$$a \overset{?}{=} b$$

Another useful symbol is one to represent the relation "approximately equal," which is symbolized as $\approx$. This symbol is often useful when it would be tedious to have to specify values precisely. Thus, 9 divided by 7 is 1.285714. . . . Instead of writing the answer out to many places to the right of the decimal point we could write

$$\frac{9}{7} \approx 1.3$$

SIGNED NUMBERS
AND ABSOLUTE VALUES

In algebra, numbers are usually allowed to be both positive and negative in sign. The easiest way to understand the sign of numbers is to picture them as specifying particular locations on an axis. An axis is devised by drawing a line and designating a point on the line as the zero point:

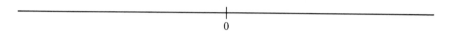

Units of equal size are then marked off to the right and to the left of the zero point:

These units are numbered by counting them off starting from the zero point. By convention, if we are counting off to the right, the units are assigned numbers positive in sign, and if we are counting off to the left, the units are assigned numbers negative in sign:

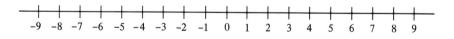

Theoretically, we could continue to count off units to the right and to the left indefinitely. But usually we arbitrarily stop marking off and counting units whenever we have enough to suit our purposes.

A positive number, then, represents a point on the axis to the right of the zero point. A positive number is denoted by prefacing the number with a + sign. Negative numbers represent points to the left of the zero point and are denoted by prefacing the numbers with a − sign. The numbers +8 and −3 specify the following points on the axis:

To simplify the writing of numbers, any number that is not prefaced by a sign is taken to be a positive number. Thus, +8 and 8 are taken to mean the same thing and represent the same point on an axis. This implies that only when a number is prefaced by − is it to be taken as a negative number.

An *integer* is a number that corresponds to a point on an axis which lies a whole number of units from the zero point. Thus 8 is an integer because it can be represented by a point that lies 8 whole units from the zero point. A *fraction* is a number that corresponds to a point whose distance from the zero point cannot be given in terms of a whole number of units. Thus, the number $11/4$ corresponds to a point that lies 2 whole units plus $3/4$ of another unit from the zero point. Decimal numbers also correspond to points that lie between whole numbers on an axis. Here are some examples of the location of fractions and decimals on an axis:

The *absolute value* of a positive number or zero is the number itself; the absolute value of a negative number is the number with its sign changed. The operation of taking the absolute value is indicated by placing the number between two vertical lines:

$$|6| = 6$$
$$|-6| = 6$$

If a is a *positive* number, then $|a| = a$ and $|-a| = a$. But if a is itself a *negative* number, then $|a| = -a$ and $|-a| = -a$. What this means is that $|a|$ cannot be simplified unless it is known whether a is positive or negative.

Signed numbers and absolute values can be interpreted in terms of the location of points on an axis. Signed numbers indicate a particular point to the right of or to the left of the zero point according to whether the number is positive or negative in sign. Thus, the numbers 3 and −3 indicate different points on an axis, the first number representing a point 3 units to the right of the zero point and the second indicating a point 3 units to the left of the zero point. But these two points have something in common — they both lie a distance equal to 3 units from the zero point. This fact is given by the absolute values:

$$|3| = 3$$
$$|-3| = 3$$

Since both of the points have the same absolute value, they both lie the same distance from the zero point on an axis. An absolute value may be interpreted, therefore, as the distance of a point on an axis to the zero point, disregarding direction.

BASIC ARITHMETIC
OPERATIONS ON SIGNED NUMBERS

Since the same number with different signs indicates the location of different points on an axis, it is necessary to take the signs of numbers into account when they are manipulated by an arithmetic operation. The four most basic arithmetic operations are addition, subtraction, multiplication, and division.

The operations of addition and subtraction are indicated by the symbols $+$ and $-$. Since these symbols are also used to represent the signs of numbers, it is necessary to distinguish when the symbols are being used to indicate the arithmetic operations of addition and subtraction and when they are being used to represent the signs of numbers involved in an addition or subtraction operation. To help make the distinction, the sign of a positive number is omitted. Thus, in the expression $3 + 4$, the $+$ indicates that the operation of addition is to be performed on the positive numbers 3 and 4, and the expression $3 - 4$ indicates that the operation of subtraction is to be performed on the positive numbers 3 and 4. If a number involved in an arithmetic operation is negative in sign and its sign may create confusion about the nature of the operation, this is clarified by placing the number and its sign within parentheses: $3 + (-4)$ or $3 - (-4)$.

The operations of addition and subtraction can be visualized as movements along an axis. Addition and subtraction can be viewed as defining the direction of movements on the axis. Suppose that we wished to perform the operation indicated by $3 + 4$. We begin by locating the point on the axis that represents the first number in the expression. To *add* the second number to it, we move from the location on the axis of the first number (3) a number of units equal to the value of the second number *in the same direction indicated by the sign of the second number*. Since the second number is 4 and is positive in sign, we move 4 units to the right (the positive direction) to add it to 3 and obtain the answer, 7:

To perform the operation $3 + (-4)$ we would move in the direction indicated by the sign of the second number from the location of the first number (3). Since the sign of the second number is negative, we move 4 units to the left (the negative direction) to add -4 to 3 and obtain the answer, -1:

To perform the operation $3 - 4$ we again start by locating the point on the axis that represents the first number. Since the indicated operation is *subtraction,* we move *in the direction opposite to that indicated by the sign of the second number.* Note that the minus sign indicates the operation of subtraction and that the sign of the second number is positive. Therefore, the movement on the axis is

And the answer is -1. Finally, we have the operation $3 - (-4)$. Since the sign of the second number is negative and the operation is subtraction, the movement will be

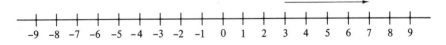

The answer is 7.

It is, of course, possible to chain together a series of addition and subtraction operations. For example,

$$4 - 3 + (-1) - (-5) = 5$$

The movements on an axis would be

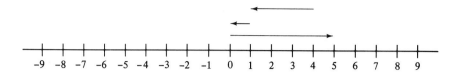

From the preceding examples, it can be seen that the expressions $3 + 4$ and $3 - (-4)$ yield the same result. This is also true of the expressions $3 + (-4)$ and $3 - 4$. This points out that addition and subtraction are the same operation except for direction. This fact can be used to simplify arithmetic expressions. Since *addition* requires moving in the *same* direction as the sign of the following number and *subtraction* requires moving in the direction *opposite* to that of the sign of the following number, expressions can be simplified by changing the operation *and* the sign of the following number. Thus, the expression $3 - (-4)$ is equivalent to $3 + 4$, and the expression $3 + (-4)$ is equivalent to $3 - 4$. (Note that in both of the original expressions 4 was negative in sign and that it becomes positive in sign when the expressions are simplified by changing the operation.) Therefore,

$$2 + (-5) - (-6) + 3 = 2 - 5 + 6 + 3 = 6$$

A number of ways are used to indicate the operation of multiplication. They can be illustrated as follows:

$$2 \times 3 = 6$$
$$2 \cdot 3 = 6$$
$$(2)(3) = 6$$

In this book we shall indicate that two (or more) terms are to be multiplied together by placing each term within parentheses. An important exception is when some of the terms involve symbols; thus,

$$4X = (4)(X)$$
$$XY = (X)(Y)$$

Since symbols such as X and Y are used to denote quantities, no confusion results when we place them next to another number or symbol to indicate that the operation of multiplication is to be performed. When one term in a multiplication is a number and the other a symbol, it is conventional to write the number first. Thus, to indicate that the terms 12 and Z are to be multiplied, the operation is written as $12Z$ rather than as $Z12$.

Again, because the numbers in algebra are signed, it is necessary to pay attention to the signs of numbers in multiplication. For multiplication involving two terms the following rules concerning the signs of the terms apply in determining the sign of the product:

1. $+$ times $+$ is $+$.
2. $+$ times $-$ is $-$.
3. $-$ times $-$ is $+$.

When more than two terms are involved in a multiplication, the following rule applies: If the number of terms that are minus in sign is odd, the product will be minus in sign; if the number of terms that are minus in sign is even, the product will be plus in sign. If all the terms are positive in sign, the product will also be positive. The following examples illustrate the rules:

$$(2)(-2)(3) = -12$$
$$(-6)(8) = -48$$
$$(-X)(-Y) = XY$$
$$(-2)(3)(-4) = 24$$
$$(-1)(3)(-2)(-4) = -24$$
$$(5X)(3)(-2) = -30X$$

Division is defined as the inverse of multiplication. That is, to divide a number a by another number, b, is defined as finding some number q for which $a = (b)(q)$. The term a is called the *dividend*, b the *divisor*, and q the *quotient*. The operation of division may be indicated in a number of ways:

$$a \div b \qquad \frac{a}{b} \qquad \text{or} \qquad a/b$$

Thus,

$$8 \div 2 = 4$$

$$\frac{8}{2} = 4$$

$$8/2 = 4$$

Note that 4 is the quotient (answer) in these illustrations *because* $8 = (2)(4)$.

In division, if the numbers have the same sign, the result is a plus; if the numbers are of unlike signs, the result is a minus.

$$\frac{-6}{-3} = 2$$

$$\frac{-8}{4} = -2$$

$$\frac{4}{2} = 2$$

$$\frac{20X}{-5} = -4X$$

A series of different arithmetic operations are often chained together. When this occurs, signs of aggregation — parentheses, brackets, and braces — are used to group operations, and work proceeds from the innermost set, the parentheses, ( ), outward to the brackets, [ ], and then to the braces, { }:

$$(3)(2 + 5) = (3)(7) = 21$$

$$(4)(5 - 2 + 1 - 7) = (4)(-3) = -12$$

$$(2)\left\{5 - [(-6)(2)] + \left(\frac{-20}{2}\right)\right\} = (2)[5 - (-12) - 10]$$

$$= (2)(5 + 12 - 10) = (2)(7) = 14$$

Question 3–1. Find the answers to the following problems:

a. $[(2 - 3)(4)] + (5)(6 - 3)$ b. $(-2)(3)(-5)(2)(-4)$

c. $\left[\left(\frac{12}{-6}\right)(2)\right] + (5)\left(\frac{-4}{2}\right)$ d. $(5X)(3) + 4X + (2)(3X - 5X)$

e. $|5 - 9 + 2|$ f. $\left|\left(\frac{7}{2}\right) - (3)(4 + 2)\right|$

It may be recalled that division involving a zero in either the numerator or denominator causes difficulties. This stems from the requirement that the result of every algebraic operation on real numbers be another uniquely determined number. As was seen, division was defined as the inverse of multiplication. Thus, if

$$\frac{a}{b} = q$$

then q must be a number for which $a = bq$. For example,

$$\frac{6}{3} = 2 \quad \text{since} \quad 6 = (3)(2)$$

But if, in a division, the numerator, a, is any nonzero number and the denominator, b, is zero, then there exists no number, q, which when multiplied by zero will yield a. This is because any number multiplied by zero always results in zero. Thus,

$$\frac{6}{0} = \text{"does not exist"}$$

since there does not exist any number q for which zero times q equals 6. Because no such number exists, the division of any nonzero number by zero is said to be undefined:

$$\frac{8}{0} = \text{undefined}$$

$$\frac{X}{0} = \text{undefined}$$

If, in a division, both the numerator and denominator equal zero, there will exist many numbers for q which will satisfy the above criterion. In other words, since $0 = 0q$, when q is *any* real number, the corresponding quotient, $\frac{0}{0}$, does not represent a unique real number. Thus, regardless of the value of the numerator, division by zero is not defined.

If the denominator in a division is a nonzero real number and the numerator is zero, the result is zero. That is, if we have

$$\frac{0}{b} = q$$

where b is any nonzero number, then q is always equal to zero, because $0 = bq$ only if q is zero. Consequently, the result of zero divided by any nonzero number is zero.

$$\frac{0}{8} = 0$$

$$\frac{0}{-8} = 0$$

$$\frac{0}{5X} = 0 \qquad (X \neq 0)$$

FRACTIONS

The manipulation of fractions frequently occurs in statistics. Many of the mistakes made in manipulating numbers occurs in the handling of fractions, so special care should be paid to them. Addition and subtraction can only be done when the denominators of the fractions involved are the same.

$$\frac{1}{2} + \frac{1}{3} \text{ is } not \text{ equal to } \frac{2}{5}$$

What is needed here is a *common denominator* in both fractions.

$$\frac{1}{2} = \frac{3}{6}$$

$$\frac{1}{3} = \frac{2}{6}$$

Therefore,

$$\frac{1}{2} + \frac{1}{3} = \frac{3}{6} + \frac{2}{6} = \frac{5}{6}$$

Here is another example:

$$\frac{1}{2} + \frac{2}{3} - \frac{3}{4} = \frac{6}{12} + \frac{8}{12} - \frac{9}{12} = \frac{5}{12}$$

The same rule holds true for variables. Take, for example, the following problem:

$$\frac{1}{a} - \frac{1}{b} = ?$$

These fractions cannot be subtracted because the denominators are different. The solution to the problem lies in recalling that the value of a fraction is not changed if both the numerator and denominator are multiplied by the same (nonzero) number. For example,

$$\frac{1}{2} = \frac{(4)(1)}{(4)(2)} = \frac{4}{8} = \frac{1}{2}$$

$$\frac{1}{2} = \frac{(25)(1)}{(25)(2)} = \frac{25}{50} = \frac{1}{2}$$

The same process can be done with variables. In the problem above, the numerator and denominator of the first fraction can both be multiplied by b. This will not change the value of the fraction, regardless of what b is equal to (as long as b is not equal to zero).

$$\frac{1}{a} = \frac{b(1)}{b(a)} = \frac{b}{ba} = \frac{b}{ab}$$

The order of multiplication is irrelevant [$(2)(3) = (3)(2)$], so $ba = ab$. The numerator and denominator of the second fraction can both be multiplied by a.

$$\frac{1}{b} = \frac{a(1)}{a(b)} = \frac{a}{ab}$$

Now the two fractions have the same denominator, and subtraction may be performed.

$$\frac{b}{ab} - \frac{a}{ab} = \frac{b-a}{ab} = \frac{-a+b}{ab}$$

(The placement of terms in subtraction and addition is irrelevant, so $b - a = -a + b$.) Here is a more complicated problem, which is done in exactly the same way:

$$\frac{5a}{3bc} + \frac{2d}{7ef} = \frac{7ef\,(5a)}{7ef\,(3bc)} + \frac{3bc\,(2d)}{3bc\,(7ef)}$$

$$= \frac{35aef}{21bcef} + \frac{6bcd}{21bcef} = \frac{35aef + 6bcd}{21bcef}$$

Without knowing the values of the letters, it is impossible to simplify the fraction further.

The multiplication of fractions is fairly simple. The numerators are multiplied together and then the denominators are multiplied.

$$\left(\frac{2}{4}\right)\left(\frac{3}{6}\right) = \frac{6}{24} = \frac{1}{4}$$

$$\left(\frac{1}{2}\right)\left(\frac{2}{3}\right)\left(\frac{3}{5}\right) = \frac{6}{30} = \frac{1}{5}$$

$$\left(\frac{5X}{3}\right)\left(\frac{2}{5}\right) = \frac{10X}{15} = \frac{2X}{3}$$

When fractions are divided the fraction in the denominator is inverted and multiplication performed.

$$\frac{^1\!/_2}{^3\!/_4} = \left(\frac{1}{2}\right)\left(\frac{4}{3}\right) = \frac{4}{6} = \frac{2}{3}$$

Sometimes the division of fractions is written as follows:

$$\frac{10}{15} \div \frac{3}{6}$$

This means that the first fraction is to be divided by the second. Again the process is the same.

$$\frac{10}{15} \div \frac{3}{6} = \left(\frac{10}{15}\right)\left(\frac{6}{3}\right) = \frac{60}{45} = \frac{4}{3}$$

$$\frac{8X/3}{2X/3} = \left(\frac{8X}{3}\right)\left(\frac{3}{2X}\right) = \frac{24X}{6X} = 4$$

Question 3–2. Find the answers to the following problems:

a. $\dfrac{2}{3} - \dfrac{4}{5}$

b. $\dfrac{3}{2X} + \dfrac{5}{4Y}$

c. $\left(\dfrac{af}{dg}\right)\left(\dfrac{2c}{3b}\right)$

d. $\dfrac{\frac{5}{6}}{\frac{8}{3}}$

e. $\dfrac{3X/4Z}{6X/2Z}$

_____

FACTORIALS

Taking the factorial of a positive integer is the operation of multiplying out the value of the number, say X, times $X - 1$, times $X - 2$, and so on until the last number in the multiplication process is equal to 1. The operation of factorial is indicated by an exclamation sign, !.

$$4! = (4)(3)(2)(1) = 24$$

$$5! = (5)(4)(3)(2)(1) = 120$$

$$3! = (3)(2)(1) = 6$$

Note that

$$5! = (5)(4!) = (5)(24) = 120$$

or

$$5! = (5)(4)(3!) = (5)(4)(6) = 120$$

This can save a lot of work when it is necessary to divide factorials.

$$\frac{7!}{5!} = \frac{(7)(6)(5)(4)(3)(2)(1)}{(5)(4)(3)(2)(1)} = \frac{5040}{120} = 42$$

$$= \frac{(7)(6)(5!)}{(5!)} = 42$$

Zero factorial is defined to be equal to 1 $(0! = 1)$.

SQUARES
AND SQUARE ROOTS

It is often necessary to take the square and square roots of numbers in statistics. It is important to be able to find them quickly and accurately. The square of a number is the product that results when the number is multiplied by itself. The operation of squaring is indicated by a 2 super-scribed to the right of the number.

$$3^2 = (3)(3) = 9$$

$$10^2 = (10)(10) = 100$$

Note that

$$(-3)^2 = (-3)(-3) = 9$$

$$(-10)^2 = (-10)(-10) = 100$$

The square root of a nonnegative number is defined as another number which when multiplied by itself will yield the original number. The operation of taking the square root is indicated by placing the number under a radical sign, $\sqrt{}$

Note also that

$$\sqrt{9} = 3 \qquad (3)(3) = 9$$

$$\sqrt{9} = -3 \qquad (-3)(-3) = 9$$

Any positive number has both a positive and a negative square root. Statistics makes use only of positive square roots, so the negative ones will be ignored.

$$\sqrt{100} = 10 \qquad (10)(10) = 100$$

$$\sqrt{X^2} = X \qquad (X)(X) = X^2 \qquad \text{(for } X \geq 0)$$

It is possible to perform arithmetic operations within the radical sign.

$$\sqrt{\frac{20 + 44}{4}} = \sqrt{\frac{64}{4}} = \sqrt{16} = 4$$

Note that

$$(\sqrt{3})(\sqrt{3}) = 3$$

$$(\sqrt{Z})(\sqrt{Z}) = Z \qquad \text{(for } Z \geq 0)$$

Sometimes it is necessary to square an algebraic expression such as $(a + b)$. This is not difficult if it is remembered that to square a number means to multiply it by itself.

$$(a + b)^2 = (a + b)(a + b)$$

The multiplication can be done as follows:

$$
\begin{array}{r}
a + b \\
a + b \\
\hline
a^2 + ab \\
ab + b^2 \\
\hline
a^2 + 2ab + b^2
\end{array}
$$

Thus, $(a + b)^2 = a^2 + 2ab + b^2$.

It is useful to know that when variables are multiplied their powers (the superscripts) can be added, and when variables are divided their powers can be subtracted.

$$(X)(X^2) = X^3$$

$$(X^2)(X^3) = X^5$$

$$\frac{X^2}{X} = X$$

$$\frac{X^3}{X^2} = X$$

Finding the square roots of numbers can be tedious. Table A-2 in the Appendix can be used to find the squares and square roots of numbers from 1 to 999. It is useful, however, to have a general notion of the range in which the square root of a number should fall. It is easy to remember the square roots of the following numbers:

$\sqrt{9} = 3$	$\sqrt{100} = 10$
$\sqrt{16} = 4$	$\sqrt{121} = 11$
$\sqrt{25} = 5$	$\sqrt{144} = 12$
$\sqrt{36} = 6$	$\sqrt{169} = 13$
$\sqrt{49} = 7$	$\sqrt{196} = 14$
$\sqrt{64} = 8$	$\sqrt{225} = 15$
$\sqrt{81} = 9$	$\sqrt{400} = 20$

If the square root of 75 were desired, it can be seen that its value must lie between 8 and 9. This must be so because 75 is bracketed by 64, whose square root is 8, and 81, whose square root is 9. By looking up 75 in Table A-2 we do indeed find that its square root is between 8 and 9:

$$\sqrt{75} = 8.6603$$

Reasoning this way provides a check on the value of the square root and helps in the placement of the decimal point. The best check, however, is to multiply what is thought to be the square root by itself to see if the original number results.

$$\sqrt{75} = 8.6603 \qquad (8.6603)(8.6603) = 75$$

A common point of confusion is taking the square roots of decimal numbers smaller than 1.0. Such numbers have square roots whose value is larger than the original number.

$$\sqrt{.64} = .80 \qquad (.80)(.80) = .64$$

$$\sqrt{.81} = .90 \qquad (.90)(.90) = .81$$

$$\sqrt{.16} = .40 \qquad (.40)(.40) = .16$$

The square roots of decimal numbers can be found using Table A–2, although it is a little tricky. What has to be done is to find a number from 1 to 999 which when the decimal point is moved *two places to the left* equals the number whose square root is to be taken. When this is done, the answer is found by moving the decimal point of the number in the square-root column in Table A–2 *one place to the left*. Suppose the square root of 2.1 were desired. There are two numbers in Table A–2 that would seem to be connected in some way, 21 and 210. Which number to use can be determined by moving the decimal point two places to the left. For 21 the value obtained by doing this is .21, which is *not* the number whose square root is desired. For 210, however, moving the decimal point two places to the left yields 2.10, which is the number sought since the trailing zero does not affect the value. Therefore, 210 is looked up in Table A–2 and its square root is found to be 14.4914. By moving the decimal point one place to the left, the square root of 2.1 is found.

$$\sqrt{210} = 14.4914$$

$$\sqrt{2.10} = 1.44914$$

The square root of decimal numbers less than 1.0 can be found in the same way. Suppose the square root of .50 were desired. Table A–2 does not give values for numbers less than 1.0. The number 50, though, is in the table, and if its decimal is moved two places to the *left*, we have the number whose square root we are seeking, .50. According to Table A–2, the square root of 50 is 7.0711. To obtain the square root of .50, the decimal point in the answer for the square root of 50 is moved *one* place to the *left*.

$$\sqrt{50} = 7.0711$$

$$\sqrt{.50} = .70711$$

In general, then, the square roots for numbers less than 1.0 can be obtained by finding a number in the table which when the decimal point is moved an *even* number of places to the *left* results in the number whose square root is sought. The square root of this number is obtained and the decimal point in this square root is moved *half* the number of places to the *left* as the decimal point in the original number was moved to obtain the value of the square root being sought.

Square roots for numbers larger than 999 can also be obtained from Table A–2, but now the decimal point is moved to the right. Suppose that we wished to find the square root of 2100. The trick now is to find a number in the table which when the decimal point is moved two places to the right equals 2100. Again, the numbers 21 and 210 seem relevant. Moving the decimal point for the number 210 two places to the right yields 21,000, which is not the number in which we are interested. Moving the decimal point for the number 21 two places to the right, though, yields our number, 2100. Therefore, 21 is looked up in Table A–2 and its square root is found to be 4.5826. By moving the decimal point one place to the right, the square root of 2100 is found.

$$\sqrt{21} = 4.5826$$

$$\sqrt{2100} = 45.826$$

In general, then, the square roots of numbers whose values fall outside the range of Table A–2 (1–999) can be found in the following way. Find the number in the table which when the decimal point is moved an *even* number of places to the right or left yields the number whose square root is sought. Then move the decimal point of the number in the square-root column half the number of these places in the same direction as the decimal point was moved in getting the original number. The result is the square root of the number whose value lay outside the range of the table.

Question 3–3. Find the answers to the following problems:

a. $|-15| + \dfrac{4!}{2!} - \sqrt{111}$ b. $\left|\dfrac{-20}{5}\right| + \dfrac{(5!)(3!)}{(3!)(4!)} - \sqrt{2.25}$

c. $\sqrt{65,700}$ d. $\sqrt{7900}$

e. $\sqrt{.07}$ f. $\sqrt{.70}$

PROPORTIONS, PERCENTAGES, AND RATIOS

Quite frequently in social science, situations occur where it is necessary to communicate information about the size of a *subclass* of objects. The "objects" that are involved could be anything. Although it would be simple to specify the number of objects or the absolute size of the subclass, this often is not what is of interest. Suppose, for example, that we suspected that a particular company discriminated against women in hiring. The company tells us that it has 200 women employees. This information tells us the size of the subclass of concern but does not provide us with enough information to even begin to make an evaluation of whether the company discriminates against women in hiring employees. What is obviously needed is information on the size of the subclass relative to the total class of all employees.

This type of information can be given in a number of forms. One form is in terms of a proportion. A *proportion* is found by dividing the number of objects in a subclass by the total number of objects in the class. Thus, if the company has 500 employees, the proportion of women employed by the company is

$$\text{proportion of women employees} = \frac{\text{number of women employees}}{\text{total number of employees}}$$

$$= \frac{200}{500}$$

$$= .40$$

A similar procedure will give the proportion of male employees:

$$\text{proportion of men employees} = \frac{\text{number of men employees}}{\text{total number of employees}}$$

$$= \frac{300}{500}$$

$$= .60$$

An important point to note about proportions is that they vary in value between 0 and 1.0.

Many people dislike working with decimal numbers smaller than 1.0. They feel that it is easier to understand numbers that do not begin with a decimal point. Fortunately, for such people proportions can be converted to *percentages* that vary between 0 and 100. *Any* proportion can be converted to a percentage by multiplying the proportion by 100. Therefore,

$$\text{percentage} = (\text{proportion})(100)$$

Since the proportion of women employed by the company is .40, the percentage of women employees is

$$\text{percentage of women employees} = (.40)(100)$$
$$= 40\%$$

If we had started from the actual numbers of men and women employed, the calculation would be

$$\text{percentage of women employees} = \frac{\text{number of women employees}}{\text{total number of employees}} (100)$$

$$= \left(\frac{200}{500}\right)(100)$$

$$= (.40)(100)$$

$$= 40\%$$

Percentages and proportions communicate exactly the same information. Percentages do it on a scale from 0 to 100; proportions do it on a scale from 0 to 1.0. Percentages, of course, will not always get rid of places to the right of the decimal point. If the numbers of Democrats, Independents, and Republicans among the 500 employees of the company are, respectively, 192, 101, and 207, the proportions would be $\frac{192}{500} = .384$, $\frac{101}{500} = .202$, and $\frac{207}{500} = .414$. The corresponding percentages are 38.4%, 20.2%, and 41.4%.

A *ratio* evaluates the size of one subclass relative to another subclass. The number of men and the number of women employees each constitute a subclass of the total class of all company employees. A ratio gives the magnitude of one subclass relative to another. A ratio is formed by dividing the number of elements in one subclass by the number of elements in the other. The ratio of male to female employees in the company is

$$\text{ratio men to women} = \frac{\text{number of men employees}}{\text{number of women employees}}$$

$$= \frac{300}{200}$$

$$= 1.5$$

This tells us that among the company's employees, there are 1.5 men for each woman.

Question 3–4. At a particular college, there are 1200 Protestant students, 900 Catholic students, 500 Jewish students, and 400 students of other religious preferences.

a. Find the proportion of each category of religious preference among students at the college.
b. Find the percentage of each category of religious preference among students at the college.
c. At this college, what is the ratio of Protestant to Catholic students?
d. What is the ratio of Catholic to Jewish students?

e. What is the ratio of students who are Catholic, Protestant, or Jewish to students of other religious preferences?

SUMMATION SIGNS

The need to sum (add together) a set of values occurs repeatedly in statistics. In computing the average weight of the players on a baseball team, for example, the weights of all the players must be added together and the total sum of these weights divided by the number of players. In statistics it is conventional to symbolize the *score* of an object on a variable as X. The score of one particular object in a set of objects is distinguished by a subscript that gives a distinct number to each object in the set. Thus, the score of the first object in the set is symbolized as X_1, the score of the second object as X_2, and so on. The capital letter N is used to symbolize the number of objects in the set so that the score of the last object is symbolized as X_N.

In the case of finding the average weight of the baseball team (or the average score of any set of objects) it is necessary to sum the individual scores. This operation could be symbolized as follows:

$$\text{sum} = X_1 + X_2 + X_3 + X_4 + X_5 + X_6 + X_7 + X_8 + X_9$$

This notation is obviously very clumsy. To simplify matters, statistics makes use of a special sign, Σ, which is called the summation sign (the Greek letter, capital sigma). When the summation sign appears, it means that all the numbers symbolized by letters that appear to its right should be summed. Since X symbolizes the scores in a set of data, the *sum* of the weights of the players on the baseball team can be represented by ΣX. Thus,

$$\Sigma X = X_1 + X_2 + X_3 + X_4 + X_5 + X_6 + X_7 + X_8 + X_9$$

For any set of objects the sum of their scores on a variable is

$$\Sigma X = X_1 + X_2 + X_3 + \cdots + X_N$$

Sometimes it is necessary to be precise about which objects are involved in a summation. This is done by a subscript notation involving the letter i.

$$\sum_{i=1}^{N} X_i = X_1 + X_2 + X_3 + \cdots + X_N$$

The subscript i symbolizes the number for any object in the set. The values i is to take on for a particular summation process are indicated by the numbers above and below the summation sign. The points scored in a game by five members of a basketball team can be used to illustrate this style of notation.

Player Number	Points Scored
1	10
2	5
3	25
4	30
5	15

$\sum_{i=1}^{N} X_i$ would symbolize the sum of the points scored by all the players on the team. Its value is determined by

$$\sum_{i=1}^{5} X_i = X_1 + X_2 + X_3 + X_4 + X_5$$

$$= 10 + 5 + 25 + 30 + 15$$

$$= 85$$

The sum of the points scored by the first four players is represented as

$$\sum_{i=1}^{4} X_i = X_1 + X_2 + X_3 + X_4$$

$$= 10 + 5 + 25 + 30$$

$$= 70$$

The sum of the points scored by the last three players is represented as

$$\sum_{i=3}^{5} X_i = X_3 + X_4 + X_5$$

$$= 25 + 30 + 15$$

$$= 70$$

The use of summation notation allows the representation of the sum of the scores of a set of objects when the number of objects may not be known. When it is clear that summations are to be made over all the objects in a set, the subscript notation is often dropped and the sum represented by ΣX.

Question 3-5. The incomes of six people are:

Person Number	Income
1	$10,000
2	12,000
3	8,000
4	15,000
5	0
6	20,000

a. What is N?   b. What is X_3?
c. What is X_N?   d. Find ΣX.

e. Find $\displaystyle\sum_{i=4}^{5} X_i$.   f. Find $\displaystyle\sum_{i=2}^{5} X_i$.   _____

OPERATIONS
WITH SUMMATION SIGNS

Summations are always made algebraically (taking into account the signs of numbers). If $X_1 = 3$, $X_2 = -7$, and $X_3 = 2$, then

$$\Sigma X = X_1 + X_2 + X_3$$
$$= (3) + (-7) + (2)$$
$$= -2$$

If $X_1 = -5$, $X_2 = 0$, $X_3 = 3$, $X_4 = 0$, and $X_5 = 6$, then

$$\Sigma X = X_1 + X_2 + X_3 + X_4 + X_5$$
$$= (-5) + (0) + (3) + (0) + (6)$$
$$= 4$$

The manipulation of summation signs often looks clumsy but is usually fairly simple. Any constant inside a summation sign (to its right) may be brought outside the summation sign (to its left). Where k represents a constant,

$$\Sigma kX = k\Sigma X$$

If $k = 5$, and $X_1 = 3$, $X_2 = -2$, and $X_3 = 5$,

$$\Sigma kX = k(X_1) + k(X_2) + k(X_3)$$
$$= (5)(3) + (5)(-2) + (5)(5)$$
$$= 15 - 10 + 25$$
$$= 30$$

and

$$k\Sigma X = k(X_1 + X_2 + X_3)$$
$$= (5)[(3) + (-2) + (5)]$$
$$= (5)(6)$$
$$= 30$$

Where only a constant (k) is summed over a set of objects, the expression may be simplified.

$$\sum_{i=1}^{N} k = Nk$$

If $N = 3$ and $k = 7$,

$$\sum_{i=1}^{3} 7 = 7 + 7 + 7$$
$$= 21$$

Thus, $\sum_{i=1}^{N} k = Nk$, since obviously $Nk = (3)(7) = 21$.

Operations inside a summation sign are performed before summation takes place.

$$\sum_{i=1}^{N} X_i^2 = X_1^2 + X_2^2 + X_3^2 + \cdots + X_N^2$$

This is different from

$$\left(\sum_{i=1}^{N} X_i\right)^2 = (X_1 + X_2 + X_3 + \cdots + X_N)^2$$

Note that in the first equation the X's are squared first and then summed; in the second equation the X's are summed first and it is the sum that is squared. Since these operations produce different results with the same numbers, care must be taken to distinguish between

$$\sum_{i=1}^{N} X_i^2 \quad \text{and} \quad \left(\sum_{i=1}^{N} X_i\right)^2$$

The first of these quantities is spoken of as the "sum of the X's squared"; the second is referred to as the "sum of the X's *quantity* squared."

The summation sign itself may be *distributed* across terms. What this means is illustrated by the following equality:

$$\sum_{i=1}^{N} (X_i + Y_i) = \sum_{i=1}^{N} X_i + \sum_{i=1}^{N} Y_i$$

The equality is easily demonstrated.

$$\sum_{i=1}^{N} (X_i + Y_i) = (X_1 + Y_1) + (X_2 + Y_2) + (X_3 + Y_3) + \cdots + (X_N + Y_N)$$

Now the parentheses are removed and the X's and Y's are grouped separately.

$$\sum_{i=1}^{N} (X_i + Y_i) = \underbrace{X_1 + X_2 + X_3 + \cdots + X_N}_{\text{first group}} + \underbrace{Y_1 + Y_2 + Y_3 + \cdots + Y_N}_{\text{second group}}$$

The first group is $\sum_{i=1}^{N} X_i$ and the second group $\sum_{i=1}^{N} Y_i$. Therefore,

$$\sum_{i=1}^{N} (X_i + Y_i) = \sum_{i=1}^{N} X_i + \sum_{i=1}^{N} Y_i$$

The summation sign may be distributed across any number of terms that are being added or subtracted.

EQUATIONS

Although in this book relatively little use is made of formulas for purposes of proofs and derivations, it is necessary to be able to "follow" formulas and to perform elementary operations on them. In the broadest sense a formula is a specification of a "rule" to be followed. We shall be especially interested in formulas that involve an equals sign.

We shall begin the discussion of formulas that involve an equals sign with equalities. An equality is a statement that two expressions represent the same number, such as

$$2 + 2 = 4$$

The two expressions on each side of the equals sign are called *members* or *sides* of the equality. Equalities in which both sides *must* represent the same number are called *identities*. Some examples of identities are:

$$1 + 3 = 4$$
$$(a)(b + c) = ab + ac$$
$$(XY)(a + b + c) = aXY + bXY + cXY$$

These equalities must always be true regardless of what numerical values the symbols represent.

If it is possible to find some values for the symbols which will make the sides of an equality unequal, the equality is called a conditional equality or *equation*. Consider

$$X - 3 = 2$$

In this equation the two sides are unequal ($\neq$) if $X = 1$.

$$1 - 3 \overset{?}{=} 2$$

$$-2 \neq 2$$

They are unequal if $X = 2$:

$$2 - 3 \overset{?}{=} 2$$

$$-1 \neq 2$$

The equation is an equality only when $X = 5$.

$$5 - 3 = 2$$

$$2 = 2$$

Usually, it is desired to find the values of the variable in an equation that will make it an equality (both sides represent the same number). Finding these values is called *solving the equation*. In the following equation how can a value of X be found that will make the equation an equality?

$$5X + 8 = 32 - 7X$$

There are two rules for manipulating equations which help to solve simple equations:

1. Adding or subtracting the same quantity to both sides of an equation does not affect the equality of the equation.
2. Multiplying or dividing both sides of an equation by the same (nonzero) quantity does not alter the equality of the equation.

Consider

$$6 = 6$$

If a constant, say 2, is added to both sides, the equality of the expression is retained.

$$6 + 2 = 6 + 2$$
$$8 = 8$$

Subtracting 3,

$$6 - 3 = 6 - 3$$
$$3 = 3$$

Multiplying by 8,

$$(6)(8) = (6)(8)$$
$$48 = 48$$

Dividing by 4,

$$\frac{6}{4} = \frac{6}{4}$$

$$1\frac{1}{2} = 1\frac{1}{2}$$

These rules also hold true for symbols. Given $X = X$, add a:

$$X + a = X + a$$

Subtract b:

$$X - b = X - b$$

Multiply by c:

$$X(c) = X(c)$$

Divide by d:

$$\frac{X}{d} = \frac{X}{d} \qquad (d \neq 0)$$

The important point is that in all these operations the equality of the expressions still holds.

The strategy used in solving simple equations is to use the two rules to isolate the unknown variable (X) on one side of the equation and the known values (the numbers) on the other. When this is done, the value of the unknown variable that will make the equation an equality is obtained. The equation being considered is

$$5X + 8 = 32 - 7X$$

The strategy here will be to get all the unknowns (terms involving X) on the left side and all the numbers on the right. To get the term $7X$ to the left side, $7X$ can be *added* to both sides of the equation.

$$5X + 8 + 7X = 32 - 7X + 7X$$
$$12X + 8 = 32$$

The term 8 can be moved to the right side of the equation by *subtracting* 8 from both sides:

$$12X + 8 - 8 = 32 - 8$$
$$12X = 24$$

The value of X that will make the equation an equality can be obtained by dividing both sides of the equation by 12.

$$\frac{12X}{12} = \frac{24}{12}$$

$$X = 2$$

To see that the equation really is an equality when the value of X is equal to 2, the value of 2 is substituted for X wherever it occurs in the original equation.

original equation: $\qquad 5X + 8 = 32 - 7X$

substituting 2 for X: $\quad (5)(2) + 8 = 32 - (7)(2)$
$$10 + 8 = 32 - 14$$
$$18 = 18$$

Both sides are equal, so the value of 2 really does make the equation an equality. The solution to the equation, then, is $X = 2$.

Question 3–6. Solve the following equations:

a. $12X - 17 = 8X + 3$.

b. $(2)(3X + 5) + \dfrac{8X}{2} = (20)(2X - 4)$.

_____

A number of equations in statistics involve two variables. An example of such an equation is

$$Y = 5X + 1$$

Fortunately, it is usually not necessary to have to solve the equation for both unknowns $(X$ and $Y)$. It is necessary, however, to be able to find the value of Y given the value of X. Thus, if $X = 3$,

$$Y = 5X + 1$$
$$= (5)(3) + 1$$
$$= 15 + 1$$
$$= 16$$

If $X = 7$, then

$$Y = (5)(7) + 1$$
$$= 35 + 1$$
$$= 36$$

Question 3–7. Find the value of Y in the following equations; given $X = 4$:

a. $Y = 5(X + 3) + 50$.
b. $Y = 10X^2 + 3X + 7$.

_____

GRAPHS

Graphs are the basis for all pictorial representations in mathematics and also in statistics. An understanding of graphs will facilitate the comprehension of these representations. Graphs are drawn using a system of rectangular coordinates. The basis for this system is a set of axes. In two-dimensional graphs such as in Figure 3–1, the horizontal axis is called the *abscissa* or *X axis*. The vertical axis is called the *ordinate* or *Y axis*. The point at which the axes cross is called the *origin.*

Graphs usually involve two variables. One variable is represented on the abscissa and is usually symbolized by X, and the other variable is represented on the ordinate and is usually symbolized by Y. Numerical intervals are marked out on both axes from the origin as in Figure 3–2. By marking out positive values to the right on the X axis and up on the Y axis and negative values to the left on the X axis and down on the Y axis, both positive and negative numbers can be represented on both axes. Theoretically, the values extend on indefinitely in all four directions, although in practice only enough values are actually marked off on the axes that are necessary to represent a problem.

Points are plotted on a graph by giving their *coordinates* on the two axes. By working backward for the moment, the meaning of the coordinates of a point can be given. If any point on the graph is picked, such as

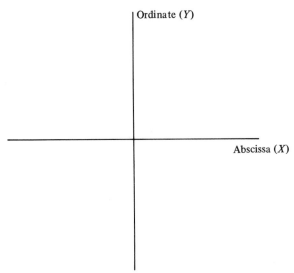

Ordinate (Y)

Abscissa (X)

Figure 3–1

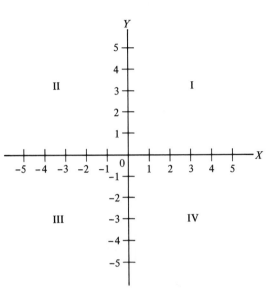

Figure 3–2

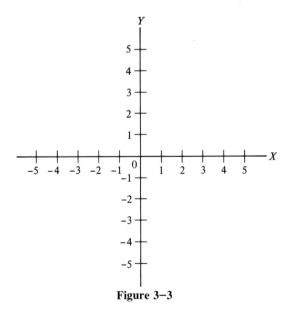

Figure 3–3

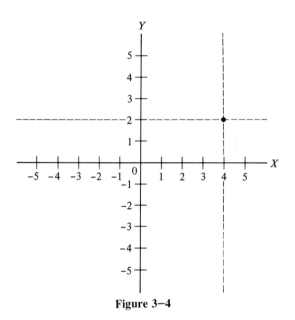

Figure 3-4

the one in Figure 3-3, its coordinates are found by drawing a vertical line perpendicular to the X axis through the point and a horizontal line perpendicular to the Y axis through it as in Figure 3-4. The value on the X axis where the vertical line intercepts it is the value of the point's X coordinate, in this case 4. And the value on the Y axis where the horizontal line intercepts it is the value of the point's Y coordinate, here 2. It is customary to specify coordinates by enclosing the values in parentheses giving the value for the X coordinate first. Thus, the coordinates for the point in Figure 3-4 are specified as (4,2). This procedure is reversed when the numerical values of the coordinates of a point are given and it is to be placed (plotted) on a graph.

The axes of a two-dimensional graph divide it into four quadrants. Conventionally, these quadrants are numbered as in Figure 3-2. It can be seen from Figure 3-2 that quadrant I consists of points whose X and Y coordinates must both be positive. Quadrant II consists of points whose X coordinate is negative and whose Y coordinate is positive. In quadrant III the X and Y coordinates must both be negative, and in quadrant IV the X coordinate is positive and the Y coordinate negative. Frequently, it is not necessary to use all four quadrants in a graph. It often happens that the points to be plotted have X and Y values that are all positive. In such cases it is only necessary that quadrant I be represented; the other three quadrants can be omitted.

Question 3–8. Plot the following points on the graph:

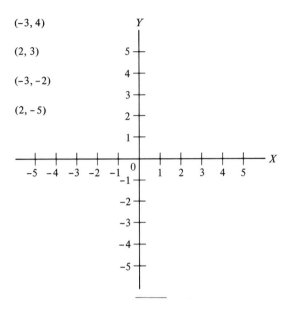

(-3, 4)

(2, 3)

(-3, -2)

(2, -5)

GRAPHING
A STRAIGHT LINE

A straight line is one of a large number of figures whose points, and therefore form, can be specified by a mathematical equation. The statistical technique of regression (Chapter 11) takes advantage of such equations to describe the nature of the relationship between two or more variables. The equation for a straight line has the form

$$Y = a + bX$$

In this equation X and Y are variables; they can take on different values, while a and b are constants for a particular line. By substituting actual numerical values for a and b, the equation precisely describes one particular straight line. Consider, for example, the equation

$$Y = 3 + 2X$$

In this equation $a = 3$ and $b = 2$ and the equation itself specifies one and only one straight line. From the equation it is possible to obtain the co-ordinates of various points that lie on the line. These points can then be plotted, and thus a graph of the line can be made.

The graph of any equation consists of all those points whose coordinates satisfy the equation; turn the equation into an equality. The coordinates of points that lie on the line can be obtained in the following way. Any arbitrary value is selected and substituted for X in the equation, and the corresponding value for Y is computed. These values for X and Y, then, are ones that make the equation into an equality. Consequently, the point whose coordinates correspond to these X and Y values must lie on the graph of the line. The process is then repeated with other arbitrarily selected values for X until the coordinates of enough points have been determined so that the line can easily be sketched in after these points have been plotted.

In Table 3–1, the first column contains arbitrarily selected numerical values that are to be substituted for X in the equation $Y = 3 + 2X$. In the second column are the computations that yield the corresponding Y value for each X value. In the third column are the resulting coordinates of points that lie on the line. In Figure 3–5, the coordinates in Table 3–1 are plotted on a graph. All the points whose coordinates have been calculated can be connected by a straight line. This is the line described by the equation $Y = 3 + 2X$.

Examination of Figure 3–5 shows that the line passes through (intercepts) the Y axis at the point +3, which is also the value of a in the equation for the line. In fact, every straight line will intercept the Y axis at the value of a in its equation. For this reason the a value in equations for straight lines is called the Y intercept. There are many lines, though, which could pass through this same Y intercept, for example, $Y = 3 - 1X$. In Figure 3–6, the line $Y = 3 + 2X$ is labeled L-1 and the line $Y = 3 - 1X$, L-2. Although the two lines pass through the same Y intercept, they are very different. In their equations, the only difference is in their values of b. For L-1, $b = 2$, and for L-2, $b = -1$. The b value is called the *slope* of the line and determines the angle of the line on the graph.

Table 3–1

X Value	Computations for Y Value from Equation: $Y = 3 + 2X$	Coordinates
1	$Y = 3 + (2)(1) = 5$	(1, 5)
2	$Y = 3 + (2)(2) = 7$	(2, 7)
3	$Y = 3 + (2)(3) = 9$	(3, 9)
4	$Y = 3 + (2)(4) = 11$	(4, 11)
-2	$Y = 3 + (2)(-2) = -1$	(-2, -1)
-3	$Y = 3 + (2)(-3) = -3$	(-3, -3)

Graphing a Straight Line

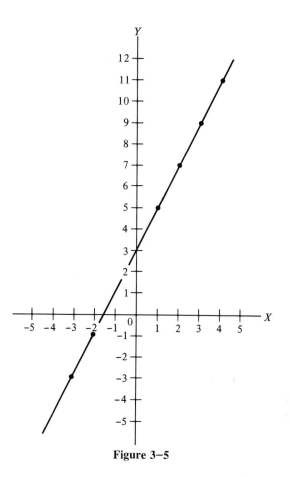

Figure 3–5

The slope of a line indicates the relation between the horizontal change (X axis) and vertical change (Y axis) that takes place when we move along the line. If on L-1 in Figure 3–6 we were to travel on the line from point (2, 7) to point (3, 9), this journey will have covered 1 unit of horizontal distance ($3 - 2 = 1$) and 2 units of vertical distance ($9 - 7 = 2$).

The horizontal change that takes place in journeying between two points on a line is represented as Δx (read "delta x") and the vertical change as Δy. The value of Δx is computed by subtracting the X coordinate of the starting point (X_{start}) from the X coordinate of the stopping point (X_{stop}) of the two points between which the journey was made.

$$\Delta x = X_{stop} - X_{start}$$

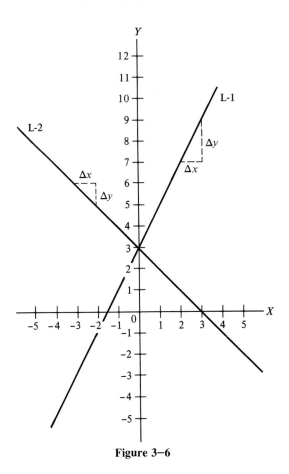

Figure 3–6

The value of Δy is computed by subtracting the Y coordinate of the starting point (Y_{start}) from the Y coordinate of the stopping point (Y_{stop}).

$$\Delta y = Y_{stop} - Y_{start}$$

For line L-1, the coordinates of the starting point were (2, 7) and the coordinates of the stopping point (3, 9). Therefore,

$$\Delta x = 3 - 2 = 1$$
$$\Delta y = 9 - 7 = 2$$

The slope of a line b is equal to $\dfrac{\Delta y}{\Delta x}$. For line L-1,

$$b = \frac{\Delta y}{\Delta x} = \frac{2}{1} = 2$$

This, of course, corresponds to the value of b in the equation for L-1.

It does not matter what two points on the line are chosen as starting and stopping points. The value of b will always be the same, regardless of what points are chosen, since a straight line does not change its "angle" anywhere along its length. The important thing about the slope is that it indicates for a line how the Y values will change for a given change in the X value. For L-1, the slope says that if the X value is increased by 1 unit the Y value will increase by 2 units.

For L-2 in Figure 3–6, the situation is quite different. If the starting point of (−3, 6) and the stopping point of (−2, 5) are chosen, then

$$\Delta x = -2 - (-3) = 1$$
$$\Delta y = 5 - 6 = -1$$

and

$$b = \frac{\Delta y}{\Delta x} = \frac{-1}{1} = -1$$

This, again, is the value of b in the equation for the line. This slope tells us that for line L-2 the relation between X and Y is such that if the X value is increased by 1 unit, the Y value will decrease by 1 unit. Obviously, the relation between X and Y values is quite different for L-1 and L-2, and it is the slope, the b value in the equation for each line, which specifies the relation.

In general, then, every line is determined by the value of its Y intercept (a) and its slope (b). Lines that differ from each other must differ in one or both of these values. Lines that have the same values for a and b are identical. If the equations for a group of lines have either the same a value *or* the same b value, they share a similarity. If the equations have the same a value, the lines all intersect the Y axis at the same point. The b value in the equation for a straight line specifies the angle of the line on the graph. Thus, all lines whose equations have the same b value are parallel.

Question 3–9. Plot the following lines on a graph:
a. $Y = -3 + X$.
b. $Y = -3 + 2X$.
c. $Y = 1 - 3X$.
d. $Y = 4 - 3X$.

NUMBER OF PLACES CARRIED
IN ARITHMETIC COMPUTATIONS

A common problem concerns how many decimal places must be carried through a set of arithmetic computations for the answer to be "accurate." Although a simple question, the answer is somewhat complex. This is because there is no single rule that is appropriate in all situations. Rather, it is necessary to understand what the numbers in each particular problem "mean."

In mathematics, numbers are symbols which get their meaning from their use in mathematical rules. For the purposes of mathematics, the numbers do not represent properties of the world. When these numbers are manipulated arithmetically, the meaning of the results comes from the rules involved. Consider the following operation:

$$\frac{7}{3} = X$$

The answer that is obtained for X gets its meaning from the following rule: X must be that number which when multiplied by 3 gives (precisely) the number 7. In terms of symbols,

$$3X = 7$$

When 7 is divided by 3, though, the result is 2.333. . . . That is, the digit to the right of the decimal point repeats itself to infinity. Now, if the question is asked how many places to the right of the decimal point must be carried, the answer is—an infinite number! Only an infinite number of places to the right of the decimal point will yield 7 precisely when multiplied by 3. Thus,

$$(3)(2.3) = 6.9$$
$$(3)(2.33) = 6.99$$
$$(3)(2.333) = 6.999$$
$$(3)(2.3333) = 6.9999$$

$$\cdot \qquad \cdot$$
$$\cdot \qquad \cdot$$
$$\cdot \qquad \cdot$$

Only

$$(3)(2.333. . .\text{infinitely}) = 7$$

What is important here is that each place to the right of the decimal point has meaning because each additional place brings the answer closer to 7. (Of course, for most practical purposes, we would probably be satisfied with an approximation to 7 of 6.99 or 6.999 and thus could stop at two or three places to the right of the decimal point.) When numbers are treated as symbols, as in mathematics, additional places to the right of a decimal point usually have meaning and it is only because of laziness or wanting to do something else in life besides arithmetic that they are not computed.

Most uses of numbers, though, treat them not just as abstract symbols but as representations of real-world properties. The number 3, for example, could stand for 3 dollars, 3 feet of height, 3 pounds of weight, 3 children, or a B in a course. When numbers are used to represent properties, there must be an assignment procedure which, in one way or another, assigns a particular number to represent a certain amount of the property. This assignment procedure is called *measurement*. If a person were 6 feet tall, the measurement procedure that assigned him the number 6 might have been having him stand next to a ruler and reading off the foot mark nearest the top of his head. *When numbers represent real-world properties they get their meaning not from their use in mathematical rules but from the measurement procedure.*

Suppose that we had a truck scale which measured the weights of

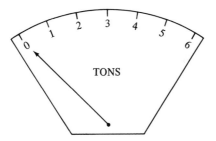

TONS

trucks in tons and that the scale looked as shown. The scale is marked off in units of tons, and let us assume that we can only determine a truck's weight to the nearest whole ton. Now suppose that a truck is driven onto the scale and the pointer on the scale is "near" 3 tons: that is, the pointer is closer to 3 than to any other number. We therefore assign a weight of 3 tons to the truck. The pointer will be closer to the number 3 than to any other number whenever the actual weight of a truck is from 2.6 tons to 3.4 tons. Thus, all trucks whose actual weight falls into the range from 2.6 tons to 3.4 tons will be assigned a scale weight of 3 tons by our measuring procedure. What we are saying here is that the measuring procedure is not sensitive enough to pick up differences of one tenth of a ton (200 pounds). If someone told us that a particular truck was weighed as 3 tons on this scale, we could not write this as 3.0 tons. To do so would imply that the truck did *not* weigh 2.6, 2.7, 2.8, 2.9, 3.1, 3.2, 3.3, or 3.4 tons but exactly 3.0 tons. This would be misleading, since each of the former weights would all yield a scale reading of 3 tons. Tacking on the zero in the tenths place implies a level of accuracy that the scale does not possess.

Now suppose that someone came in and said he could reliably "eyeball" the scale to one-tenth of a ton. Another truck is now driven on the scale and our expert scale reader says that the nearest one-tenth of a ton the scale points to is 4.3. Now what actual weights of trucks would yield a reading of 4.3?—any truck whose actual weight was anywhere in the range 4.26 to 4.34 tons, because within this range the pointer would be closer to the 4.3 tenth of a ton than to any other. Again, if we were told that a truck had a scale weight of 4.3 tons, it would be misleading to write this as 4.30 tons because the zero in the hundredths place implies a level of accuracy that the scale, even with our expert reading it, does not possess.

The point should now be clear. When numbers represent real-world properties, they get their meaning from the measurement procedure. The places in a number that have a meaning in terms of the measurement procedure are called *significant digits*. Numbers in places to the right of the significant digits are meaningless because there is no interpretation for them in terms of the measurement procedure. There is, of course, nothing to prevent us from writing out more places to the right on a number than its significant digits, but to do so is misleading because it implies a higher level of accuracy than the measuring instrument possesses.

Assume, for example, that with a ruler we could measure the length of an object to a hundredth of an inch. Using the ruler we obtain a reading of, say, 4.34 inches. The actual length of the object, though, could be anywhere in the range from 4.336 to 4.344 inches, and there is no way for us to discriminate among these possibilities using the ruler. Now say that we measure the same object using a micrometer which is accurate to the nearest ten-thousandth of an inch and obtain a reading of 4.3428 inches. If we were to report our reading with the ruler as 4.3400 inches, the last two zeros imply that the ruler is as accurate as the micrometer. And when we use the micrometer we see that this reading would be off. If we now examine the reading from the micrometer, we see that it would be obtained if the actual length of the object were anywhere from 4.34276 to 4.34284 inches. So again if we tried to add a zero to 4.3428 in the hundred-thousandth place, we would imply a level of accuracy that our measuring instrument does not have. No matter to what accuracy a measuring instrument is built, at some point the limit of its accuracy is reached. And the only numbers that are meaningful are the significant digits, which are numbers in places to which the measuring instrument gives a physical interpretation. Numbers in places beyond the significant digits are meaningless.

Let us return to weights of trucks. Suppose that we have measured four trucks to the nearest tenth of a ton and found them to weigh 3.1, 2.8, 3.4, and 3.2 tons. The average weight of the four trucks is now computed as follows:

$$\text{average} = \frac{3.1 + 2.8 + 3.4 + 3.2}{4} = \frac{12.5}{4} = 3.125$$

What has been done is perfectly correct arithmetically. But the 2 in the hundredths place and the 5 in the thousandths place are meaningless. If we took them seriously they would imply that the average is more accurate than the original numbers from which it was computed. This, of course, is nonsense. Therefore, even though it is correct arithmetically to carry out calculations far beyond the number of significant digits in the original numbers, it is useless to do so because numbers in places beyond the significant digits have no meaning.

We are now finally ready to answer the question of how many places should be carried in making arithmetic computations on numbers which represent measurements. In general, one more place than the number of significant digits of the original numbers should be carried through all computations to the end, at which time the "answer" should be rounded off to the number of significant digits of the original numbers. Thus, the average weight of the four trucks should be reported as 3.1 tons and not as 3.125.

This means that the number of places to be carried in computations will vary from problem to problem. If measurements were obtained from the micrometer (accurate to a ten-thousandth of an inch), then five places to the right of the decimal point should be carried, and at the end the answer rounded off to four. If measurements were obtained from a ruler that is accurate to a hundredth of an inch, then three places to the right of the decimal point should be carried, and at the end the answer rounded off to two places. When a particular set of numbers is given, the number of places to be carried is implied by the number of significant digits of the numbers. For the numbers 2.0, 5.0, 6.0, and 4.0, there is one significant digit to the right of the decimal point, whereas for the numbers 2, 5, 6, and 4, there are no significant digits to the right of the decimal point. For the first set, two places to the right of the decimal point must be carried; for the second, only one place to the right of the decimal point need be carried.

After setting forth these rules, we must now say that in various illustrations we will carry calculations to more places than are required. This is done primarily to show that the various arithmetic operations involved in a problem (such as addition, subtraction, multiplication, division, and taking the square root) were performed correctly. But the fact that more places are carried in order to check on the arithmetic does not make the additional places in the "answer" meaningful in terms of the original measurement procedure. The answer is not more "accurate" because additional places have been carried.

Question 3–10.

a. Calculate the average weight of the following trucks: 2.7 tons, 1.8 tons, 6.2 tons, 4.6 tons.
b. Convert the weight of each truck in part a to pounds and then calculate the average weight of the trucks in pounds (1 ton equals 2000 pounds). What digits in your answer are significant? Why?
c. Calculate the average age for the following people: 18, 20, 35, 25, 36.
d. The people in part c report their ages more precisely as: 18.4, 20.3, 35.4, 25.3, and 36.4 years. Use the more precise figures to calculate their average age.
e. Explain how reporting the average age in part c as 26.8 would be misleading.

_____

COMPUTATIONS
ON DISCRETE VARIABLES

There is a type of variable for which an "answer" may properly have more places than do the original numbers from which it is computed. This occurs when the numbers represent "counts." If, for example, we were to survey families and inquire as to the number of children in each family, we would obtain numbers such as 0, 1, 2, 3, and 4, which represent the number of children in each family. We would never obtain a number such as 1.5 or 2.78. This could not happen because there is no such thing as .5 or .78 of a child; the variable number of children can only take on whole or integer values which result from counting the number of children in each family. These counts are perfectly accurate. That is, if, say, a family is reported as having 3 children, there is no question as to whether this number represents 2.9, 2.99, 3.1, or 3.05 children. Thus the question as to what range of values would give rise to observing a particular measured value, such as in the case of weights of trucks discussed earlier, does not arise in this situation because it is impossible, even theoretically, for the variable number of children to take on fractional values.

A variable which cannot take on all possible values within a certain range, such as number of children per family, is called a *discrete* variable. While variables which, at least theoretically, can take on all possible values in a certain range are called *continuous* variables. Examples of continuous variables would be the weight of trucks or the height of people. For both variables there are practical limits outside of which we would never find an actual case, such as a person being 20 feet tall or a truck weighing 1,000,000 tons. But within the practical limits of either variable it is possible to at least theoretically conceive of obtaining any value, such as a person being 5.876534 feet tall. And it is only the imperfections of our measuring instruments that prevent us from identifying any value precisely.

Numbers which arise from counting on a discrete variable, then, are perfectly accurate and contain no measurement error. Consequently, if an arithmetic operation produces a number with more places than the original numbers have, the digits in these places will be meaningful. Thus we might find that among all the families that were surveyed there is an average of 2.35 children per family. The digits 3 and 5 now have meaning even though the original numbers from which the average was calculated did not have digits in the tenths or hundredths places. This is possible only because the original numbers were counts which are perfectly accurate. It should be noted, though, that the digits 3 and 5 in this example get their meaning from the arithmetic operations involved: If the number of children in all the families is summed and the sum divided by the number of families, the result is 2.35. It is, of course, impossible for any family to have 2.35 children. The digits in the "extra" places tell us that the distribution of children per family is such that a number slightly larger than 2 results when the average is computed. This implies something about the distribution of the variable number of children per family, not that the "typical" or "average" family has 2.35 children.

EXERCISES FOR CHAPTER 3

In Exercises 1 through 13, perform the indicated operations.

1. $4 + (-3) + 5 - 2$

2. $\left(\dfrac{3}{2}\right) \div \left(\dfrac{2}{5}\right)$

3. $\left(\dfrac{1}{7}\right)\left(\dfrac{3}{5}\right) \div \left(\dfrac{5}{7}\right)$

4. $a - (b - c)$

5. $(-2)(7 - X)$

6. $\sqrt{|-16|}$

7. $\left|-\dfrac{7!}{5!}\right|$

8. $\dfrac{(10!)(5!)}{(8!)(4!)}$

9. $\sqrt{\left(\dfrac{2}{7}\right)\left(\dfrac{2}{7}\right) + \left(\dfrac{5}{49}\right)}$

10. $\sqrt{\dfrac{1}{4} + \dfrac{4}{9}}$

11. $\sqrt{8.5}$

12. $\sqrt{.85}$

13. $\sqrt{1.36}$

14. The number of suicides in seven cities are: city 1, 5; city 2, 10; city 3, 3; city 4, 15; city 5, 7; city 6, 25; city 7, 30. Find the values of the following quantities:

    a. N

    b. ΣX

    c. $\sum\limits_{i=1}^{N} X_i$

    d. X_3

    e. X_N

    f. $\sum\limits_{i=3}^{6} X_i$

    g. $\sum\limits_{i=4}^{N} X_i$

15. If $X_1 = 2$, $X_2 = 4$, $X_3 = 1$, $X_4 = 5$, and $k = 3$, find $\sum\limits_{i=1}^{N} kX_i$

    In Exercises 16 through 20, solve the equations for X.

16. $X - 5 = 9$

17. $3X - 5 = 2X + 5$

18. $5X + 2(3X - 1) = -1 - 5(X - 3)$

19. $12(\tfrac{1}{3}X - 14) + 18 = 10(\tfrac{4}{5}X - 5)$

20. $4X + \tfrac{1}{2}(X + 1) - 2(X + 2) = 24$

21. Plot the following points on a graph:

    $(-4, 8)$ $(1\tfrac{1}{2}, 5\tfrac{1}{2})$ $(-2, 0)$

    $(-2, 9)$ $(1, 5)$ $(-2, -3)$

    $(0, 9\tfrac{1}{2})$ $(-1, 4\tfrac{1}{2})$ $(2, 7)$

$$(1, 9) \qquad (-2, 3) \qquad (-2, 1)$$
$$(2, 8) \qquad (-2, 2)$$

In Exercises 22 through 26, plot the lines on a graph.

22. $Y = 4 + \frac{3}{2}X$
23. $Y = 4 - \frac{3}{2}X$
24. $Y = -3 + 2X$
25. $Y = 2 + 2X$
26. $Y = 2 - 2X$
27. Calculate the average weight of the following people: 156, 138, 119, 162, 143, 175.
28. In a 100-yard race, the five contestants had the following times: 9.9, 10.4, 10.6, 10.7, 11.2. Calculate the average time of the runners.
29. In a survey of 500 people, 50 classified themselves as belonging to the upper class, 75 to the upper-middle class, 200 to the middle class, and 175 to the working class. What proportion of people in the survey are in each class? What percentage of people in the survey are in each class? What is the ratio of upper-class to working-class people? Of middle-class to upper-class people?

ANSWERS TO
QUESTIONS IN CHAPTER 3

3–1: a. 11   b. −240
       c. −14   d. 15X
       e. 2   f. 14.5

3–2: a. $-\dfrac{2}{15}$ b. $\dfrac{5X + 6Y}{4XY}$

       c. $\dfrac{2acf}{3bdg}$ d. $\dfrac{5}{16}$

       e. $\dfrac{1}{4}$

3–3: a. 10.46   b. 7.5
       c. 256.32   d. 88.882
       e. .26457   f. .83667

3–4: a. .40, .30, .167, .133   b. 40%, 30%, 16.7%, 13.3%
       c. 1.33   d. 1.8
       e. 6.5

3–5: a. 6   b. 8000
       c. 20,000   d. 65,000
       e. 35,000   f. 35,000

3–6: a. $X = 5$ b. $X = 3$

3–7: a. 85   b. 179

3–8:

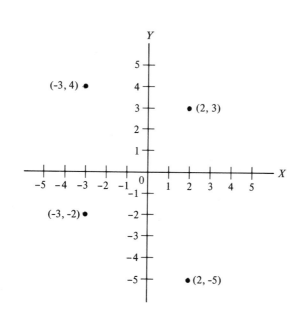

3–9:

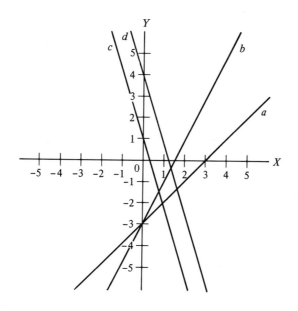

3–10: a. 3.8 tons, not 3.825 tons.
b. 7600 pounds, not 7650 pounds.
c. 27, not 26.8.

d. 27.2, not 27.16.

e. Part d shows how the people in part c could have an average age of 27.2 years if they report their age to one-tenth of a year. Thus, to report an average age of 26.8 for them in part c would be incorrect.

4
Measuring
Central Tendency

Data are usually collected with an eye toward investigating a problem the researcher has in mind. Collecting the data is a very important process and much care must be devoted to it. It is, though, only the first step. Data cannot speak by themselves. The data must be manipulated, analyzed, before they will yield the information they contain on the problem in a comprehensible form. As we have seen, summarizing the information in a set of data so that it is comprehensible is the function of descriptive statistics.

Which descriptive statistics are to be calculated depends upon what characteristics of the data the researcher is interested in and the level of measurement of the data. Because measurements at the nominal and ordinal levels do not contain as much information as measurements at the interval or ratio levels, there are different descriptive techniques at each level so as to maximize the information that is contained in the descriptive statistic. Some techniques for summarizing nominal and ordinal data will be taken up in Chapters 9 and 10. The problems of interval and ratio level data will be the subject of the next few chapters.

Usually analysis of data begins by obtaining a description of some basic characteristics of the data. At the interval and ratio levels of measurement, where it is known that the units of measurement are equal in size, the basic characteristics involve the relationship between the data and the scale of measurement. It will be recalled that the scale of measurement refers to the way in which numerical values are attached to a variable. If, for example, length is the variable, then inches might be the units of the scale of measurement and a ruler might be used to assign numbers to objects. When we have a description of the relationship between the data and its scale of measurement, a good idea of the nature of the data is obtained.

The relationship between a set of data and its scale of measurement is in part described by two important characteristics. The first is where the data tend to be located on the scale of measurement. If a researcher obtained the age of workers at an automobile assembly plant (a ratio level variable), he would want to know whether the workers tend to be older, say around 50, or younger, say around 30. Finding where the data tend to be located on the scale of measurement is called *finding the central tendency* of the data. The second characteristic is to understand how the data are spread out across the scale of measurement. What is desired here is to know how much variation there is in the data. In the case of the age of the workers, little variation would mean that the ages of the workers all tend to be lumped closely around the central tendency in the data; the age of all workers tends to be close to one value. A great deal of variation would mean that the ages of the workers tend to be widely spread out on the scale of measurement and are not all close to one value. The variation in a set of data is called *dispersion*. The characteristics of central tendency and dispersion are important both as summarizing measures that describe a set of data and make it comprehensible and as building blocks for more sophisticated statistical techniques. In this chapter we take up the problem of measuring central tendency. Measures of dispersion are considered in Chapter 5.

The property on which information is sought is called the *variable*. The objects that are studied, workers in the case of the assembly plant, are called *observations* in statistics. We have seen that statistics always attaches numbers to its variables. In the case of the age of the workers, numbers are already attached, the years of age. The numerical value of an observation on a variable is called a *score*. If a worker was 45 years old, his score on the variable of age would be 45.

Assume, now, that we are interested in investigating the education of the workers at the automobile assembly plant. By questioning each of the 1000 workers we could obtain the number of years of education for each. The numerical value of years of education for a worker will be his score on the variable of education. This procedure would give us a collection of 1000 numbers. It is unlikely that we could obtain any understanding of the nature of education among the workers by examining the 1000 scores individually. Some way is needed to summarize the information present among all 1000 scores. One way to do this is to draw a "picture" of the location of the scores on the scale of measurement.

FREQUENCY DISTRIBUTIONS
AND GRAPHICAL PRESENTATIONS

In order to draw a picture, it is necessary first to tabulate the number of times each value of education occurs in the data. It would be cumbersome to have to deal with 1000 scores, so a sample of 50 scores will be selected for illustrative purposes. The scores for 50 workers are tabulated in Table 4 – 1. This table of tabulations is called a *frequency distribu-*

Table 4-1
Frequency Distribution for 50 Worker's
Scores on Education

Years of Education	Frequency	Years of Education	Frequency
0	0	8	9
1	1	9	7
2	0	10	3
3	1	11	6
4	0	12	4
5	3	13	3
6	4	14	2
7	6	15	1

tion. In such a table, "frequency" refers to the *number of times* that a particular value of the variable occurs in the data. In Table 4-1 each possible value of education is listed and next to each value is its frequency, the number of times the value occurs in the set of 50 scores. From this table it can be seen, for example, that none of the 50 workers had 4 years of education while 3 workers had 13 years of education. From a frequency distribution a special type of graph, called a *histogram,* can be drawn which pictorially presents this information.

A histogram is constructed by placing the variable on the horizontal or X axis (the abscissa) of the graph and counting off how many occurrences there are of a particular value of the variable, the frequency, on the vertical or Y axis (the ordinate) of the graph. Notice that in Figure 4-1, the axes are labeled, the X axis as "Years of Education" and the Y axis as "Frequency." And the graph as a whole is given a title. In any table or graph everything ought to be clearly labeled and a title provided to specify exactly what the contents of the table or graph are. These details are important so that an unfamiliar reader can figure out what the table or graph is about. Over each value of the variable a vertical bar is drawn whose height corresponds to the frequency of occurrence of that value among the scores. The histogram of the scores for the 50 workers on the variable of education is shown in Figure 4-1.

The variable "Years of Education" in Figure 4-1 has been treated as if it were a *discrete* variable. A discrete variable is one in which all possible numerical values within a given range of values cannot occur. Families, for example, can have 0, 1, 2, 3, or more *integer* number of children. A family cannot have 1.34 children. Thus, the nature of some variables is such that it is nonsensical for them to take on all possible numerical values. In contrast, *continuous* variables can take on all possi-

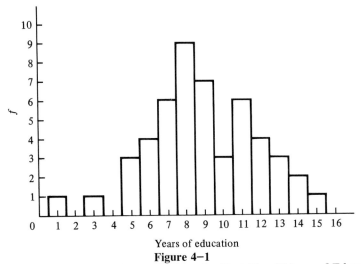

Years of education
Figure 4–1
Histogram of a Sample of 50 Workers on the Variable of Years of Education

ble values within a certain range. Age is a continuous variable because it makes some sense to think of someone as, say, 20.576843 years old, although we are unlikely to ever have any reason to specify a person's age so precisely.

In the case of the example in Figure 4–1, it may be possible for someone to have 11.4 or even 11.75 years of education. But we will assume that when the workers report their education, they round off to the nearest year. Thus, two workers who had actually completed 10.25 and 9.75 years of education would both report 10 years of education. Since the actual data reported by the workers is in integer years of education, we can select midpoints between integer years as limits for the bars on the graph (i.e., .5, 1.5, 2.5, 3.5, and so on). It is customary to choose as the limits of intervals numerical values that do not actually occur in the data. This ensures that each score can be uniquely placed in only one interval.

Since education was reported to the nearest year, all scores that appear in a particular interval are treated as having the same value and plotted on the same place on the histogram. Thus, the bar drawn above 8 years of education represents the reported scores of nine workers whose actual years of education may really not be precisely the same but which may vary within the range from 7.5 to 8.5. What is happening here is that the workers in reporting their years of education are grouping the data for us into 1-year intervals. In dealing with continuous variables, some sort of grouping of scores into intervals is usually necessary.

Otherwise, it is at least theoretically possible to so finely determine scores that the score of each observation would be different, and each would require its own distinct bar of one unit height on the graph. Grouping scores makes it possible for the height of the bars to present differences in the frequency of occurrence of scores that fall into particular ranges.

One is free to choose intervals of any size in making a frequency distribution or histogram. But two fundamental rules must be observed in constructing intervals. The first is that the intervals must be *mutually exclusive*. This means that intervals must not overlap. Two intervals that went from 10 to 11.2 and from 10.8 to 12.5 overlap by half a unit. A score such as 11 would be counted twice, once in the 10-to-11.2 interval and once in the 10.8-to-12.5 interval. Such a procedure would misrepresent the data, since some scores are given more "weight" than others. Any measurement procedure, not only frequency distributions and histograms, must ensure that its categories are mutually exclusive.

The second rule is that the intervals must be *exhaustive*. This means that *all* the scores will be presented somewhere in the intervals. If this rule is violated, certain scores are being ignored or discarded.

Within these rules, intervals of any size may be chosen. The scores in Table 4–1, for example, could be grouped into three intervals, as shown in Table 4–2. The corresponding histogram for this frequency distribution is presented in Figure 4–2. The first thing to note about these intervals is that they are not of equal size. Unequal intervals can greatly distort the nature of data. The age of adults, for example, might be grouped into three intervals, called young, medium, and old. If the young interval went from 21 to 50, medium from 51 to 60, and old from 61 to 80, there would appear to be an unusually large number of young adults. On the other hand, if the intervals went young 21 to 30, medium 31 to 40, and old 41 to 80, there would appear to be an unusually large number of older people.

Sometimes the variable itself presents "natural" limits for the selec-

Table 4–2
Frequency Distribution for Observations
in Figure 4–1, Grouped into Three Intervals

Interval (years)	Frequency
.5 – 8.5	24
8.5 – 12.5	20
12.5 – 15.5	6

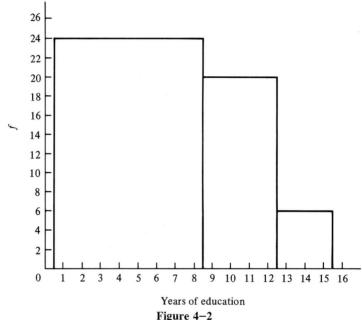

Years of education

Figure 4–2

Histogram of 50 Workers on Education Grouped into Three Intervals with the Height of the Bars Proportional to Frequencies

tion of intervals. In the case of our present example, it may make sense to distinguish between workers who had a grade school, high school, or some college education. This led to the construction of the unequal intervals of .5 to 8.5, 8.5 to 12.5, and 12.5 to 15.5. If a variable does not present such "natural" limits for intervals, it is usually better to stick to intervals of equal size.

There are a number of problems in grouping data into intervals. The most serious is that whenever data are grouped, information is lost. And the larger the intervals into which the data are grouped, the more information is lost. An example of this is seen in Figure 4–2. Looking at this histogram it can be seen that there are 24 workers who had between .5 and 8.5 years of education. But from this histogram it cannot be determined whether these workers as a group tended to have 6, 7, or 8 years of education or whether they tended to have only 1, 2, or 3 years of education. The information that was available to answer this question was lost when the data were grouped into larger intervals. In Figure 4–1 this information is preserved, and it can be seen that the 24 workers tended to have 6, 7, or 8 years of education and not 1, 2, or 3. Thus, whenever scores are grouped into larger intervals, information is lost about the data, since every score in an interval is treated in exactly the same way regardless of its location within the interval.

A related problem is deciding on how many intervals there shall be. The fewer the number of intervals, the larger each interval must be and the more information is lost. Many intervals, while preserving more information, may leave gaps in the histogram, where few or no scores occurred in the data and present so many bars on the graph that it is difficult to decipher. It is necessary to arrive at a compromise between these competing problems.

The last problem in grouping scores is the decision to let the frequency of an interval be represented by the height of the bar in the histogram. If the intervals are not of equal size, the size of the bars can be misleading to the reader. The size of the first bar in Figure 4–2, for example, is more than twice the size of the second bar, even though the frequency for the first interval is 24 and the second, 20. This problem can be overcome by making the *area* of the bars proportional to frequencies rather than their heights. This is done in Figure 4–3 for the frequency distribution in Table 4–2.

Since the first interval covers a range of 8 units (from .5 to 8.5) and there are 24 workers in this interval, the height of the bar is 3.0 (24 divided by 8). This procedure makes the area of each bar represent the frequency of the scores occurring in each interval.

The selection of the number of intervals, their limits, and whether

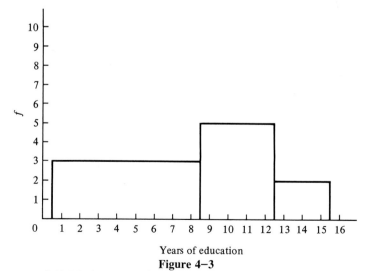

Years of education
Figure 4–3
Histogram of 50 Workers on Education Grouped into Three Intervals with the Area of the Bars Proportional to Frequencies

height or areas should represent frequencies in a histogram are ultimately arbitrary decisions. There are no firm rules about these things, and the best guide is common sense combined with knowledge about the nature of the variable. The point of frequency distributions and histograms is to communicate information about the nature of the variable. Decisions should be made which allow an effective presentation and do not distort the data.

An important point about histograms which will be of great use later is that it is possible to let the areas of all the bars in a histogram be equal to 1.0. This is true only for histograms where the height of the bars represents the frequencies. In Figure 4–1, the height of the bar for 5 years of education is 3. Since the total number of scores represented in the histogram is 50, the area of this bar is $^3/_{50}$, or .06 of the total area of all the bars. The height of the bar for 8 years of education is 9. Thus its proportion of the total area is $^9/_{50}$, or .18. If the proportional area for each bar in the histogram is computed in this way, we would find that the sum of all these proportional areas would be 1.0. Thus, the total area of the bars in a histogram can be set equal to 1.0 and the height of each bar represented as a proportion of the total area.

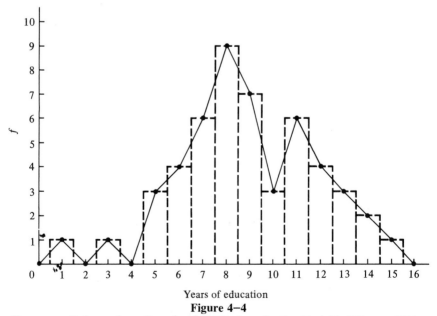

Years of education

Figure 4–4

Frequency Polygon for a Sample of 50 Workers for the Variable "Years of Education"

Another way of graphically presenting a frequency distribution is through a *frequency polygon*. A frequency polygon is constructed from a histogram in which height (not area) represents frequency. A frequency polygon is constructed by connecting the midpoint at the top of each bar in a histogram by a continuous line. The height of the line at any point then represents the frequency of occurrence of the value of the variable at that point. Figure 4–4 shows how the histogram of Figure 4–1 is made into a frequency polygon.

The area under a frequency polygon can also be treated as equal to 1.0. In Figure 4–4, some areas of the bars are cut off by the line and lie outside the frequency polygon. These areas, however, are compensated for by the areas inside the line which are not part of the bars. Since the area of the bars in a histogram can be treated as equal to 1.0, the area under a frequency polygon can also be treated as approximating 1.0. Graphs of similar logic to the frequency polygon will be used later to pictorially represent the distribution of large populations of scores.

Histograms and frequency polygons give some idea of where and how scores are located on the scale of measurement. They indicate where scores tend to be centrally located and give a rough indication of the dispersion in the scores. From these graphs a better understanding of the data can be obtained than by inspecting each score individually. But graphical presentations do not provide measures of central tendency and dispersion which can be mathematically manipulated. Such measures are needed to precisely summarize and compare characteristics of data both for descriptive purposes and as information for more advanced statistical techniques.

THE MEAN
AS A MEASURE OF
CENTRAL TENDENCY

One familiar measure of central tendency is the arithmetic *mean*. Sometimes this measure is called the *average*, but the term mean will be used here. The mean is simply the algebraic sum of all scores divided by the number of observations. It is customary in statistics to use the symbol X to represent the *scores* of observations on a variable. This is also referred to as the *raw score*. The mean is symbolized by the Greek letter μ ("mu," pronounced "mew") for the case of a population. If a sample is being dealt with, the mean is symbolized by $\overline{X}$. Since it is rare to directly calculate the mean of a population, the formula for the mean will be stated in terms of the symbol for a sample, although the same arithmetic operations are involved in both cases.* Thus,

$$\bar{X} = \frac{\Sigma X}{N}$$

This formula is easily memorized. But it is essential not only to know formulas but to understand what it is they are supposed to do. Formulas indicate what operations are to be performed upon a set of scores. This formula indicates that if we wish to calculate the mean for a sample, the scores for the observations are to be summed (ΣX) and then divided by the total number of observations in the sample.

Assume that we wish to know the mean winning percentage of the popular vote for presidential candidates in the United States in the twentieth century. The variable here is percentage of total popular vote. The scores are the percentages of popular votes obtained by a particular winner in a presidential election.

Observation	Percentage of Popular Vote (X)
1900 (McKinley)	51.7
1904 (T. Roosevelt)	56.4
1908 (Taft)	51.6
1912 (Wilson)	41.9
1916 (Wilson)	49.3
1920 (Harding)	60.3
1924 (Coolidge)	54.0
1928 (Hoover)	58.2
1932 (F. Roosevelt)	57.4
1936 (F. Roosevelt)	60.8
1940 (F. Roosevelt)	54.7
1944 (F. Roosevelt)	53.4
1948 (Truman)	49.6
1952 (Eisenhower)	55.1
1956 (Eisenhower)	57.4
1960 (Kennedy)	49.7
1964 (Johnson)	61.1
1968 (Nixon)	43.4
1972 (Nixon)	60.7
	$\Sigma X = 1026.7$

*See pages 54 to 58 for an explanation of the use of summation notation. An understanding of summations is essential before proceeding.

By adding together the winning percentage of the popular vote for each observation, we find that $\Sigma X = 1026.7$. To find the mean, the sum of the raw scores (ΣX) is divided by the number of observations (N).

$$\bar{X} = \frac{\Sigma X}{N} = \frac{1026.7}{19} = 54.04$$

This mean shows that the "typical" winning percentage of the popular vote in presidential elections is about 54%. It also indicates where the raw scores are located on the variable of "percentage of popular vote." The mean tells us that the raw scores are located not around 65% nor around 45% but at about 54%. Thus, we have located the central tendency of this distribution of scores on the scale of measurement.

Question 4–1. The amounts of money that a sample of people contributed to political campaigns in the last election were, in dollars: 1, 2, 5, 25, 10, 0, 2, 0, 5, 10.

a. What is the total sum of money contributed by these people, ΣX?
b. What is the money contributed by the seventh person, X_7, in the sample?
c. Find N for the sample.
d. Find $\bar{X}$.

The mean has the very nice property that the deviations of the scores from the mean algebraically sum to zero. A deviation is the difference obtained when the mean is subtracted from a raw score. The lowercase letter x is used to represent a deviation. Thus,

$$x_i = X_i - \bar{X}$$

where X_i is the raw score for the ith observation and x_i is the corresponding deviation for the ith observation. Since it is usually pretty clear which deviation goes with which raw score, the subscript is usually omitted.

Assuming the following ages for five people, the calculation of their mean age is as follows:

Person	Age (X)
1	25
2	30
3	20
4	35
5	40
	$\Sigma X = 150$

$$\overline{X} = \frac{\Sigma X}{N} = \frac{150}{5} = 30$$

Therefore, for observation 1:

$$x = X - \overline{X} = 25 - 30 = -5$$

The deviation for the first observation is -5. In the same way, the deviations for the other observations can be calculated and the *deviations* summed. The summing of deviations is represented by Σx.

Person	Age (X)	Mean ($\overline{X}$)	$x = X - \overline{X}$
1	25	30	−5
2	30	30	0
3	20	30	−10
4	35	30	5
5	40	30	10
	$\Sigma X = 150$		$\Sigma x = 0$

Notice how the various pieces of information are laid out and clearly labeled. This may seem somewhat pedantic, but it is useful to lay out all problems in a similar way so that other people can see what manipulations are taking place. (This is especially important on examinations.)

Each deviation (x) was obtained by algebraically subtracting the mean $(\overline{X})$ from the raw score for each observation (X). When these deviations are algebraically summed, the result is zero. This is a mathematical property of the mean and *must* be true if the mean has been calculated correctly. It is sometimes useful to calculate Σx to see if it comes out to be zero as a check on whether the mean has been calculated correctly.

Question 4–2. Use the mean you calculated in Question 4–1 to compute the x for each observation and check to see if the Σx is equal to zero.

It is useful to note some equivalent notational forms. Since $x = X - \overline{X}$, then

$$\Sigma x = \Sigma(X - \overline{X})$$

The substitution of x for $(X - \overline{X})$, or vice versa, will prove to be useful.

Another mathematical property of the mean is that the sum of the squared deviations from the mean (Σx^2) is less than the sum of the squared deviations about any other number. For the example of the ages of five people:

Person	Age (X)	Mean ($\bar{X}$)	$x = X - \bar{X}$	x^2
1	25	30	−5	25
2	30	30	0	0
3	20	30	−10	100
4	35	30	5	25
5	40	30	10	100
	$\Sigma X = 150$		$\Sigma x = 0$	$\Sigma x^2 = 250$

If we choose any number besides the mean, the Σx^2 will *always* be larger than 250. If, for example, the deviations had been calculated about 20 instead of the mean of 30, we would have:

Person	Age (X)	20 (not $\bar{X}$)	$x = X - 20$	x^2
1	25	20	5	25
2	30	20	10	100
3	20	20	0	0
4	35	20	15	225
5	40	20	20	400
	$\Sigma X = 150$			$\Sigma x^2 = 750$

Note again that $\Sigma x^2 = \Sigma (X - \bar{X})^2$.

Question 4–3. Find the Σx^2 for Question 4–2. What is $\Sigma (X - \bar{X})$? What is $\Sigma (X - \bar{X})^2$?

OTHER MEASURES
OF CENTRAL TENDENCY

The mean is the most commonly used measure of central tendency and the one that is most frequently employed in the development of various statistical techniques. In some situations, however, the mean can be a misleading indicator of central tendency because of its sensitivity to extreme values. This can be illustrated in the example of the ages of five

persons. If the age of the last person is changed from 40 to 90, the calculation for the mean will proceed as follows:

Person	Age (X)
1	25
2	30
3	20
4	35
5	90
	$\Sigma X = 200$

$$\bar{X} = \frac{\Sigma X}{N} = \frac{200}{5} = 40$$

In the original example, the age for observation number 5 was 40 and
the mean was 30. By changing the value of the last observation to an
extreme score the mean as a measure of central tendency shifted by 10
years! If the ages are plotted on a graph, it can be seen that the new
mean misrepresents the location of the bulk of the observations. In Figure 4–5, 4 of 5 of the observations lie to the left of the mean and only
one to the right. That is hardly the "central location" of the observations.

The *median* is a measure of central tendency that overcomes the
problem of extreme values. The median is defined as the value in a distribution below which and above which equal numbers of *observations*
fall. If there are an odd number of observations in the distribution, the
score of the observation in the middle is defined as the median. If there
are an even number of observations, the median is defined as the arithmetic mean of the values of the two middle observations.

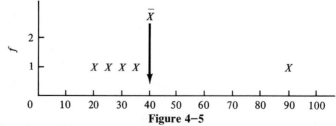

Figure 4–5
Location of Mean of Age of Five People with One Extreme Score

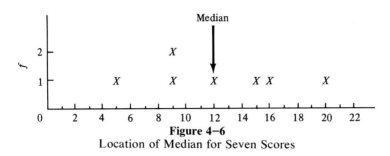

Figure 4–6
Location of Median for Seven Scores

To compute the median, it is first necessary to order the scores in ascending order from the smallest to the largest value. Once this is done, we count sequentially, starting from the smallest value, the occurrence of *observations* (not values) until the middle observation is reached – the one below which and above which are equal numbers of observations. Consider the following seven numbers: 5, 9, 12, 16, 9, 15, 20. These numbers can be ordered from smallest to largest by plotting them on a graph as in Figure 4–6. There are seven numbers here, so the fourth number will be the middle one. By counting off the occurrence of the observations on the graph, we find that the fourth observation has the *value* of 12. Equal numbers of observations lie on either side of this value: 3 and 3. Here are 10 numbers that are plotted in Figure 4–7: 20, 6, 7, 3, 19, 14, 3, 14, 12, 10. Because there are an even number of observations, 10, the median will be the arithmetic mean of the values of the two middle observations. In this case the middle observations will be the fifth and sixth ones. When observations are counted off starting from the one with the smallest value, these middle observations are found to have the values of 10 and 12. When added together and divided by 2, we obtain a value for the median of 11. Note again that equal numbers of observations fall above and below this point.

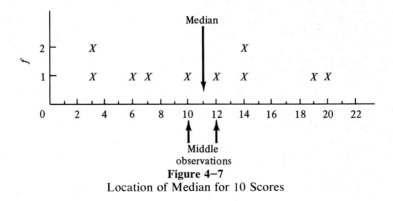

Figure 4–7
Location of Median for 10 Scores

Question 4–4. Compute the median for the sample of people who contributed money to political campaigns in Question 4–1. These values were: 1, 2, 5, 25, 10, 0, 2, 0, 5, 10. _____

Let us now compare what happens to the median in the case of the original example of the ages of five persons and when the value of age for the last person is changed to 90. In Figure 4–8 both situations are graphed. In both cases there are five observations, so the median will be equal to the value of the third observation. In both situations this turns out to be 30. Thus, we see that the median is relatively (and in this particular case, perfectly) unaffected by one observation taking on an extreme value.

A third measure of central tendency, the *mode* is frequently mentioned, although it is not used as much as the mean or the median. The mode is simply the most frequently occurring score among a distribution of scores. In Figure 4–6 the mode is 9 because that value occurs twice and all other values occur only once or not at all. In Figure 4–7 the distribution is *bimodal;* that is, there are two peak values which occur with equal frequency, 3 and 14. The number of peaks in a distribution, the

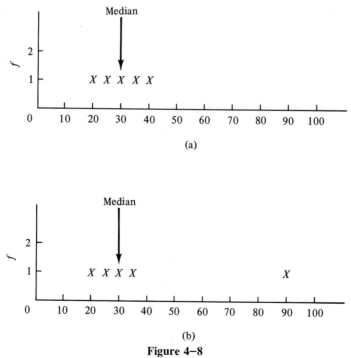

(a)

(b)

Figure 4–8
Effect upon the Median of an Extreme Score

points at which observations occur most frequently, determine whether the distribution is unimodal, bimodal, trimodal, or multimodal, according to whether there are one, two, three, or more peaks in the distribution.

The mode is useful when information about the peak frequency of occurrence is needed. In determining the number of seats in a concert auditorium, for example, it may be better to plan for the peak frequency of concert attendance rather than for the mean or median attendance. It is also useful to report the modes when a distribution has two or more modes. In such situations the mode loses effectiveness as a measure of central tendency since there are two or more values for it. But the mean and the median would also lose effectiveness as measures of central tendency in such situations, and reporting the modes would signal this fact. In general, though, the mean or median is a more useful measure of central tendency if a distribution is unimodal.

Of the other two measures of central tendency, the mean and the median, neither can be said to be "the best" as a descriptive measure of central tendency. There are times when the mean is inappropriate and the median should be used. The mean, however, takes into account more information about the data than the median. The mean incorporates information about the distance between scores; the median does not. Therefore, where *appropriate,* the mean is the preferred measure of central tendency.

Generally, the mean is appropriate for unimodal distributions that are symmetric in shape. Sometimes nonunimodal distributions, such as bimodal or rectangular distributions, occur in which the idea of central tendency does not have a clear meaning. Fortunately, almost all the distributions that we shall encounter are unimodal in shape.

Many unimodal distributions, however, are not symmetric but are "skewed" in shape. A skewed distribution is one that has considerably more extreme scores in one direction than in the other. Figure 4–9 shows the difference between a symmetric unimodal distribution and skewed unimodal distributions. In skewed distributions the direction of the *skewness* refers to the direction toward which the long tail is extended. If this is the direction in which increasing values occur (usually to the right), the distribution is said to be *positively skewed.* If it is the direction in which decreasing values occur (usually to the left), the distribution is said to be *negatively skewed.*

If a distribution is highly skewed, the median is a more appropriate measure of central tendency than the mean. This is because the median is less affected by extreme scores than is the mean. In Figure 4–9 it can be seen that in a symmetric distribution the mean and the median will be located at the same point. But in the skewed distributions, the mean is pulled in the direction of skewness away from the middle of the distribu-

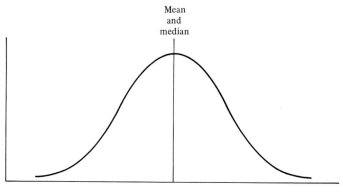

(a)

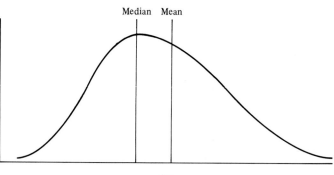

(b)

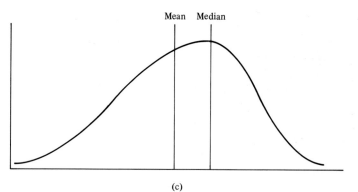

(c)

Figure 4–9
Location of Mean and Median in (a) Symmetric, (b) Positively Skewed, and (c) Negatively Skewed Distributions

tion. In such situations the median will lie closer to the central tendency than the mean. The distribution of physicians' incomes, for example, is skewed to the right. That is, there is a small proportion of doctors whose income exceeds $100,000 or even $200,000 per year. If the mean were reported as the measure of central tendency for doctors' incomes, these extreme values would pull the mean to a misleadingly high value; most doctors do not have incomes anywhere near the mean. Therefore, the median would be the appropriate measure of central tendency to report.

Statistics makes greater use of the mean than of the median. There are two reasons for this. The first has already been mentioned, the fact that the mean incorporates more information about the data than does the median. The second has to do with "sample stability." In a unimodal symmetric distribution (a population), the mean (μ) and the population median will have the same value. If, however, repeated random samples are drawn from this population, the sample means $(\bar{X}$'s) will show less fluctuation among their values for the samples than will the sample medians. This means that any single sample mean is likely to be a more accurate estimator of the population mean than any sample median would be of the population median. Since unimodal symmetric distributions occur quite frequently in statistics, this difference in sample stability is a major reason for using the mean in inferential statistics instead of the median.

EXERCISES FOR CHAPTER 4

1. The following numbers are suicide rates per 100,000 persons for various metropolitan areas in the United States:

8.4	7.9	10.1	11.8	13.2
13.0	7.2	16.1	10.8	11.1
15.9	8.7	4.5	9.6	10.3
11.2	6.0	10.6	13.2	1.5
21.2	7.4	17.1	7.8	6.9
9.5	14.4	6.9	10.2	11.0
7.5	8.2	6.7	21.4	7.6
9.5	15.4	12.2	3.8	10.1
13.8	5.3	7.2	7.9	17.2
7.4	21.5	9.7	7.6	8.8

Construct a frequency distribution by grouping the data into intervals 1 unit in size. Construct a histogram and a frequency polygon from this frequency distribution.

2. Use the data in Exercise 1 to construct a frequency distribution by grouping the data into intervals 2 units in size. Construct a histogram and a frequency polygon from this frequency distribution.

3. The per capita gross national product for 60 countries is given in the following descending array:

2900	550	272	160	88
1947	490	263	144	78
1388	478	239	142	73
1316	467	225	135	70
1196	423	220	129	70
1130	395	219	120	64
943	377	194	108	60
836	362	189	105	60
726	357	178	100	57
670	340	173	99	55
610	316	172	98	50
572	293	161	94	50

Use 10 intervals to construct a frequency distribution, histogram, and frequency polygon for these data.

4. Find the mean, median, and mode for the following numbers: 18, 13, 23, 19, 9, 13, 7, 15, 36.

5. Find the mean, median, and mode for the following numbers: 2, 12, 20, 7, 15, 7, 10, 45, 7, 15.

6. In a class consisting of 10 girls and 20 boys, the average score on an

examination for the girls is 80 and the average score for the boys is 70. What is the average score on the examination for the whole class?

7. A politician charges the opposing political party with spending an average of over $100,000 for its candidates from the state and that this is an outrageous sum for a party to spend on its candidates, especially on candidates for state senator and state representative. The campaign spending figures for the party are:

Office	Number of Candidates	Average Amount Spent
U.S. senator	2	$1,000,000
U.S. congressman	16	400,000
Governor	1	800,000
State senator	50	35,000
State representative	50	23,000

Calculate the mean and median for campaign spending by the party. Is the politician's criticism valid? Explain.

8. Using the numbers 4, 3, 2, 1, and 0 to represent the grades of A, B, C, D, and F, the grades of 10 students in a class are:

Student	Grade
1	2
2	4
3	0
4	1
5	1
6	3
7	2
8	3
9	2
10	2

a. Find N.

b. Find ΣX.

c. Find $\sum_{i=3}^{N} X_i$.

d. Find $\bar{X}$.

e. Find x_2, x_3, x_6, and x_N.

f. Find Σx^2.

ANSWERS TO
QUESTIONS IN CHAPTER 4

4-1: a. $\Sigma X = 60$.   b. $X_7 = 2$.
      c. $N = 10$.   d. $\overline{X} = 6.0$.

4-2: $\overline{X} = 6.0$, $\Sigma x = 0$, where

$x_1 = -5$ $x_6 = -6$
$x_2 = -4$ $x_7 = -4$
$x_3 = -1$ $x_8 = -6$
$x_4 = 19$ $x_9 = -1$
$x_5 = 4$ $x_{10} = 4$

4-3: $\Sigma x^2 = 524$, where

$x_1^2 = 25$ $x_6^2 = 36$
$x_2^2 = 16$ $x_7^2 = 16$
$x_3^2 = 1$ $x_8^2 = 36$
$x_4^2 = 361$ $x_9^2 = 1$
$x_5^2 = 16$ $x_{10}^2 = 16$
$\Sigma(X - \overline{X}) = 0$ $\Sigma(X - \overline{X})^2 = 524$

4-4: Median $= 3.5$.

5

Measures of Dispersion

The mean describes one important characteristic of interval or ratio scores, the central tendency of the distribution of scores on the scale of measurement. Another important characteristic of a distribution of scores is the spread or *dispersion* among the scores, the tendency of the scores to deviate from the central tendency. If the central tendency is thought of as the point on the scale of measurement that best represents a "typical" score in the distribution, the dispersion presents the other side of the coin: it reflects the "goodness" or "poorness" of the central tendency as a representation of all the scores in a distribution. Dispersion, then, is the degree to which scores deviate from the central tendency of the distribution.

Dispersion is an important property of distributions because it is possible for different distributions to be the same in that they have the same central tendency, as, say, measured by the mean, but to be different in their dispersions. Figure 5–1 presents the histograms for three distributions which have the same mean but different dispersions. In Figure 5–1 the distribution in (a) has the smallest dispersion because a great proportion of its scores are clustered close to the mean; the distribution in (c) has the greatest dispersion because a great proportion of its scores lie relatively far from the mean.

As in the case of central tendency, a way is needed to calculate a measure of the degree of dispersion in a set of scores. And as in the consideration of measures of central tendency, a measure of dispersion ought to allow us to make accurate inferences concerning the population dispersion from a measure of a sample's dispersion. This goal can be maximized by taking into account information about all the scores in a sample and not just a few.

It is for this reason that the most obvious indicator of dispersion, the *range,* is usually inadequate. The range is the difference between the largest and smallest scores in a distribution. What it measures is the range of possible values within which scores actually occur. Because the only information that the range considers are the values of the largest

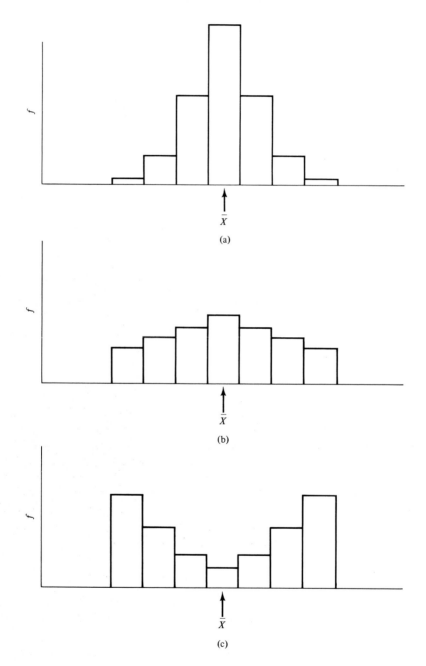

Figure 5–1
Histograms of Three Distributions with the Same Mean but Different Dispersions

and smallest scores, it cannot tell us whether the other scores are all lumped together in the middle or are evenly distributed across the range.

A different approach to dispersion is to obtain a measure of how far any randomly selected score in the distribution is likely to deviate from the central tendency of the distribution. A measure of this sort will be a type of average; that is, it would be calculated by taking into account the deviation of each score from the central tendency of the distribution and dividing the sum of these deviations by the number of observations. The result would be a measure of the typical deviation to be found in the distribution: the distance that any randomly selected score could be expected to lie from the central tendency.

Measures of dispersion constructed in this way usually employ the mean as the measure of central tendency. Since for unimodal symmetric distributions the mean is less susceptible to sampling fluctuations than the median, any measure of dispersion would possess more sampling stability when based upon the mean than upon the median. And since unimodal symmetric distributions, or approximations to them, occur quite frequently in statistics, it is advantageous to work with a measure of dispersion based upon the mean.

It might appear that a straightforward measure of the typical deviation could be constructed by finding the deviation from the mean for each observation, summing these deviations, and dividing the sum by the number of observations. These calculations are performed on the example of the ages of five persons discussed in Chapter 4.

Person	Age (X)	Mean $(\overline{X})$	$x = X - \overline{X}$
1	25	30	-5
2	30	30	0
3	20	30	-10
4	35	30	5
5	40	30	10
	$\Sigma X = 150$		$\Sigma x = 0$

$$\overline{X} = \frac{\Sigma X}{N} = \frac{150}{5} = 30$$

We see that when the deviations are added together their sum equals zero. As was stated in Chapter 4, this must be so because it is a mathematical property of the mean. Consequently, a measure of dispersion cannot be constructed on the basis of a simple summing of deviations, because this sum will always equal zero.

THE AVERAGE DEVIATION

There are two ways to get around this problem. The simplest results in a measure of dispersion called the *average deviation*. The average deviation is constructed by summing the absolute values of the deviations and then dividing by the number of observations. The absolute value of any number, say k, is represented by the symbol $|k|$ and is equal to the numerical value of the number disregarding its sign (positive or negative). In the example above, the average deviation would be calculated as follows:

Person	Age (X)	Mean $(\bar{X})$	$x = X - \bar{X}$	$\|x\|$
1	25	30	−5	5
2	30	30	0	0
3	20	30	−10	10
4	35	30	5	5
5	40	30	10	10
	$\Sigma X = 150$			$\Sigma\|x\| = 30$

$$\text{average deviation} = \frac{\Sigma|x|}{N} = \frac{30}{5} = 6$$

Since

$$\Sigma|x| = \Sigma|X - \bar{X}|$$

the formula for the average deviation could be written as

$$\text{average deviation} = \frac{\Sigma|X - \bar{X}|}{N}$$

By taking the absolute value of each deviation *before* summing, the average deviation gets around the problem that deviations around the mean must always sum to zero and results in an easily interpretable measure of dispersion. The average deviation tells us the distance which a randomly selected score will typically deviate from the mean.

Question 5−1. The number of terms that five randomly selected congressmen have served in the U. S. House of Representatives are: 3, 10, 12, 7, 8. Find the average deviation of these scores.

THE STANDARD
DEVIATION AND VARIANCE

For descriptive purposes, for describing the degree of dispersion in a distribution of scores, the average deviation is an adequate and easily interpretable measure. Unfortunately, the mathematical properties of the average deviation are such that it does not meet the needs of advanced mathematics and cannot be used in the development of statistical theory. Consequently, the development of most important statistical techniques does not make use of the average deviation but of a different measure of dispersion, called the *standard deviation.*

The standard deviation makes use of a different method to get around the problem that $\Sigma x = 0$. This method squares each deviation and then adds the squared deviations together. If x is used to represent a deviation, x^2 will represent a squared deviation and Σx^2 represents the sum of the squared deviations. It will be recalled that to square a number means to multiply that number by itself. If the sign of the number is positive, squaring it will result in a positive number. If the sign of the number is negative, squaring it will also result in a positive number, because the product of two negative numbers is positive. Thus, no matter what sign a deviation has, the square of the deviation will always be positive and the sum of the squared deviations (Σx^2) will be a positive, non-zero number, even though the sum of the deviations (Σx) must equal zero. The sum of the squared deviations (Σx^2) occurs so frequently in statistical calculations that it is given a special name: the *sum of squares.*

Population characteristics are distinguished from sample characteristics by using Greek letters to represent population characteristics and English letters to represent the corresponding sample characteristics. In the case of the mean, the symbol used to represent a population mean is μ and the symbol used to represent a sample mean is $\overline{X}$. Similarly, in the case of the standard deviation, the Greek letter σ (lowercase "sigma") is used to represent the standard deviation of a population and s is used to represent the standard deviation of a sample.

The formula for a population standard deviation then may be written as follows:

$$\sigma = \sqrt{\frac{\Sigma x^2}{N}}$$

Thus, the population standard deviation is equal to the square root of the quantity of the sum of squares divided by the number of observations. The new operation here is the square-root sign. By squaring the deviations the unit of measurement of the variable being dealt with is also squared. If we were dealing with the weights of people, squaring the deviations turns "pounds" into "pounds squared." But "pounds squared" is not the unit of measurement about which we are concerned. To get back to good old ordinary pounds, the square root must be taken.

When the square root is computed it can have either a positive or negative value. For example, the square root of 9 ($\sqrt{9}$) can be either +3 or −3, since either one when multiplied by itself yields a value of +9. But substantively, it does not make sense to have a minus or negative standard deviation. That would be saying that the dispersion in a distribution is less than zero, which is impossible. Thus, only the positive value of the square root is used when computing the standard deviation.

In finding the standard deviation, the quantity composed of the sum of squares (Σx^2) divided by the number of observations (N) must be calculated. This quantity is functionally related to the standard deviation — it is equal to the square of the standard deviation. Consequently, this quantity can also serve as a measure of the dispersion in a distribution and it is called the *variance*. Statistical theory, in fact, makes more use of the variance than the standard deviation as a measure of dispersion. Although the units of measurement of the variance are squared and thus makes the variance difficult to interpret in practical terms, this causes no difficulties in manipulating it mathematically. We will often make use of the variance. Since the variance is equal to the square of the standard deviation, the variance for a population is symbolized by σ^2 and its formula is

$$\text{population variance} = \sigma^2 = \frac{\Sigma x^2}{N}$$

Consequently,

$$\text{population standard deviation} = \sigma = \sqrt{\sigma^2} = \sqrt{\frac{\Sigma x^2}{N}}$$

The definition of a deviation is the difference obtained from subtracting the mean from the score for an observation. In the case of a population, the mean is symbolized by μ, a raw score is represented by X, and a deviation x equals $(X - \mu)$. From this it follows that a squared deviation x^2 must be equal to $(X - \mu)^2$. Therefore, the formulas for the variance and the standard deviation of a population can also be written as

$$\sigma^2 = \frac{\Sigma(X - \mu)^2}{N}$$

$$\sigma = \sqrt{\frac{\Sigma(X - \mu)^2}{N}}$$

Exactly the same type of procedures are used to find a sample standard deviation and a sample variance, which are symbolized by s and s^2.

$$\text{sample standard deviation:} \quad s = \sqrt{\frac{\Sigma x^2}{N}}$$

$$\text{sample variance:} \quad s^2 = \frac{\Sigma x^2}{N}$$

Similarly, the following relationships hold:

$$s = \sqrt{s^2} = \sqrt{\frac{\Sigma x^2}{N}}$$

Because it is a sample that is being dealt with now the mean is represented as $\overline{X}$ and $x = (X - \overline{X})$ and $x^2 = (X - \overline{X})^2$. Therefore,

$$s^2 = \frac{\Sigma(X - \overline{X})^2}{N}$$

$$s = \sqrt{\frac{\Sigma(X - \overline{X})^2}{N}}$$

The calculation of the standard deviation for the example of the ages of five people proceeds as follows:

Person	Age (X)	Mean ($\overline{X}$)	$x = X - \overline{X}$	x^2
1	25	30	−5	25
2	30	30	0	0
3	20	30	−10	100
4	35	30	5	25
5	40	30	10	100
	$\Sigma X = 150$			$\Sigma x^2 = 250$

$$\overline{X} = \frac{150}{5} = 30$$

$$s = \sqrt{\frac{\Sigma x^2}{N}} = \sqrt{\frac{250}{5}}$$

$$= \sqrt{50} = 7.07$$

It can be seen that the standard deviation for this example (7.07) is a little larger than the average deviation was for the same example (6).

Like the average deviation, the standard deviation provides a mea-

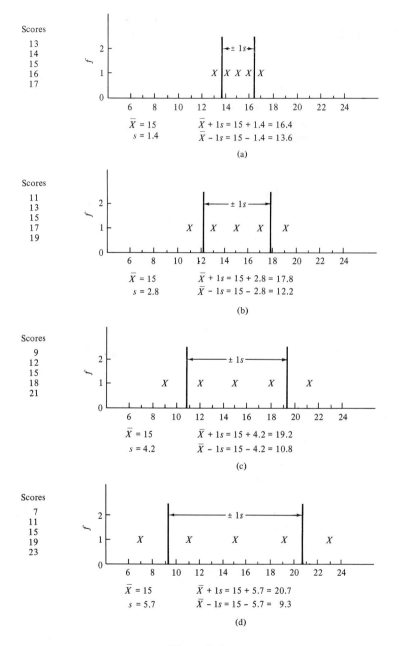

Figure 5–2
Examples of the Standard Deviation as a Measure of Dispersion

sure of the dispersion of the scores in a distribution. Unfortunately, the interpretation of a numerical value for a standard deviation is not as easy as for an average deviation. An average deviation represents the typical deviation of a score from the mean of a distribution. A more complicated statement is necessary for interpreting a standard deviation. If a distribution of scores is normal in shape, or is a close approximation to a normal distribution, 68.26% of the scores will fall within ± 1 standard deviation of the mean, and, consequently, 31.74% of the scores will lie further from the mean than a distance equal to 1 standard deviation.

Before pursuing this interpretation further, let us examine how the standard deviation measures dispersion. The basic idea is that the value of the standard deviation gets larger as the degree of spread or dispersion in a distribution increases. In Figure 5–2 four distributions of five scores each are plotted. The mean of each distribution is 15, but the degree of spread among the scores increases from distribution (a) through distribution (d). The standard deviation for each distribution is calculated and the interval of ±1 standard deviation about the mean is plotted on each graph. By examining these figures it can be seen how the standard deviation measures the spread or dispersion in a distribution of scores.

Question 5–2. Calculate the standard deviation for the sample in Question 5–1.

_____

INTERPRETING
THE STANDARD DEVIATION

In order to deal with a small set of numbers, the illustrations in Figure 5–2 used only five scores and did not represent normal distributions. Suppose, now, that we had obtained the weights of all professional football players in the United States and found these weights to have a normal distribution with a mean of 225 pounds and a standard deviation of 15 pounds. We will consider these weights to be a population. Therefore, $\mu = 225$ and $\sigma = 15$. It was said earlier that in a normal distribution 68.26% of the scores lie within ±1 standard deviation of the mean. Since the standard deviation in the distribution of football players' weights is equal to 15 pounds, this means that 68.26% of the players' weights lie between 210 pounds ($\mu - 1\sigma = 225 - 15 = 210$) and 240 pounds ($\mu + 1\sigma = 225 + 15 = 240$). If there are, say, 3000 scores in the distribution, then about 2048 players [$(.6826)(3000) = 2047.8$] have weights between 210 and 240 pounds.

If, instead of 1 standard deviation, we had taken an interval of 2

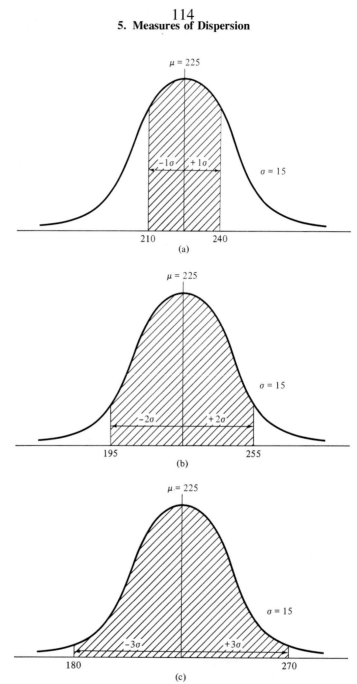

Figure 5–3
Intervals about the Mean in a Normal Distribution of the Weights of Football
Players: (a) 1 Standard Deviation (68% of the Scores); (b) 2 Standard Deviations
(95.46%); (c) 3 Standard Deviations (99.73%)

standard deviations above and below the mean, 95.46% of the players would have weights that lie in the interval. Since 1 standard deviation in this distribution equals 15 pounds, 2 standard deviations equals 30 pounds. Therefore, the limits of this interval are 195 pounds ($\mu - 2\sigma = 225 - 30 = 195$) and 255 pounds ($\mu + 2\sigma = 225 + 30 = 255$). For a distance of 3 standard deviations about the mean, 99.73% of the scores are contained in the interval in a normal distribution. Three standard deviations equals 45 pounds in this example, so the limits of this interval

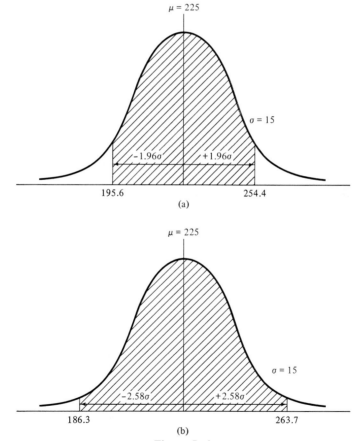

Figure 5–4

Intervals about the Mean in a Normal Distribution of the Weights of Football Players: (a) 1.96 Standard Deviations (95% of the Scores); (b) 2.58 Standard Deviations (99%).

are 180 pounds ($\mu - 3\sigma = 225 - 45 = 180$) and 270 pounds ($\mu + 3\sigma = 225 + 45 = 270$). These intervals are presented in Figure 5–3.

Quite frequently in statistics it happens that we wish to know the limits of intervals that contain exactly 95% or 99% of the scores in some normal distribution. It works out that an interval which covers a distance of 1.96 standard deviations above and below the mean contains 95% of the scores. And an interval that covers a distance of 2.58 standard deviations above and below the mean contains 99% of the scores. For the distribution of football players' weights, 1 standard deviation equals 15 pounds and 1.96 standard deviations equal 29.4 pounds [(1.96) (15) = 29.4]. Therefore, 95% of the players have weights between 195.6 pounds [$\mu - 1.96\sigma = 225 - (1.96)(15) = 225 - 29.4 = 195.6$] and 254.4 pounds [$\mu + 1.96\sigma = 225 + (1.96)(15) = 225 + 29.4 = 254.4$]. And 99% of the players have weights between 186.3 pounds [$\mu - 2.58\sigma = 225 - (2.58)(15) = 225 - 38.7 = 186.3$] and 263.7 pounds [$\mu + 2.58\sigma = 225 + (2.58)(15) = 225 + 38.7 = 263.7$]. These intervals are shown in Figure 5–4.

A point of which great use will be made is that statements about what percentage of scores lie within a certain number of standard deviations of the mean are true for *any* normal distribution. Thus regardless of what the scores in a distribution represent and regardless of the values of the mean and the standard deviation of the distribution, it is always true that, for example, 95% of the scores lie within a distance of 1.96 standard deviations about the mean. Or that 99% of the scores lie within a distance of 2.58 standard deviations about the mean. Thus if we were told that on a college entrance examination, the scores were normally distributed with a mean of 330 and a standard deviation of 40, it must be the case that 95% of the students had scores between 251.6 [$330 - (1.96)(40) = 330 - 78.4 = 251.6$] and 408.4 [$330 + (1.96)(40) = 330 + 78.4 = 408.4$].

Question 5–3. The ages of the 435 U.S. congressmen are normally distributed with a mean of 55 and a standard deviation of 8.

a. What are the values for an interval of 1 standard deviation above and below the mean?
b. What are the limits of an interval about the mean within which the ages of 95% of the congressmen fall?
c. What are the limits of an interval about the mean within which the ages of 99% of the congressmen fall?
d. About how many of the congressmen have ages that fall into the interval constructed in part c? _____

COMPUTATIONAL FORMULAS FOR
THE VARIANCE AND STANDARD DEVIATION

The calculation of the variance and the standard deviation can be laborious, especially when a large number of observations is involved. Some time can be saved by using "computational formulas." Starting from its definitional formula, a computational formula for the variance will be derived. (Since the standard deviation can be found by taking the square root of the variance, a computational formula for the standard deviation can be obtained by simply taking the square root of the computational formula for the variance.) To arrive at this formula, two algebraic equivalences must be kept in mind. The first is that the mean $(\overline{X})$ may be replaced by $\Sigma X/N$. The second involves the process of the summation of a constant. When ΣX is written, it says to sum together the values of X for N observations. Because X is a variable, X will take on different values during the summation process. A constant is a number whose value does not change, at least within the execution of a particular problem. Therefore, Σk, where k is a constant, says to add together N times the value of k. But since the value of k is a constant and thus the same in each case, Σk is equal to Nk. (Adding the same number, say 5 times, is equal to 5 times that number.) The value of $\overline{X}$ is a constant for a particular problem. Therefore, $\Sigma \overline{X} = N\overline{X}$. Keeping these points in mind, the derivation of a computational formula for the variance can proceed as follows:

$$s^2 = \frac{\Sigma x^2}{N} = \frac{\Sigma (X - \overline{X})^2}{N}$$

Multiplying out the quantity $(X - \overline{X})^2$ we have

$$s^2 = \frac{\Sigma (X^2 - 2X\overline{X} + \overline{X}^2)}{N}$$

[This step should be checked by multiplying out $(X - \overline{X})(X - \overline{X})$.] The summation sign is distributed across the numerator*:

$$s^2 = \frac{\Sigma X^2 - \Sigma 2X\overline{X} + \Sigma \overline{X}^2}{N}$$

The constants inside the summation sign are moved outside and $\Sigma \overline{X}^2$ is replaced by $N\overline{X}^2$:

$$s^2 = \frac{\Sigma X^2 - 2\overline{X}\Sigma X + N\overline{X}^2}{N}$$

Replacing $\overline{X}$ by $\Sigma X/N$,

*See pages 57 to 58.

$$s^2 = \frac{\Sigma X^2 - 2\left(\frac{\Sigma X}{N}\right)(\Sigma X) + N\left(\frac{\Sigma X}{N}\right)^2}{N}$$

$$= \frac{\Sigma X^2 - 2\frac{(\Sigma X)^2}{N} + N\left(\frac{\Sigma X}{N}\right)\left(\frac{\Sigma X}{N}\right)}{N}$$

$$= \frac{\Sigma X^2 - 2\cdot\frac{(\Sigma X)^2}{N} + \frac{(\Sigma X)^2}{N}}{N}$$

$$= \frac{\Sigma X^2 - \frac{(\Sigma X)^2}{N}}{N}$$

This last formula is the computational formula for the variance. Since $s^2 = \Sigma x^2/N$, then

$$\frac{\Sigma x^2}{N} = \frac{\Sigma X^2 - \frac{(\Sigma X)^2}{N}}{N}$$

Multiplying both fractions by N yields

$$\Sigma x^2 = \Sigma X^2 - \frac{(\Sigma X)^2}{N}$$

This provides us with another way of computing the sum of squares. The computational formula for the standard deviation is

$$s = \sqrt{\frac{\Sigma X^2 - \frac{(\Sigma X)^2}{N}}{N}}$$

It is important to note that ΣX^2 is not the same thing as $(\Sigma X)^2$ in these formulas. ΣX^2 tells us to find the square of each raw score and then sum the squared scores. It is referred to as "the sum of the raw scores squared." $(\Sigma X)^2$ tells us to *first* sum up the raw scores and then square the sum. It is referred to as "the sum of the raw scores *quantity* squared."

The advantage of the computation formulas lies in the fact that the deviations for each observation from the mean do not have to be calculated. The computational formulas are especially useful when a calculator can be used as many calculators are able to compute ΣX and ΣX^2 at the same time. The calculation of the standard deviation for the example of the ages of five people through the computational formula would be as follows:

Person	Age (X)	X^2
1	25	625
2	30	900
3	20	400
4	35	1225
5	40	1600
	$\Sigma X = 150$	$\Sigma X^2 = 4750$

$(\Sigma X)^2 = 22{,}500$

$$s = \sqrt{\frac{\Sigma X^2 - [(\Sigma X)^2/N]}{N}} = \sqrt{\frac{4750 - (22{,}500/5)}{5}}$$

$$= \sqrt{\frac{4750 - 4500}{5}} = \sqrt{\frac{250}{5}}$$

$$= \sqrt{50} = 7.07$$

This result checks perfectly with the value obtained for the standard deviation by the formula $\sqrt{\Sigma x^2/N}$ on page 111, as it should since the two formulas are mathematically equivalent.

Question 5–4. Using the computational formula, recalculate the standard deviation for the terms of congressmen in Question 5–1.

ESTIMATING
POPULATION PARAMETERS

Many subjects are complicated, and statistics is no exception. We must now face up to one complication about the standard deviation. In the case of the mean, the sample mean $(\overline{X})$ is an unbiased estimator of the population mean (μ). The term *unbiased estimator* in a special way means that $\overline{X}$ accurately estimates μ. That this is a peculiar use of "accurate" is readily apparent when you consider that it is unlikely that the $\overline{X}$ of any randomly selected sample is going to have the same value as the μ of the population from which the sample is drawn.

Suppose that we have a large population of high school boys whose average weight is 140 pounds. Thus, μ equals 140. A random sample of 50 boys is drawn from this population. The mean of this sample is calculated and found to be equal to 130. Since this is the first sample drawn from the population, we will indicate its mean as $\overline{X}_1$. Therefore, $\overline{X}_1 = 130$. Now we return these boys to the population and randomly select a second sample and find that $\overline{X}_2 = 150$. We continue this process for a long time, recording the value of the sample mean each time. Suppose that the values for the means of the first 20 samples looked like this:

$$\overline{X}_1 = 130 \qquad \overline{X}_{11} = 135$$
$$\overline{X}_2 = 150 \qquad \overline{X}_{12} = 145$$
$$\overline{X}_3 = 140 \qquad \overline{X}_{13} = 140$$
$$\overline{X}_4 = 135 \qquad \overline{X}_{14} = 140$$
$$\overline{X}_5 = 125 \qquad \overline{X}_{15} = 145$$
$$\overline{X}_6 = 140 \qquad \overline{X}_{16} = 135$$
$$\overline{X}_7 = 155 \qquad \overline{X}_{17} = 130$$
$$\overline{X}_8 = 150 \qquad \overline{X}_{18} = 145$$
$$\overline{X}_9 = 140 \qquad \overline{X}_{19} = 140$$
$$\overline{X}_{10} = 145 \qquad \overline{X}_{20} = 135$$

Remember that *each* of these values represents the mean weight for a random sample of 50 boys. These sample means can be treated as observations and can be plotted on a graph in the same way that has been done with other data. The 20 means are plotted in Figure 5–5.

The type of distribution shown in Figure 5–5 is called a *sampling distribution of the mean,* since it is the means of samples that are plotted. It must be kept clear that this distribution is different from the distribution composed of the weights of the individual boys. The latter distribution will be referred to as the *original distribution.*

Now we could continue indefinitely drawing random samples from the original distribution and plotting the mean of each sample in the sampling distribution until the number of scores in the sampling distribution became so large that it could be considered to be a population. But an important fact is already apparent from the means of the 20 samples plotted in Figure 5–5. That is that the mean of the sampling distribution is equal to 140 pounds. But this is also the value of μ for the original population! In other words, the mean of the "sampling distribution of the

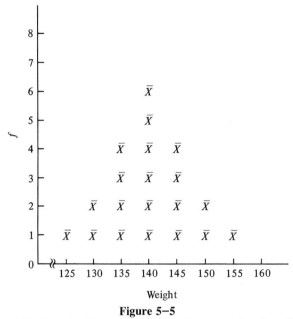

Figure 5–5
Distribution of 20 Sample Means, Each Based upon a Random Sample of 50
High School Boys

means" will equal the mean of the original population. This will always hold true in the long run. It is this property that is referred to when we say that $\overline{X}$ is an unbiased estimator of μ: A sample statistic is an *unbiased estimator* of its corresponding population parameter, if the mean of the sampling distribution of the sample statistic is equal to the value of that population parameter. Sampling distributions can be constructed for many different characteristics of samples. We shall have cause to refer to a number of different kinds of sampling distributions later. But the logic of their construction is the same. And the test of whether a sample statistic is an unbiased estimator of the corresponding population parameter is whether the mean of the sampling distribution for the sample statistic is equal to the population parameter.

Now the variance for a population is $\sigma^2 = \Sigma x^2/N$, and for a sample it is $s^2 = \Sigma x^2/N$. But it happens, strange as it may seem, that the mean of the sampling distribution for s^2 does not equal σ^2, even though the formulas are the same. Thus, s^2 is a *biased* estimator of σ^2.

Where a description of the degree of dispersion in a sample is desired, s^2 is a perfectly adequate measure. But when it becomes necessary to estimate σ^2, s^2, because it is biased, is inadequate. What is needed is an unbiased estimator of σ^2. We shall represent an unbiased estimator of

σ^2 by the symbol $\hat{s}^2$ (read s-hat squared).* Fortunately, the calculation of $\hat{s}^2$ is only slightly different from that for s^2. It is

$$\hat{s}^2 = \frac{\Sigma x^2}{N-1}$$

The difference between s^2 and $\hat{s}^2$ is that the sum of squares is divided by N in the case of s^2 and by $N-1$ in the case of $\hat{s}^2$. Since the numerator or sum of squares part of the formula is not altered, the computational formula for the unbiased estimator for the variance is

$$\hat{s}^2 = \frac{\Sigma X^2 - \dfrac{(\Sigma X)^2}{N}}{N-1}$$

This solves the problem of obtaining an unbiased estimator of the population variance and gives us an unbiased statistic to measure dispersion. Since the standard deviation is equal to the square root of the variance, we will define the statistic $\hat{s}$ as

$$\hat{s} = \sqrt{\hat{s}^2} = \sqrt{\frac{\Sigma x^2}{N-1}} = \sqrt{\frac{\Sigma X^2 - \dfrac{(\Sigma X)^2}{N}}{N-1}}$$

It might be thought that since $\hat{s}^2$ is an unbiased estimator of σ^2, $\hat{s}$ (as defined above) is an unbiased estimator of σ. This, strangely enough, is not true. However, an unbiased estimator of σ is not needed, since when we want an unbiased measure of dispersion, we can use $\hat{s}^2$. But to keep the convenient correspondence that the standard deviation is the square root of the variance, we will use $\hat{s}$ as meaning $\sqrt{\hat{s}^2}$, even though $\hat{s}$ is not an unbiased estimator.

We now have two sets of formulas for the sample variance and standard deviation. It is necessary to carefully distinguish their use. The formulas for s^2 and s are to be used when a *description* of the dispersion in a sample is needed. The formula for $\hat{s}^2$ (and its square root $\hat{s}$) is used when an *estimate* of the dispersion in a population is needed.

How much difference does it make whether the sum of squares is divided by N or $N-1$? When the sample size is small, it can make a great deal of difference. For $N=10$, the difference is about 10%; for $N=20$, about 5%. As the sample size gets larger, the difference continues to get smaller.

Question 5–5. Compute an unbiased estimate of σ^2 for the terms of congressmen from the sample in Question 5-1.

---

*Many statistics books use the symbol $\hat{\sigma}^2$ to represent an unbiased estimator of the variance. $\hat{s}^2$ is used here to retain the distinction that English letters are used to represent characteristics of samples and Greek letters to represent characteristics of populations. It may also be noted that some elementary books do not bother to distinguish between s^2 and $\hat{s}^2$.

EXERCISES FOR CHAPTER 5

1. Calculate the average deviation for the following set of scores: 8, 2, 0, 10, 4, 5, 6, 9, 2, 4. Give the numerical values for a range of ± 1 average deviation about the mean. How many of the original scores lie inside this range? How many lie outside it?

2. Calculate the sample variance and standard deviation (s^2 and s) for the scores in Exercise 1.

3. A researcher interested in problems of city government selects a random sample of 12 cities. He finds that the ages of the mayors of these cities are: 25, 50, 52, 40, 60, 45, 68, 72, 35, 48, 35, 70. Find the variance and standard deviation to describe the degree of dispersion in the data.

4. Use the computational formulas to find the variance and the standard deviation for the data in Exercise 3. What is the value of the quantity $\Sigma X^2 - [(\Sigma X)^2/N]$? Is this equal to the Σx^2 found in Exercise 3?

5. A researcher studying homicide problems in cities finds the following homicide rates per 100,000 population in a random sample of 8 cities: 8, 10, 15, 7, 12, 5, 8, 14. Calculate an estimate of the variance and the standard deviation ($\hat{s}^2$ and $\hat{s}$) in the population.

6. If the mean age of a sample of college students is 18 and the standard deviation is 2, what will the standard deviation of the sample be 20 years later? What will the mean be? Explain why adding 20 to the value of each score changed the mean but not the standard deviation.

7. On an exam the scores of 500 students are normally distributed with a mean of 62 and a standard deviation of 12. Find the limits of an interval about the mean within which 68.26% of the scores lie; in which 95% of the scores lie; in which 99% of the scores lie. About how many of the students fall into the 95% interval? The 99% interval?

8. Assume that 300 runners finish a marathon race in which their times are normally distributed with a mean of 160 minutes and a standard deviation of 10 minutes. Find an interval about the mean within which the times of 95% of the runners lie; in which 99% of the runners' times lie. About how many runners had times which fall into the 95% interval? The 99% interval?

ANSWERS TO
QUESTIONS IN CHAPTER 5

5–1: $\Sigma X = 40$, $\overline{X} = 8.0$.

$\lvert x_1 \rvert = 5$	$\lvert x_4 \rvert = 1$
$\lvert x_2 \rvert = 2$	$\lvert x_5 \rvert = 0$
$\lvert x_3 \rvert = 4$	$\Sigma \lvert x \rvert = 12$

Average deviation $= 2.4$.

5-2: $\Sigma X = 40$, $\bar{X} = 8.0$, $s = 3.03$, where

$$x_1^2 = 25 \qquad x_4^2 = 1$$
$$x_2^2 = 4 \qquad x_5^2 = 0$$
$$x_3^2 = 16$$
$$\Sigma x^2 = 46.$$

5-3: a. 47, 63.   b. 39.32, 70.68.
     c. 34.36, 75.64.   d. About 431.

5-4: $\Sigma X = 40$, $\Sigma X^2 = 366$, $s = 3.03$, where

$$\Sigma X^2 - \frac{(\Sigma X)^2}{N} = 366 - \frac{1600}{5} = 366 - 320 = 46$$

5-5: $\Sigma x^2 = 46$, $\hat{s}^2 = 11.5$.

6
Probability I

We have seen that statistics has two major purposes: to provide comprehensible descriptions about a body of data and to estimate the accuracy of inferences made from information about a sample to the characteristics of a population. The notion of probability is the basis for the achievement of the second goal. When random sampling is used, an estimate of the accuracy of a sample statistic as a measure of its corresponding population parameter can be calculated through probability theory. This is the only known method of estimating the accuracy of sample information without actually knowing the values of the population parameters. This chapter discusses the basic rules of probability that are involved in making such estimates.

In Chapter 2 the probability of an event was defined as a relative frequency: the ratio of the number of successes to the total number of trials in the long run. It is rare, however, that probabilities are computed empirically by observing the outcomes of a repeated series of trials over the long run or even over a very large number of trials. It is more usual to compute probabilities by constructing idealized experiments and analyzing their possible outcomes. In flipping a coin, for example, the probability of the coin coming up heads is not computed from observing the outcome of thousands of flips (trials) and calculating the ratio of the number of heads to the total number of flips. Rather an idealized experiment is constructed: The coin is imagined to be perfectly balanced (honest), the possible outcomes of a single flip (trial) are stipulated to be only two (heads and tails), and a flip is assumed to be a perfectly random process. In this idealized experiment, the outcomes of each flip, heads or tails, are equally likely to occur, and no other outcomes are possible. In such a situation, the probability of an event can be found by calculating the ratio of the number of ways the event of interest can occur to the total number of possible outcomes to the experiment. Where P stands for probability and E for the event of interest, then

$$P_E = \frac{\text{number of ways event of interest can occur}}{\text{total number of possible outcomes to the experiment}}$$

In the idealized coin-flipping experiment, there is one way in which heads can occur (the event of interest) and two possible outcomes to the experiment. Therefore,

$$P_H = \frac{1}{2} = .50$$

It is by reasoning in this way that we knew that the probability of a head in flipping a coin is .50 without ever having flipped a coin at all.

Another example of this method of computing probabilities is to ask what the probability is of obtaining a 3 in rolling a die. A die has six faces, only one of which can come up at a time. In an ideal experiment, all six faces are assumed to be equally likely to appear. Since only one of these faces is a 3, there is only one way a 3 can appear and a total of six equally likely possible outcomes to a single roll. Therefore,

$$P_3 = \frac{1}{6} = .167$$

Thus probabilities are calculated through a "counting" process: An idealized experiment is imagined, the number of possible ways the event of interest could occur in the experiment is counted, the number of total possible ways the experiment could turn out is counted, and the ratio of these two counts is the probability of the event. An important qualification here is that the various "ways" the experiment could turn out are all equally likely. This is true of a great variety of "experiments." But where the likelihood of the various outcomes is not equally likely, such as with a dishonest coin or die, the differences in likelihood would have to be taken into account.

The counting process can also be used to calculate the probability of more complex events. We might ask, for example, what the probability is of obtaining two heads when two coins are flipped. In order to perform the counting it is necessary to list all possible outcomes to the experiment. This can be done as follows:

Outcome Number	Coin 1	Coin 2
1	H	H
2	H	T
3	T	H
4	T	T

The listed outcomes exhaust all possible outcomes to the experiment; there are no other possible outcomes to flipping two coins besides the four in the list. Examination of these possible outcomes shows that there is only one way in which the event of interest, two heads, can occur. Therefore,

$$P_{2H} = \frac{1}{4} = .25$$

In the same experiment, the probability of obtaining a head and a tail might be desired. Examination of the possible outcomes shows that this event could occur in two ways. Thus,

$$P_{(H \text{ and } T)} = \frac{2}{4} = .50$$

Question 6–1. Calculate the probability of two heads and one tail in flipping three coins. _____

PERMUTATIONS
AND COMBINATIONS

The number of possible outcomes to an experiment becomes very large as the number of objects involved in the experiment increases. For example, experiments involving flipping three coins have 8 possible outcomes, four coins 16 outcomes, five coins 32 outcomes, six coins 64 outcomes, and seven coins 128 outcomes. For dice, we know that in rolling a pair of dice there are 36 possible outcomes, in rolling three dice there are 216 possible outcomes, and for four dice 1296 possible outcomes. Obviously, it would be quite tiresome to calculate probabilities for experiments involving a large number of objects if, before counting could take place, it were necessary to list all the possible outcomes. Fortunately, there are various "counting rules" that can be used to expedite this process. We shall consider only two such counting rules here. These two rules deal with determining the number of ways it is possible to select K objects from a total set of N objects.

Suppose that four objects labeled A, B, C, and D are randomly mixed together. A blindfolded person is told to pick out one object with each hand, proceeding by drawing one object at a time, the first with his right hand and the second with his left. Our problem is to calculate the probability that the person will be holding object A in his right hand and B in his left at the end of the experiment.

In computing probabilities, particular attention must be paid to the statement of the problem. In this case the problem is stated in such a way as to indicate that the order in which the objects are selected is important: what is of concern is that object A is selected first and object

B second. If in selecting among a set of objects the order of selection is relevant, the selection is called a *permutation*. If for some reason the order in which objects are selected is not of concern, the selection is called a *combination*. Stated another way, a permutation involves selecting the precise order of objects in the outcome of an experiment; a combination involves selecting what set of objects there will be in an outcome but not their order.

Our problem, then, is one of permutations: What is the probability of selecting *A* first and *B* second? This probability can be calculated by listing all the outcomes and using the counting process. The possible ways of selecting two objects one at a time from a set of four are

$$
\begin{array}{cccc}
AB & BA & CA & DA \\
AC & BC & CB & DB \\
AD & BD & CD & DC
\end{array}
$$

Out of these 12 possible outcomes, there is only one that yields the event of concern, *AB*. Thus, the probability of obtaining the permutation *AB* is

$$ P_{AB} = \frac{1}{12} = .083 $$

Instead of listing all the permutations, the number of permutations can be calculated from a permutations formula. If *N* equals the total number of objects in a set and *K* equals the number of objects to be selected from the set, the number of permutations of *K* objects it is possible to select from a set of *N* objects is

$$ P_K^N = \frac{N!}{(N-K)!} $$

In our present example $N = 4$ and $K = 2$. Therefore,

$$ P_2^4 = \frac{4!}{(4-2)!} = \frac{4!}{2!} = \frac{(4)(3)(2)(1)}{(2)(1)} $$

$$ = \frac{24}{2} = 12 $$

This is the same number of permutations that was obtained when all the permutations were listed out and counted.

Question 6–2. Taking into account the order of selection, how many ways are there to select three objects from a set of seven?

In many situations, the order in which objects are selected is irrelevant and our interest would then be in combinations. Among the objects

A, B, C, and D, what is the probability of holding A and B at the end of the experiment? In this situation, whether A was selected first and B second or B first and A second is irrelevant. Consequently, AB and BA are considered to be the same outcome – the combination of A and B. The possible combinations for selecting two objects from the set of A, B, C, and D are

$$
\begin{array}{ll}
AB & BC \\
AC & BD \\
AD & CD
\end{array}
$$

Instead of listing the possible combinations, the number of combinations can be obtained from a combinations formula. The number of possible combinations of K objects that can be selected is given by the formula

$$C_K^N = \frac{N!}{K!(N-K)!}$$

where N equals the total number of objects in the set and K the number of objects to be selected. For $N = 4$ and $K = 2$,

$$C_2^4 = \frac{4!}{2!(4-2)!} = \frac{4!}{2!(2!)}$$

$$= \frac{(4)(3)(2)(1)}{(2)(1)(2)(1)} = \frac{24}{4}$$

$$= 6$$

Since there are six possible outcomes to this experiment and only one way the event of interest can occur, the probability of selecting the combination of A and B is

$$P_{(A \text{ and } B)} = \frac{1}{6} = .167$$

Question 6–3

a. At a party there are seven people and two chairs. How many different pairs of people may sit at the party?
b. If one chair is a rocker and the other an armchair, how many different arrangements of "sitters" is it possible to form from the seven people if we distinguish arrangements not only by who is sitting but in what chair they sit?

The counting rules for permutations and combinations can also be applied to experiments that consist of a series of events where each event can result in the same two outcomes. The two outcomes for each event are termed "success" and "failure." In flipping a number of coins, for example, each event is the flipping of a coin. If we are interested in seeing if the coins come up heads, then an event, the flipping of a particular coin, is termed a success if it does turn up a head and a failure if it turns up a tail. If we were interested in knowing the number of ways, say, that four heads could occur in flipping five coins, this is analogous to asking the number of ways it is possible to select four objects from a set of five objects, and we can apply the permutations and combinations formulas.

Now suppose that five honest coins were flipped and the probability of obtaining four heads were desired. We do not care about the order in which heads appear, so the concern here is with combinations. To calculate the probability, we need the number of ways the event of interest, four heads, can occur and the total number of possible outcomes to the experiment. The number of ways the event of four heads can occur can be obtained by substituting into the combinations formula:

$$C_4^5 = \frac{5!}{4!(5-4)!} = \frac{5!}{4!(1!)} = \frac{(5)(\cancel{4!})}{(\cancel{4!})(1)}$$

$$= 5$$

This calculation tells us that there are five different ways in which it is possible to obtain four heads when five coins are flipped. It is important to understand in these problems that the same combination may occur in more than one way. The five ways that the combination of four heads can occur are:

Way Number	Coin 1	Coin 2	Coin 3	Coin 4	Coin 5
1	H	H	H	H	T
2	H	H	H	T	H
3	H	H	T	H	H
4	H	T	H	H	H
5	T	H	H	H	H

Next, it is necessary to determine the total number of possible outcomes to the experiment. It is quite possible, of course, that in flipping five coins, none of them would come up heads. It is also possible that the experiment might result in only one head appearing, or two, or three, or four, or five. Consequently, the number of ways in which *each* of these possible combinations could occur must be determined so that the total number of possible outcomes to the experiment can be calculated. This can be done as follows:

$$C_0^5 = \frac{5!}{0!(5-0)!} = \frac{120}{120} = 1$$

$$C_1^5 = \frac{5!}{1!(5-1)!} = \frac{120}{24} = 5$$

$$C_2^5 = \frac{5!}{2!(5-2)!} = \frac{120}{12} = 10$$

$$C_3^5 = \frac{5!}{3!(5-3)!} = \frac{120}{12} = 10$$

$$C_4^5 = \frac{5!}{4!(5-4)!} = \frac{120}{24} = 5$$

$$C_5^5 = \frac{5!}{5!(5-5)!} = \frac{120}{120} = 1$$

total number of possible
outcomes to the experiment = 32

These calculations show that there is one way in which no heads could occur, five ways in which one head could occur, 10 ways in which two heads could occur, and so on. Consequently, there are a total of 32 different possible outcomes to the experiment of flipping five coins. The probability of four heads appearing in flipping five coins, then, is

$$P_{4H} = \frac{5}{32} = .156$$

Since the major calculations have already been done, let us also calculate the probabilities for the other possible combinations of heads that could appear. This is done by dividing the number of ways a particular combination of heads could appear by the total number of outcomes to the experiment:

Combination	Number of Ways a Combination Could Appear	Probability
0 Heads	1	$\frac{1}{32} = .03125$
1 Head	5	$\frac{5}{32} = .15625$

2 Heads	10	$\frac{10}{32} =$	.31250
3 Heads	10	$\frac{10}{32} =$	.31250
4 Heads	5	$\frac{5}{32} =$	.15625
5 Heads	$\frac{1}{32}$	$\frac{1}{32} =$	.03125
			1.00000

Notice that the probabilities of all the combinations add up to 1.00. This makes sense intuitively since the listed combinations exhaust all the possible outcomes to the experiment: nothing else can happen in the experiment except one of the listed combinations and thus there is no probability of anything else happening. This property is true in general: The probabilities of all the posssible outcomes to an experiment must add up to 1.0.

Question 6–4. Bob and Tom are playing a game where they bet on the outcome of flipping a coin. They have agreed that Bob will always call heads and Tom tails. In the first six flips the coin has come up tails each time. What is the probability of this happening by chance, that is, that the coin is honest? Bob thinks he should continue to play because he feels that luck has to turn in his favor after all these tails. Do you think that Bob should continue to play the game? Why?

RULES OF PROBABILITY

Ordinarily, a researcher does not compute probabilities himself. Rather, he will make use of various tables of probability that have been constructed by statisticians. In this section some of the basic mathematical rules of probabilities are developed. The purpose is to give some insight into the process of how probabilities are manipulated to develop the various kinds of probability tables.

SOME TERMINOLOGY

To symbolize the probability of an event, P is used to represent probability, and the event whose probability is of concern is subscribed to P.

Thus, the statement: "The probability of event A occurring is" is symbolized as

$$P_A =$$

When two separate events occur together, we speak of this as a *joint event*. The statement that "the probability of the joint occurrence of the separate events A and B is" is symbolized as

$$P_{(A \text{ and } B)} =$$

For example, if one event is rolling a single die and obtaining a 3 and another event is flipping a coin and obtaining a head, then to have the two events occur together is a joint event. The probability of this joint event might be symbolized as

$$P_{(3 \text{ on die and H on coin})}$$

Two events are said to be *mutually exclusive* if they cannot possibly occur together in the same experiment. In flipping a coin the events of a head and a tail are mutually exclusive. The occurrence of the event of a head precludes the occurrence of the event of a tail, and vice versa.

A *conditional probability* is the probability of one event occurring given that some other event has already occurred. To say that "given that event A has occurred, what is the probability that event B will occur" is a statement of conditional probability and is symbolized as

$$P_{(B|A)} =$$

This symbol is read as "the probability of event B given that event A has occurred is."

Question 6–5. In each of the following situations, determine whether it is an example of the probability of a joint event, mutually exclusive events, or of conditional probability.

a. The probability of a person having red hair and freckles.
b. The probability of a voter casting his ballot for either the Democratic, Republican, or Socialist party.
c. The probability of getting a deuce on the second draw from a deck of cards.
d. The probability of it being a sunny day tomorrow if it is a sunny day today.
e. The probability of a congressman voting yes, no, or abstaining on a particular bill.
f. The probability that Congress will pass a bill and that the president will sign it.

_____

THE RULE
OF INDEPENDENCE

Two events are said to be independent if the occurrence of one event in no way affects the *probability* of occurrence of the other. If A and B are independent events, their independence is symbolized by the following two equations:

$$P_{(A|B)} = P_A$$

$$P_{(B|A)} = P_B$$

These equations say that the probability of either of the events is unaffected by the occurrence of the other event. We know, for example, that the events of flipping two coins are independent events. The outcome of flipping the first coin in no way affects the probabilities of the outcomes of flipping the second coin, and vice versa.

One implication of events being independent is that the knowledge of the outcome of one event is of no aid in predicting the outcome of the second event. The colors and face values of a deck of cards are independent events: knowing that a card is black does not help to predict its face value and knowing the face value of a card does not help to predict its color. This commonsense fact can easily be illustrated mathematically. The probability of selecting an ace from a deck is $\frac{4}{52}$ since there are 4 aces in a 52-card deck. Thus,

$$P_{ace} = \frac{4}{52}$$

The probability of getting an ace given that the card is black is $\frac{2}{26}$ since there are two black aces and 26 black cards in a deck. Thus,

$$P_{(ace|black)} = \frac{2}{26}$$

Consequently,

$$P_{ace} = P_{(ace|black)}$$

because

$$\frac{2}{26} = \frac{4}{52}$$

These computations show that knowing that the card is black does not change the probability of a card being an ace. The color of the card does not help to predict whether or not it is an ace. Similar calculations would show that knowing that the card is an ace does not change the probability of its color.

Question 6–6. Identify each of the following as independent or noninde-pendent events.

a. Getting a head on the first toss of a coin and a tail on the second.
b. Flipping a weighted coin and getting a head on the first toss and a head on the second.
c. A person being both tall and heavy.
d. The amount of time a person studies and his grade in a course.
e. A coin coming up heads after it has come up heads on all five pre-vious flips.

————

THE MULTIPLICATION RULE

The multiplication rule is used to compute the probability of the joint occurrence of two events. We shall first consider a simple form of the multiplication rule which applies to independent events. *If A and B are independent events, the probability of the joint occurrence of A and B is equal to the product of their separate probabilities.* Or

$$P_{(A \text{ and } B)} = (P_A)(P_B)$$

Since flipping two coins are independent events, the probability of ob-taining the joint event of a head on coin 1 and a head on coin 2 can be found through the multiplication rule:

$$P_{(H_1 \text{ and } H_2)} = (P_{H_1})(P_{H_2})$$
$$= (.5)(.5)$$
$$= .25$$

This rule can be extended to any number of events as long as they are all independent of each other. The probability for the joint occurrence of five heads in flipping five coins is

$$P_{(H_1 \text{ and } H_2 \text{ and } H_3 \text{ and } H_4 \text{ and } H_5)} = (P_{H_1})(P_{H_2})(P_{H_3})(P_{H_4})(P_{H_5})$$
$$= (.5)(.5)(.5)(.5)(.5)$$
$$= (.5)^5$$
$$= .03125$$

If the probability of the joint occurrence of *non*independent events is desired, it is necessary to use a more complex form of the multiplica-tion rule. Nonindependence means that the occurrence of one event affects the probability of the occurrence of the other event. Mathemati-cally, this means that if A and B are nonindependent events, then

$$P_{(A|B)} \neq P_A$$

and/or

$$P_{(B|A)} \neq P_B$$

The calculation of the joint occurrence of the two events must take their nonindependence into account. If we want to know the probability of drawing two aces from a deck of cards, what happens on the first draw (ace or not-ace) will affect the probability of getting an ace on the second card. If an ace is drawn on the first card, it means that at the time of the second draw there are 51 cards left, three of which are aces. In this situation the probability of an ace on the second draw is $\frac{3}{51}$. If an ace was not obtained on the first draw, the probability of an ace on the second draw would be $\frac{4}{51}$. Because what happens on the first draw affects the probability of obtaining an ace on the second draw, the two events are not independent.

The probability of obtaining an ace on the first draw is $\frac{4}{52}$. Since we are interested in the probability of drawing two aces, the probability of getting an ace on the first draw (P_{ace_1}) is multiplied by the probability of getting an ace on the second draw, assuming that an ace was obtained on the first draw $(P_{(ace_2|ace_1)})$. Thus,

$$P_{(ace_1 \text{ and } ace_2)} = (P_{ace_1})(P_{(ace_2|ace_1)})$$

$$= \left(\frac{4}{52}\right)\left(\frac{3}{51}\right) = \frac{12}{2652}$$

$$= .0045$$

From this example it can be seen that the general form of the multiplication rule is: *If A and B are any two events (not necessarily independent), the probability of the joint occurrence of A and B is the product of the probability of getting one of these events times the conditional probability of the other given that the first event has occurred.* This rule is symbolized by

$$P_{(A \text{ and } B)} = (P_A)(P_{(B|A)}) = (P_B)(P_{(A|B)})$$

Question 6–7. In the following situations, determine whether the events are independent or nonindependent and calculate the joint probability.

a. The probability of drawing a picture card (jack, queen, or king) on the first draw and a picture card on the second draw from a deck of cards.
b. The probability of selecting two gold balls, one from an urn randomly filled with 80 gold balls and 20 silver balls and one from an urn randomly filled with 30 gold balls and 70 silver balls.
c. The probability of selecting two gold balls, one at a time, without replacement, from an urn filled with 40 gold balls and 60 silver balls.
d. At a college where 60% of the students are male and 40% are female,

66% of the students are Republicans. However, 90% of the males are Republican and 30% of the females are Republican. Use the multiplication rule to find the probability that a student selected at random would be male and Republican. Female and Republican.

THE ADDITION RULE

The multiplication rule gives the probability that two events will occur together (joint occurrence). Often we wish to know the probability of obtaining *either* of two events. This probability can be determined through the addition rule. When the two events are mutually exclusive (they cannot occur together) the addition rule takes on a simple form: *If events A and B are mutually exclusive, the probability of getting either A or B is equal to the probability of A plus the probability of B.* In symbols,

$$P_{(A \text{ or } B)} = P_A + P_B$$

If one card were drawn from a deck, the probability of its being *either* a jack or a queen can be calculated through this form of the addition rule as it is obviously impossible for both a jack and a queen to occur on the same draw. Since there are four jacks and four queens in a deck, the probability of drawing a jack is $\frac{4}{52}$ and of drawing a queen also $\frac{4}{52}$. Therefore,

$$P_{(\text{jack or queen})} = \frac{4}{52} + \frac{4}{52}$$

$$= \frac{8}{52}$$

$$= .154$$

This form of the addition rule can be extended to cover more than two events as long as they are all mutually exclusive. The probability of drawing a card higher than a 10 from a deck is

$$P_{(\text{jack or queen or king or ace})} = P_{\text{jack}} + P_{\text{queen}} + P_{\text{king}} + P_{\text{ace}}$$

$$= \frac{4}{52} + \frac{4}{52} + \frac{4}{52} + \frac{4}{52}$$

$$= \frac{16}{52}$$

$$= .308$$

A more complex form of the addition rule must be used whenever events are not mutually exclusive. The problem with such events can best be seen through a diagram such as Figure 6–1. If events *A* and *B* are mutually exclusive, no point exists where the two events have a chance to occur together. A diagram of mutually exclusive events would look as in Figure 6–1a, where the areas of *A* and *B*, which represent the probabilities of these events, do not overlap.

If *A* and *B* were *not* mutually exclusive, this would mean that there is at least one point, and perhaps more, where the two events could occur together. This situation is depicted in Figure 6–1b, where the points at which the two events could occur together are represented as the overlapping areas of the circles representing the probabilities of *A* and *B*.

In a situation where *A* and *B* are not mutually exclusive it can be seen that if we were to simply add together the individual probabilities of event *A* and event *B*, the overlapping area, the shaded part in Figure 6–1b, would be counted twice: once as part of the probability of event *A* and once as part of the probability of event *B*. Such an addition process

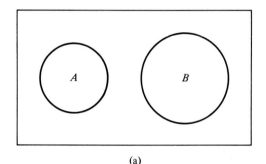

(a)

(b)

Figure 6–1
Graphical Representation of the Probabilities that Events *A* and *B* Will Be (a) Mutually Exclusive; (b) Not Mutually Exclusive

would yield a probability area larger than that which is actually covered by the probabilities of events A and B. To correct the double count of the overlapping areas, it is necessary to subtract the shaded area from the sum obtained by adding the probability of A and the probability of B. Since the shaded area is equal to the probability that A and B will occur together, it is equal to their joint probability, $P_{(A \text{ and } B)}$. The more general form of the addition rule can now be stated: *If A and B are any events (not necessarily mutually exclusive), the probability of getting either A or B is equal to the sum of their separate probabilities minus their joint probability.* Or

$$P_{(A \text{ or } B)} = P_A + P_B - P_{(A \text{ and } B)}$$

Suppose that we wish to know the probability of getting either an ace or a red card on a single draw. These are not mutually exclusive events because there are two red aces, and the general form of the addition rule must be used. The probability of drawing an ace is $\frac{4}{52}$. The probability of drawing a red card is $\frac{26}{52}$. And the probability of their joint occurrence (the overlap) is $\frac{2}{52}$ (two red aces). Therefore,

$$P_{(\text{ace or red card})} = P_{\text{ace}} + P_{\text{red card}} - P_{(\text{ace and red card})}$$

$$= \frac{4}{52} + \frac{26}{52} - \frac{2}{52}$$

$$= \frac{28}{52}$$

$$= .538$$

Question 6–8. In the following situations, determine whether the events are or are not mutually exclusive and calculate the probability of obtaining either of the events.

a. The probability of rolling a 3 or 4 with a single die.
b. The probability of drawing an ace or a club from a deck of cards.
c. If 20% of the adults in a city are on welfare, 15% of the adults are unemployed, and 10% of the adults are both unemployed and on welfare, what is the probability that an adult picked at random would be either unemployed or on welfare?

THE BINOMIAL DISTRIBUTION

Through probability theory it is possible to derive some formulas to directly calculate the probabilities of experiments which are composed of a series of independent events. We shall consider only one such formula.

If an experiment is composed of a number of simple events of the same type and each simple event has only two outcomes, the probability of a particular outcome can be determined through the formula for the binomial distribution.

The binomial distribution is applicable to situations analogous to coin-flipping experiments, and its use can best be illustrated through consideration of a coin-flipping problem. In flipping seven coins, for example, what is the probability of obtaining six heads? Instead of calculating the number of ways this outcome could occur and dividing by the total number of possible outcomes to the experiment, the probability can be obtained directly from the formula for the binomial distribution. This formula is

$$P_K = C_K^N p^K q^{(N-K)}$$

where N = number of objects or trials in the experiment
$\quad K$ = number of successes whose probability is of interest
$\quad p$ = likelihood of a K-type outcome, or success, in a simple event
$\quad q$ = $1 - p$ (the likelihood of a not-K outcome, or failure, in a simple event)

In the experiment where we wish to know the probability of obtaining six heads in flipping seven coins, $N = 7$ and $K = 6$. If the coins are honest, then $p = \frac{1}{2}$ and $q = 1 - p = 1 - \frac{1}{2} = \frac{1}{2}$. Therefore,

$$P_{6H} = C_6^7 \left(\frac{1}{2}\right)^6 \left(\frac{1}{2}\right)^{7-6}$$

Since

$$C_K^N = \frac{N!}{K!(N-K)!}$$

then

$$C_6^7 = \frac{7!}{6!(7-6)!}$$

Substituting back into the binomial formula

$$P_{6H} = \frac{7!}{6!(7-6)!} \left(\frac{1}{2}\right)^6 \left(\frac{1}{2}\right)^1$$

$$= \frac{(7)\,(6!)}{6!\,(1!)} \left(\frac{1}{2}\right)^7$$

$$= (7)\left(\frac{1}{128}\right) = \frac{7}{128}$$

$$= .055$$

It is not necessary that the outcomes of the simple event be equally likely to use the binomial distribution. Suppose that at a particular college 70% of the students are male. What is the probability that if a random sample of 10 students were selected *exactly* half the sample would be males and half females? In this situation $N = 10$, $K = 5$, $p = .7$, and $q = .3$. Thus

$$P_{(5 \ males)} = C_5^{10}(.7)^5(.3)^{(10-5)}$$

$$= \frac{10!}{5!(10-5)!}(.7)^5(.3)^5$$

$$= \frac{(10)(9)(8)(7)(6)(5!)}{(5!)(5!)}(.7)^5(.3)^5$$

$$= \frac{(10)(9)(8)(7)(6)}{(5)(4)(3)(2)(1)}(.7)^5(.3)^5$$

$$= (252)(.168)(.002)$$

$$= .085$$

Question 6–9. Bob and Tom are now betting on the outcome of rolling a single die. Bob is betting that the die will come up an even number and Tom is betting that it will come up an odd number. On the first seven rolls, the die comes up even each time. Use the binomial distribution to calculate the probability that this could have occurred by chance. (Note that any number raised to the zero power is by definition equal to 1.) Since the die has come up even seven times in a row, Tom feels he will start to win soon and that he should continue to play. Should he? Why?

SAMPLING AND
THE ASSUMPTION OF INDEPENDENCE

In Chapter 2 it was indicated that some types of sampling did not precisely meet the requirements of a simple random sample. Often even seemingly "random" sampling procedures violate the independence requirement of a simple random sample. When this happens, statistical tests that assume a simple random sample, such as the ones in this book, can yield very misleading results. The definition of a simple random sample is that the sample be selected by some random-selection procedure which gives every member of the population an equal and independent chance of being selected and makes the selection of all possible samples which could be chosen equally likely. We are now in a position to discuss more precisely how some sampling procedures may violate the independence requirement.

If a sample is chosen by "randomly" selecting one person at a time from the population, the sample will not meet the independence requirement. Returning to an urn containing 1000 marbles, we saw in Chapter 2 that the probability of randomly selecting a particular marble on the first draw is 1 in 1000, 1 in 999 on the second draw, and so on. This is called *sampling without replacement*. It does not yield a random sample because what happens on the first draw affects the *probability* of what happens on the second draw, which, as we have seen in this chapter, is a situation of nonindependence.

In contrast, *sampling with replacement* does yield a simple random sample. In the case of the urn of marbles, this technique requires that each marble be returned to the urn after it is chosen and before another selection is made. In this way the probability that a particular marble would be selected on any draw is always the same; in the case of the urn, 1 in 1000.

In practice, most sampling is done without replacement. This is of little concern where the size of the sample is small compared to the size of the population. In such a situation the chances of an object being selected twice are very slight even if the object were replaced, so the probabilities involved on each draw are almost unaffected by sampling without replacement. If, however, the size of the sample approaches 20% of the population, the distortion becomes large enough to be important. It is possible to introduce a correction factor in these situations so that the statistical tests designed for a simple random sample can be used and a book on sampling or a statistician should be consulted. If sampling without replacement is used, the results of statistical tests premised on simple random samples are usually more "conservative" than they would have been with a true simple random sample. This loss of power can be regained by the introduction of the appropriate correction factor.

Although sampling without replacement does not result in crippling problems, the failure to give every possible sample an equal chance to be selected can result in a serious violation of the independence assumption. If a deck of cards were sorted into four piles according to suit (clubs, diamonds, hearts, and spades) and one pile randomly selected, every suit in the deck would have an equal chance, one in four, of being selected. But obviously it is not possible to select all possible combinations of 13 cards, such as five spades, three hearts, two diamonds, and three clubs. The results that are obtained from a sample selected in such a manner can be very misleading. Surveys that make use of cluster sampling are open to these errors.

An example of cluster sampling is to divide a city into neighborhoods, say 50, randomly select 10 neighborhoods, and then interview randomly selected people within the 10 selected neighborhoods. Since people residing within a particular neighborhood are more likely to be similar than people residing in different neighborhoods (rich – poor, white – black, low education – high education), this is similar to sorting a deck of cards by suit. Even though people are selected at random in a cluster, combinations of people living within a cluster have a better chance of being selected as a sample than combinations of people living in different clusters. Consequently, the independence assumption is violated.

The advantage of cluster sampling is that it can greatly reduce the travel time and expense associated with interviewing. If a cluster sample of a state were drawn on the basis of counties, each interviewer could be assigned 100 interviews within a county. If a simple random sample of the state's population were drawn, it would probably be necessary for each interviewer to travel across a number of counties to obtain 100 interviews.

When statistics premised on a simple random sample are used on a sample drawn by cluster methods, serious distortions can result. It may be possible to make corrections, but this can become very complicated. The savings obtainable through cluster sampling are very attractive. Consequently, cost-conscious surveyors are drawn to it. Under the proper conditions and if executed correctly, a cluster sampling procedure can yield good results. It is, however, easily abused and the results from any cluster sampling survey should be carefully scrutinized. An expert should be consulted if the findings of a cluster sampling survey are going to be put to an important use.

In general, then, it is important to pay attention to the independence assumption in selecting a sample since statistical tests are usually based on the premise of a simple random sample, and at the least it will be necessary to introduce corrections if serious deviations from independence occur. As has been seen, the idea of a random sample is not a simple thing. With surveys playing an increasing role in social science, politics, and social policy, it is important not to accept the results of "a survey" without carefully inspecting the sampling procedure.

EXERCISES FOR CHAPTER 6

1. An experiment consists of tossing four coins simultaneously and not-ing the number of heads and tails that appear. List the possible out-comes to this experiment. What is the probability of obtaining two heads and two tails? What is the probability of obtaining all tails? What is the probability of obtaining three heads and one tail?

2. If there were 10 nations engaged in negotiations, how many ways are there that these nations could form three-nation alliances?

3. If at a party convention there were seven candidates under considera-tion for nomination as president and vice-president, in how many ways could two candidates be chosen?

4. An examination consists of seven true–false questions. Using the combinations formula, compute the probability that by chance alone a student would get all seven correct. Using the same method, what is the probability that a student could get five correct by chance? What is the probability that by chance alone a student would get only one answer correct?

5. A sequence of three letters is needed to open a safe. How many three-letter sequences are possible using the English alphabet?

6. If the Internal Revenue Service randomly checks 20% of all income tax returns each year, what is the probability that a person would have his return checked two years in a row?

7. A survey of a freshman college class investigating prejudice and class background produced the following results:

		Class		
		Lower Class	Upper Class	
Prejudice	High	200	300	500
	Low	100	400	500
		300	700	1000

Suppose that individuals were selected at random from this freshman class:

a. What is the probability of getting a low-prejudiced student? What is the probability of getting a student from an upper-class family?

b. If family class and prejudice were independent events among these students, what would be the probability of the joint occurrence of upper class and low prejudice?

c. What is the actual probability of the joint occurrence of upper class and low prejudice among the students?

d. Use probabilities to show whether or not family class and prejudice are independent events among the students.

8. In a group of 100 people, 50 are Protestant, 30 are Catholic, and 20 are Jewish. Also 40 of the 100 are Democrats, 30 are Republicans, and 30 are independents. It is found that for these people religion and party identification are independent events. What is the probability that if a person were selected at random from the 100 people:

   a. He would be both Protestant and Republican?
   b. He would be both Jewish and Democratic?
   c. He would be either Protestant or Republican?
   d. He would be either Jewish or Democratic?

9. Joan and Bob are betting on the outcome of rolling a single die. Joan claims that she is a very lucky person, so it is decided that Joan will win if the die comes up 1 or 2, and Bob will win if it comes up 3, 4, 5, or 6. Joan wins the first four rolls of the die. Use the formula for the binomial distribution to calculate the probability of this result occurring by chance alone. Based upon your calculations, do you think that Bob should continue to play? Why?

ANSWERS TO
QUESTIONS IN CHAPTER 6

6–1:

Outcome Number	Coin 1	Coin 2	Coin 3
1	H	H	H
2	H	H	T
3	H	T	H
4	H	T	T
5	T	H	H
6	T	H	T
7	T	T	H
8	T	T	T

$$P_{(2H \text{ and } 1T)} = \frac{3}{8} = .375$$

6-2:

$$P_3^7 = \frac{7!}{(7-3)!} = \frac{(7)(6)(5)(4!)}{4!} = 210$$

6-3:

a. $C_2^7 = \frac{7!}{2!(7-2)!} = \frac{(7)(6)(5!)}{(2!)(5!)} = \frac{42}{2} = 21$

b. $P_2^7 = \frac{7!}{(7-2)!} = \frac{(7)(6)(5!)}{5!} = 42$

6-4:

$$C_0^6 = \frac{6!}{0!(6-0)!} = 1$$

$$C_1^6 = \frac{6!}{1!(6-1)!} = 6$$

$$C_2^6 = \frac{6!}{2!(6-2)!} = 15$$

$$C_3^6 = \frac{6!}{3!(6-3)!} = 20$$

$$C_4^6 = \frac{6!}{4!(6-4)!} = 15$$

$$C_5^6 = \frac{6!}{5!(6-5)!} = 6$$

$$C_6^6 = \frac{6!}{6!(6-6)!} = \frac{1}{64}$$

$$P_{6T} = \frac{1}{64} = .016$$

No. If the coin were honest, the probability that the result of the first six flips would all be tails is about two times out of 100. Since the probability that six tails would have occurred by chance is so small, Bob should conclude that they did *not* occur by chance — that the coin is dishonest (weighted) and he had better quit.

6-5: a. Joint.
   b. Mutually exclusive.
   c. Conditional.
   d. Conditional.
   e. Mutually exclusive.
   f. Joint.

6-6: a. Independent.

b. Independent.
c. Nonindependent.
d. Nonindependent.
e. Independent.

6–7: a. Nonindependent;

$$P_{(2 \text{ pictures})} = \left(\frac{12}{52}\right)\left(\frac{11}{51}\right) = \frac{132}{2652} = .05$$

b. Independent;

$$P_{(2 \text{ gold balls})} = \left(\frac{80}{100}\right)\left(\frac{30}{100}\right) = \frac{2400}{10,000}$$

$$= .24$$

c. Nonindependent;

$$P_{(2 \text{ gold balls})} = \left(\frac{40}{100}\right)\left(\frac{39}{99}\right)$$

$$= \frac{1560}{9900} = .157$$

d. Male and Republican: a nonindependent event:

$$P_{(\text{male and Republican})} = (.60)(.90) = .54$$

Female and Republican: a nonindependent event:

$$P_{(\text{female and Republican})} = (.40)(.30) = .12$$

6–8: a. Mutually exclusive:

$$P_{(3 \text{ or } 4)} = \frac{1}{6} + \frac{1}{6} = \frac{2}{6} = .333$$

b. Not mutually exclusive:

$$P_{(\text{ace or club})} = \frac{4}{52} + \frac{13}{52} - \frac{1}{52}$$

$$= \frac{16}{52} = .308$$

c. Not mutually exclusive:

$$P_{(\text{unemployed or on welfare})} = (.15) + (.20) - (.10)$$

$$= .25$$

6–9:

$$P_{(7 \text{ even})} = C_7^7(.5)^7(.5)^{(7-7)}$$

$$= \frac{7!}{7!(7-7)!} (.5)^7 (.5)^0$$

$$= \frac{7!}{(7!)(0!)} (.008)(1)$$

$$= .008$$

No. The probability that the die would come up even seven times in a row by chance alone is about 8 in 1000. This chance is so small that it can be safely concluded that the outcome is not due to chance but to something else — that the die is weighted.

7
Probability II

Flipping coins, rolling dice, and playing cards are interesting pastimes and provide clear examples of the basic operations of probability. Games of chance have even stimulated work in certain areas of probability theory. But these games are much too simple to represent the probabilities of complex social phenomena. For such events more sophisticated models of probability are needed. Statistics meets this need by developing various mathematical models of probability. These models are derived from probability theory and are purely mathematical conceptions. Their utility lies in the fact that the probabilities of many real-world events have been shown to closely approximate the mathematical models.

THE NORMAL CURVE

One of the most useful models in statistics is the normal curve. It is important to keep in mind that the normal curve exists at two levels: At the mathematical level it is an idealized model, while at the empirical level it describes the shape of the distributions of many real variables, such as height, weight, and IQ. It is not somehow necessary that variables have a normally shaped distribution—indeed, there are many that do not—but it works out that many variables do. It should also be noted that the word "normal" is not meant to imply anything about a distribution or that distributions with a different shape are somehow "abnormal." The word "normal" is used as a name which has become associated with this particular type of distribution. Before considering the normal curve as a mathematical model of probability, we shall investigate some of its properties as an empirical distribution.

What distinguishes the normal from other distributions is its shape. Figure 7–1 presents an example of a normally shaped curve. As can be seen, the normal curve is symmetric and bell-shaped. Because it is unimodal and symmetric, the mean, median, and mode of a normal distribu-

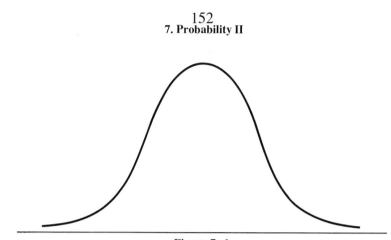

Figure 7–1
The Normal Curve

tion all coincide. Theoretically, the tails of the curve extend out to infinity in both directions without touching the horizontal axis. It is unlikely, of course, that real variables have instances of values which extend to infinity, so in practice the curve can be ended at definite points.

A normal curve as a description of the shape of the distribution of some variable is generated by constructing a frequency distribution and then a frequency polygon from a set of data. Suppose that we were able to obtain the weights of all male college students in the United States. The frequency for each value of weight that occurred in the data could be counted and plotted on a graph as was done for a histogram in Chapter 4. Then a smooth curve can be passed through the points that represent the frequency for each value of weight. The shape of this curve is determined empirically by the way the data are. Past experience has shown that for weight the result will be a normally shaped curve such as that in Figure 7–2. It would be possible by use of the formulas developed in Chapters 4 and 5 to calculate the numerical values for the mean and standard deviation for the data. With thousands of observations involved, this would be a tedious process to say the least, so we will assume without actually doing them that the calculations result in a mean of 150 pounds and a standard deviation of 10 pounds. Since a population is being dealt with, these characteristics are labeled μ and σ in Figure 7–2.

Because the curve in Figure 7–2 was generated by fitting it to the data, it encompasses 100% of the observations. And since all the observations are given equal amounts of space on the graph, there exists a correspondence between various areas under the curve and the proportion of observations in these areas. For example, the mean divides the area of the curve in half: 50% of the total *area* under the curve, *what-*

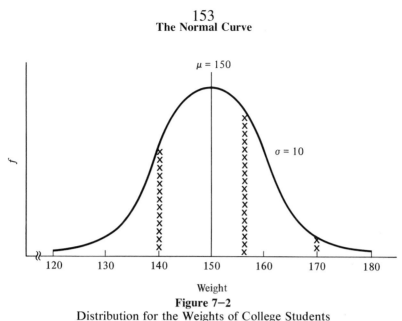

Weight

Figure 7–2
Distribution for the Weights of College Students

ever that might be, lies on either side of the mean. Correspondingly, 50% of the *observations* also lie on either side of the mean. If *any* slice of the distribution were taken, the proportion of the area under the curve in that slice to the total area under the curve will be equal to the proportion of the number of observations in the slice to the total number of observations in the distribution. This means that the proportion of observations that fall into any interval in the distribution can be determined by calculating the area of that interval under the curve.

It is relatively easy to find areas of squares, rectangles, circles, and triangles. But it is difficult to find the area of a figure that is in part bounded by an uneven curve. Fortunately, some properties of the standard deviation can help us with this problem.

It will be recalled from Chapter 5 that 68.26% of the observations in a normal distribution fall within ±1 standard deviation of the mean. This interval is plotted in Figure 7–3 for the distribution of the weights of male college students. Suppose that the distributions in Figures 7–2 and 7–3 were based upon 1,000,000 observations. This would mean that 682,600 college men (68.26% of 1,000,000) had weights which fall into the interval from 140 pounds ($\mu - 1\sigma = 150 - 10 = 140$) to 160 pounds ($\mu + 1\sigma = 150 + 10 = 160$). It also means that 68.26% of the area under the curve lies in the interval on the graph from 140 pounds to 160 pounds. Since 68.26% of the observations lie inside this interval, 31.74% of the observations must lie outside it. And because the curve is symmetrical, half of 31.74% of the observations lie on either side of the

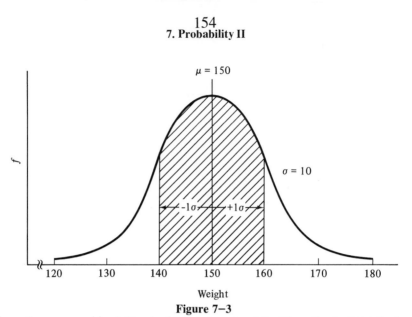

Weight

Figure 7–3
Area Encompassed by 1 Standard Deviation about the Mean for the Distribution in Figure 7–2

interval: 15.87% of college men weigh less than 140 pounds and 15.87% weigh more than 160 pounds. Similar statements hold for the corresponding *areas under the curve:* 15.87% of the *area* under the curve lies to the left of a line drawn at 140 pounds and 15.87% of the *area* under the curve lies to the right of a line drawn at 160 pounds.

Let us consider another example of areas under the normal curve and the distribution of observations. Suppose that the cumulative grade-point averages of all students at a university are obtained. Using a point scale of A $= 4$, B $= 3$, C $= 2$, D $= 1$, and F $= 0$, a normal distribution is obtained when the students' averages are graphed as in Figure 7–4. We will again assume that calculations have been done for the mean and standard deviation and result in $\mu = 2.5$ and $\sigma = .5$. By knowing that 68.26% of the area under *any* normal curve lies within ± 1 standard deviation of the mean (regardless of what the values are for the mean and standard deviation) we can immediately determine that 68.26% of the students at the university have grade-point averages that fall into the interval 2.0 to 3.0. This interval was found by simply adding and subtracting the value of the standard deviation to the mean as follows:

$$\text{value of upper limit of}$$
$$\text{a 68.26\% interval} = \mu + 1\sigma$$
$$= 2.5 + .5$$
$$= 3.0$$

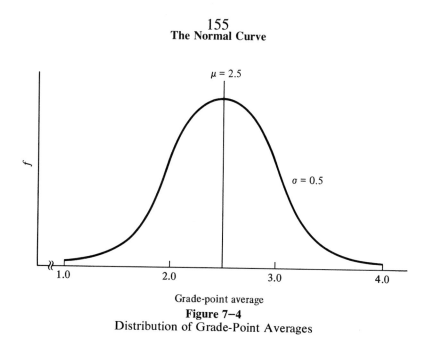

Grade-point average
Figure 7–4
Distribution of Grade-Point Averages

value of lower limit of
a 68.26% interval $= \mu - 1\sigma$
$= 2.5 - .5$
$= 2.0$

Rather than taking an interval encompassing 68.26% of the area under the curve, it is more usual to inquire about intervals encompassing 95% and 99% of the area under the curve. It turns out that 95% of the area under a normal curve is encompassed by an interval of 1.96 standard deviations about the mean and 99% is encompassed by an interval of 2.58 standard deviations about the mean. Thus if someone asked within what range do the middle 95% of all students' grade-point averages fall, the answer could be found as follows:

value of upper limit of
a 95% interval $= \mu + 1.96\sigma$
$= 2.5 + (1.96)(.5)$
$= 2.5 + .98$
$= 3.48$
value of lower limit of
a 95% interval $= \mu - 1.96\sigma$
$= 2.5 - (1.96)(.5)$
$= 2.5 - .98$
$= 1.52$

Thus 95% of the students have grade-point averages which fall into the range 1.52 to 3.48.

If a 99% interval were asked for, it would be found in the same way:

value of upper limit of
$$\text{a } 99\% \text{ interval} = \mu + 2.58\sigma$$
$$= 2.5 + (2.58)(.5)$$
$$= 2.5 + 1.29$$
$$= 3.79$$

value of lower limit of
$$\text{a } 99\% \text{ interval} = \mu - 2.58\sigma$$
$$= 2.5 - (2.58)(.5)$$
$$= 2.5 - 1.29$$
$$= 1.21$$

A range from 1.21 to 3.79 contains the middle 99% of all student grade-point averages.

We can use knowledge of areas under the normal curve to answer questions concerning the location of an observation in a distribution. For example, if a student has a 3.48 grade-point average, where would he rank compared to all the other students at the university? From the calculations above it can be seen that a score of 3.48 is 1.96 standard deviations above the mean. The area under the curve encompassed by a distance of 1.96 standard deviations above and below the mean is 95% of the total area. Half of this area, the area under the curve from the mean to $+1.96\sigma$, is 47.5% (half of 95%). Recalling that the mean divides the area under the curve in half, we can add 47.5% (the area above the mean to $+1.96\sigma$) to 50% (the total area below the mean) and say that 97.5% of the area under the curve lies below a line drawn through the point marked by 3.48. This corresponds to a statement that 97.5% of all scores have values less than 3.48. Consequently, a student with a grade-point average of 3.48 ranks above 97.5% of all students at the university. And since the total area under the curve encompasses 100% of the observations, 2.5% (100 − 97.5) of the students have grade-point averages higher than 3.48.

Question 7–1. Using the information presented in Figure 7–3, answer the following questions.

a. What range of weight encompasses the middle 95% of the weights of all male college students?
b. What range of weight encompasses the middle 99% of the weights of all male college students?
c. If a student weighed 140 pounds, how much of the population is lighter than he is? How much is heavier?

d. If a student weighed 175.8 pounds, how much of the population is lighter and how much is heavier than he is?

STANDARD SCORES

The areas under the normal curve contained in intervals of 1.0, 1.96, and 2.58 standard deviations about the mean, although often used, may not be the only ones of interest. Also, there are many different normal curves, curves normally shaped but with different values for either the mean and/or the standard deviation. It would be convenient if there were one single way to find any areas under any normal curve. This is the purpose of standard scores: With standard scores it is possible to use a single table to find areas bounded by any values in any normal distribution.

Standard scores are derived through the use of a linear transformation. This sounds formidable but is fairly simple, and we frequently make use of such transformations in our everyday lives. A linear transformation translates the original values of a set of scores to a new set of values without changing any of the relationships within the data.

Most schools, for example, use a 4-point system for grades. Some schools, though, use a 5-point system, where an A = 5, B = 4, C = 3, D = 2, and F = 1. In order to translate the scores from a school using a 4-point system into the 5-point system we simply add the value of "1" to each score. If one student has an A average and another a C average under the 4-point system, they would also have A and C averages under the 5-point system when a 1 is added to their scores. The linear transformation, while changing the numerical values of the scores, does not change the relationship of the scores to each other.

Some schools even use an 8-point system, where A = 8, B = 6, C = 4, D = 2, and F = 0. (These schools usually give credit for "plus" grades by using the odd numbers; for example, a C+ would equal a 5 in this system.) Again the scores in the 4-point system can be translated into the 8-point system, this time by multiplying the scores of the 4-point system by 2. Thus, an A in the 4-point system, which equals a 4, would when multiplied by 2 come out to be an A in the 8-point system, an 8. And a C in the 4-point system, which equals a 2, would come out a C in the 8-point system, a 4. Again the relationship between the scores is not altered by the transformation.

An everyday example of a linear transformation is the volume control on a radio or television set. Turning the knob higher causes all the sounds to get louder. But a sound that was relatively soft when the knob was set low is still relatively soft (even though it is louder) when the knob is set high. In other words, the original relationships between loud and soft sounds is preserved despite the fact that everything is louder than it was before. This is essential if the sound is to remain intelligible. (One of the reasons that distortion occurs in the sound when the volume control is turned very high is that the power curve tails off and becomes nonlinear as the power increases.)

It is possible, then, to manipulate the scores in a set of data through a linear transformation without affecting any of the relationships within the data. And since it is usually the relationships within the data that are of concern and not the specific values of the scores, nothing is lost through this process. Any of the arithmetic operations of adding or subtracting a constant or multiplying or dividing by a constant (as long as the same constant is used on all the scores) result in a linear transformation.

The linear transformation of concern here changes the raw scores of a set of observations on a variable (symbolized by X) into what are known as standard scores (symbolized by z and often called z scores). The formula for standard scores is

$$z = \frac{\text{score} - \text{mean}}{\text{standard deviation}}$$

Verbally, this formula says that to find the corresponding z score for any score, subtract the mean of the scores (a constant) from the score and divide the result by the standard deviation of the scores (also a constant).

Figure 7–5 depicts what happens to a distribution of raw scores when it is put through this transformation. For this distribution, a score is symbolized by X, the mean as μ, and the standard deviation as σ. The formula for the transformation, then, is

$$z = \frac{X - \mu}{\sigma}$$

By subtracting the value of the mean from each raw score the location of the distribution is shifted on the scale of measurement to a point where its middle sits over the zero point. That the mean of the new distribution has to be zero if the value of the original mean is subtracted from each raw score can be seen intuitively from the fact that the result of subtracting the value of any mean from itself is zero.

The consequence of dividing by the standard deviation in the z-score formula is that the value of the standard deviation in the new distribution is equal to 1. This point is somewhat more difficult to see. The standard deviation is a measure of the amount of dispersion in a distribution. By dividing by the value of the standard deviation, the amount of

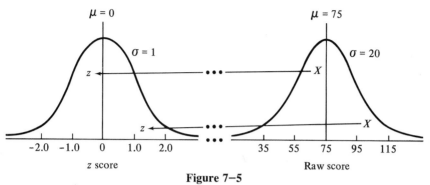

Figure 7–5

Transformation of Raw Scores in a Normal Distribution into z scores in the Standard Normal Distribution

dispersion in the new distribution is adjusted by a factor equal to that value. In Figure 7–5, the value of the standard deviation of the original distribution is 20. When the z-score formula divides by 20, the amount of dispersion in the distribution is reduced by a factor of 20. If a distribution for which a measure of dispersion equals 20 has its dispersion reduced by a factor of 20, that same measure of dispersion, when recalculated on the new scores, must now equal 1. As a result of dividing by the standard deviation in the z-score formula, the new distribution will always have a standard deviation equal to 1 regardless of what the value of the standard deviation was in the original distribution.

As a result of the z-score transformation, any original normal distribution will always yield a *standard normal distribution:* a normal distribution whose mean is equal to zero and whose standard deviation is equal to 1.

Let us illustrate how the z-score transformation preserves the relationships that exist within a set of scores. Figure 7–5 graphically shows the transformation of two observations whose raw scores are 65 and 105. Mathematically, the corresponding standard scores for these observations are found as follows:

$$z_{65} = \frac{65 - \mu}{\sigma}$$

$$= \frac{65 - 75}{20}$$

$$= \frac{-10}{20}$$

$$= -.5$$

$$z_{105} = \frac{105 - \mu}{\sigma}$$

$$= \frac{105 - 75}{20}$$

$$= \frac{30}{20}$$

$$= 1.5$$

In the original distribution, the raw score of 65 is 10 units below the mean, and 105 is 30 units above the mean. This means that 105 is 3 times as far from the mean as is 65. In the standardized distribution the z score of $-.5$ is .5 of a unit below the mean, and the z score of $+1.5$ is 1.5 units above the mean. Again one score is 3 times as far from the mean as the other, and the original relationship among the observations (their relationship to each other and to the mean) is preserved.

A very important property of the standard normal distribution is that *the units of measurement for the z scores are units of standard deviation.* A z score gives the location of an observation in terms of how many units of standard deviation the observation is above or below the mean. In the original distribution in Figure 7–5, the standard deviation is 20. The raw score of 65 is 10 units below or $-.5$ of a standard deviation $(-10/20 = -.5)$ from the mean $(\mu = 75)$. The value for the z score for this observation is $-.5$. Similarly, the raw score of 105 is 30 units above or $+1.5$ standard deviations from the mean. Again, the z score for this observation is $+1.5$. Thus, a z score tells us immediately how far a score is from the mean in terms of standard deviations.

AREAS UNDER
THE STANDARD NORMAL CURVE

By transforming raw scores in a distribution into z scores, a single table may be used to find areas in any normal distribution. Table A–3 in the Appendix gives the proportion of the total area under a standard normal curve contained in an interval from the mean to a particular positive z score. If a z score of 1.0 is used to enter the table, the body of the table gives the area under the curve between the mean and this score as .3413. This number is the *proportion of the total area under the curve* bounded by the mean and a line drawn through the point on the horizontal axis marked by a z score of 1.0.

Since the normal curve is symmetric in shape, the table only gives areas for positive z scores. Areas for negative z scores may be found by using the absolute value of a negative z score and keeping in mind that this area is located to the left of (below) the mean.

Question 7–2.

a. Find the proportion of the area under a normal curve between the mean and a z score of .5.
b. What proportion of the area under a normal curve lies to the right of a z score of .5?
c. What proportion of the area under the normal curve lies to the left of a z score of −.5?
d. What proportion of the area under the normal curve lies in the interval bounded by z scores of −1.0 and −2.0?
e. Assuming that the age of state governors in the United States is normally distributed with a mean of 56 and a standard deviation of 8, how many governors are between 60 and 70 years of age?

ADDITIONAL USES OF z SCORES

It is more usual to deal with samples in actual work than with populations. If the distribution of the sample scores is normally shaped and the sample size is fairly large, it is possible to obtain standard scores and to use Table A–3 to investigate the distribution. When a sample is involved, the values for the mean and the standard deviation for the distribution are symbolized as $\overline{X}$ and s. Consequently, the formula for z scores becomes

$$z = \frac{X - \overline{X}}{s}$$

One of the advantages of standardized scores over raw scores is that they locate a particular value relative to the other scores in a distribution. Suppose that we were interested in studying property taxes and amount of money spent per pupil for education in a sample of local communities. If we were to find that in a particular community the average local property tax was $700 per year and the per student education expenditure $200, these numbers would have little meaning for us. Is $700 a high or a low tax? Is $200 a little or a lot to spend on a student?

From the information on the communities in the sample we may find that the mean property tax for all the communities is $600 and the standard deviation $100, and the mean expenditure per pupil is $300 and the standard deviation is $50. If these variables are also normally distributed, standardized scores can now be calculated to put into perspective the figures for any one community relative to all the communities in the sample. Thus,

$$z_{700} = \frac{700 - 600}{100} = \frac{100}{100} = 1.0$$

$$z_{200} = \frac{200 - 300}{50} = \frac{-100}{50} = -2.0$$

The standardized scores immediately show that this particular community has a high property tax and a low education expenditure compared to the other communities in the sample. More specifically, we can say that this community has a property tax which is higher than about 84% of the communities in the sample and an educational expenditure that is smaller than about 98% of them. These figures were determined by using areas under the normal curve. The area between the mean and a z score of 1.0 is .3413. Added to the area below the mean (.5000) yields .8413. Since a tax of $700 per year has a z score of 1.0 in the distribution, about 84% of the observations in the sample have taxes less than this value.

For education expenditures, the area between the mean and z score of -2.0 is .4772. Added to the area under the normal curve above the mean (.5000) yields .9772. Thus, an educational expenditure of $200 per year is less than that made by about 98% of the communities in the sample. This particular community, then, seems to have a relatively low education expenditure. Obviously, this is much more informative and interesting than just knowing the community's raw scores on the variables of taxes and educational expenditure.

Although standard scores convey more information than raw scores, sometimes we want to know the raw score for some given z score. As long as the mean and standard deviation for the distribution are known, this can be done through a simple manipulation of the standard score formula. For a normally distributed population of scores,

$$z = \frac{X - \mu}{\sigma}$$

Multiplying both sides of the equation by σ,

$$(\sigma)(z) = \frac{X - \mu}{\sigma}(\sigma)$$

$$\sigma z = X - \mu$$

Adding μ to both sides,

$$\sigma z + \mu = X - \mu + \mu$$

$$\sigma z + \mu = X$$

Rewriting the equation so that X is on the left side yields

$$X = \sigma z + \mu$$

If the distribution was a large sample of scores instead of a population, we would merely substitute the appropriate symbols for the mean and the standard deviation and obtain

$$X = sz + \overline{X}$$

Suppose we are told that Mary had a z score of 2.5 on a college entrance math test for which the mean was 75 and the standard deviation 8. Her raw score on the test can be found by substituting into the formula

$$X = \sigma z + \mu$$
$$= (8)(2.5) + 75$$
$$= 20 + 75$$
$$= 95$$

If on the same test Bob had a z score of -1.75, his raw score would be

$$X = \sigma z + \mu$$
$$= (8)(-1.75) + 75$$
$$= -14 + 75$$
$$= 61$$

Question 7–3. In timing a random sample of 200 high school boys in the 100-yard dash, the distribution of times was found to be normally shaped with a mean of 13 seconds and a standard deviation of 1.2 seconds.

a. How many boys ran the 100 yards in less than 11 seconds?
b. How many boys ran it between 11 and 12 seconds?
c. One boy ran the 100 yards 1 standard deviation faster than the average time. What was his time?
d. Another boy had a time of $+2.5z$ (he ran slower than the average). What was his time?

_____

THE NORMAL CURVE
AS A PROBABILITY DISTRIBUTION

Mathematical models of chance are called *probability distributions*. A probability distribution indicates the probabilities of the various outcomes associated with a variable. The idea of a probability distribution closely parallels that of a frequency distribution. In a frequency distribution each possible outcome to a variable is paired with a frequency — the number of times the outcome occurred in a set of data. In a probability distribution each possible outcome is paired with its probability of occurrence. Since probability distributions are mathematical models, they can be described by mathematical formulas. In Chapter 6 the binomial probability distribution was discussed (although it was not then called a probability distribution), and its mathematical formula was given as

$$P_K = C_K^N p^K q^{(N-K)}$$

This formula can be used to calculate the probabilities of the outcomes of experiments consisting of a number of simple independent events where each simple event has only two outcomes. For an experiment consisting of flipping five honest coins, the probability distribution would be

$$P_{0H} = C_0^5 (.5)^0 (.5)^5 = .03125$$

$$P_{1H} = C_1^5 (.5)^1 (.5)^4 = .15625$$

$$P_{2H} = C_2^5 (.5)^2 (.5)^3 = .31250$$

$$P_{3H} = C_3^5 (.5)^3 (.5)^2 = .31250$$

$$P_{4H} = C_4^5 (.5)^4 (.5)^1 = .15625$$

$$P_{5H} = C_5^5 (.5)^5 (.5)^0 = .03125$$

(Note that these probabilities are the same as those obtained in Chapter 6 using the counting method.) As in the case of a frequency distribution, it is possible to graph a probability distribution. This is done for the probability distribution obtained from the binomial distribution for the experiment of flipping five coins in Figure 7–6.

In constructing a graph of a probability distribution, the possible outcomes are arranged on the horizontal axis. Thus the height of the point above each outcome represents its probability of occurrence as determined from the mathematical model of the binomial distribution. It is because the outcomes of flipping coins have been found to closely, and perhaps perfectly, approximate the model of the binomial distribution that we are justified in using the binomial formula to compute the probabilities of the outcomes to coin-flipping experiments.

So far the normal curve has been discussed as an empirical description of the distributions of certain variables. The normal curve can also be used as a mathematical model of probability. It is because the normal curve has been found to provide a good description of the distributions for many variables that it is useful as a mathematical model.

In contrast to the binomial distribution, which deals with discrete variables (it is logically impossible, for example, to obtain 3.6 heads as

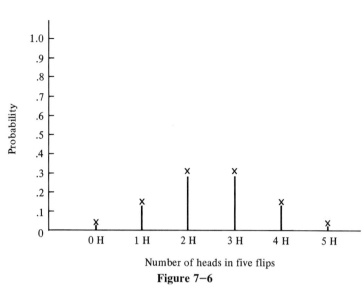

Number of heads in five flips

Figure 7–6

Graph of the Binomial Probability Distribution for the Experiment of Flipping Five Coins

the outcome to a coin-flipping experiment), the normal distribution deals with continuous variables. Since a continuous variable may logically take on all possible values within a particular range, all the values in the range may have some probability of occurring. Consequently, a graph of the normal curve as a probability distribution is a smooth curve rather than a plot of discrete points as in the case of the binomial.

It would seem that the height of the curve above the horizontal axis represents the probability of occurrence of the values of the variable— that the higher the curve over a particular point, the more likely is the outcome at that point. A technical problem arises here, however. In the case of a discrete probability distribution, only certain *exact* values, such as five heads or four heads or three heads and so on, can occur. In the case of a continuous variable, though, any values within a certain range can occur. In a distribution composed of the ages of college students we would expect many students to be 20 years old. But how many would be precisely 20.544572 years old or 20.544572983257 years old? It is possible that there would be many 20-year-olds but none who were precisely as old as either of these ages. Yet the curve would have a real height over these points, indicating some probability of their occurrence. This problem arises because technically any point is infinitely small in size. To avoid this problem the probabilities for continuous variables are not given by the height of the curve over a particular point, but by the *area* under the curve for some interval bounded by two points. This interval

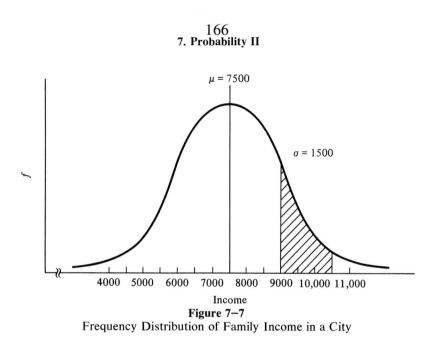

$\mu = 7500$

$\sigma = 1500$

4000 5000 6000 7000 8000 9000 10,000 11,000

Income

Figure 7–7

Frequency Distribution of Family Income in a City

can be made very small if desired. Thus instead of asking what is the probability of a person being 20.544572 years old, we can ask what is the probability that a person's age would fall into the interval from 20.5 to 20.6 years old. As a mathematical model, then, *probabilities are computed from the normal curve by finding the area under the curve for a certain interval.*

The process of finding areas under the normal curve is by now a familiar one. The normal curve is used as a probability distribution by interpreting an area as representing the probability that a variable would have an outcome in the interval which bounds that particular area. An illustration will demonstrate this process.

Suppose that the distribution of family income in a city is obtained. We will assume that in this city family income is found to be normally distributed and is graphed as in Figure 7–7. If in this particular city there are 200,000 families, we might ask how many of the families have an income between $9000 and $10,500 per year. This is a question concerning *the number of observations* that fall into a particular interval. It can be answered by finding the corresponding z scores for these incomes and the proportion of the area under the normal curve bounded by the z scores. This area will then correspond to the *proportion* of the total number of observations in the distribution which fall into the interval of concern:

$$z_{9000} = \frac{9000 - 7500}{1500} = \frac{1500}{1500} = 1.0$$

$$z_{10,500} = \frac{10{,}500 - 7500}{1500} = \frac{3000}{1500} = 2.0$$

area from mean to z score of $1.0 = .3413$

area from mean to z score of $2.0 = .4772$

area of interval bounded by z scores of 1.0 and $2.0 = .4772 - .3413$

$$= .1359$$

This area is the shaded but undashed portion of the curve in Figure 7–8. The number .1359, then, is the proportion of the total area under the normal curve bounded by z scores of 1.0 and 2.0. These z scores correspond to incomes of $9000 and $10,500 per year. To find the number of families who have an income in this range, the total number of families in the city (and in the distribution), 200,000, is multiplied by .1359. This yields 27,180 and is the number of families whose income falls into the range 9000 to 10,500.

Let us now ask a different type of question. If one family out of the city were selected at random what would the probability be that its income would fall into the range $9000 to $10,500? Since there are 200,000 families in the city, and, as we have just seen, 27,180 of them have incomes in this range, the probability would be

$$P_{(9000 \text{ to } 10{,}500)} = \frac{27{,}180}{200{,}000} = .1359$$

What we have done so far is to use the normal curve as a description of the original frequency distribution of family income. Another way to arrive at the probability in question, however, is to treat the normal curve as a probability distribution. By knowing that the distribution for family income is normally distributed with a mean of $7500 and a standard deviation of $1500, the corresponding z scores for the interval from $9000 to $10,500 can be found, and the proportion of the area under the curve bounded by those z scores found as was done in Figure 7–8. The normal curve as a probability model then tells us that the *probability* of one randomly selected family having an income between $9000 and $10,5000 is .1359 without having to know either the total number of families in the city or the number that actually falls into this range. In this situation the normal curve is being used as a model that computes probabilities rather than as a description of a particular frequency distribution. It is important to note that to use the normal curve as a mathematical model to calculate probabilities, it is only necessary to know the

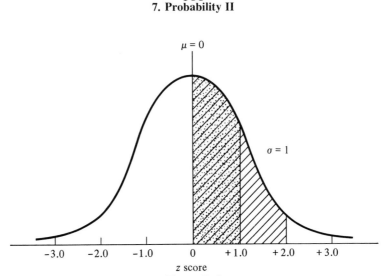

Figure 7–8
Finding the Area in the Standard Normal Distribution that Corresponds to the
Shaded Area in Figure 7–7: Dotted Area, Proportion of Area from Mean to
z score of $1.0 = .3413$; Crosshatched Area, Proportion of Area from Mean to z
score of $2.0 = .4772$

mean and standard deviation of a variable and to know the variable is
approximately normally distributed.

Question 7–4. Assuming that in a particular state the distribution for
number of years of education completed by adults is normal with a mean
of 10 and a standard deviation of 3, answer the following questions:

a. What is the probability that if an adult were picked at random, he
   would have more than 12 years of education?
b. What is the probability that an adult picked at random would have
   between five and eight years of education?
c. If a sample of 500 adults were selected at random in the state, how
   many of them could be "expected" to have between 13 and 16 years
   of education? _____

SAMPLING DISTRIBUTIONS

Suppose that 75 students in a class are each given a standard deck of 52
cards and assigned the following task: Each student, after shuffling his
deck thoroughly, deals himself 10 cards. Using a point scheme wherein
an ace equals 1, a deuce 2, a trey 3, and so on for the numbered cards,
and a jack equals 11, a queen 12, and a king 13, each student records the

point value for each of his 10 cards. Then the points are summed and the mean point value for the 10 cards is computed. For example, the following hand might have been one of the 75 dealt:

Card	Point Value
K clubs	13
J hearts	11
J diamonds	11
10 spades	10
7 spades	7
7 hearts	7
3 diamonds	3
3 clubs	3
2 hearts	2
Ace clubs	1
Total points	68

$$\overline{X} = \frac{\Sigma X}{N} = \frac{68}{10} = 6.8$$

Each hand may be considered an approximately randomly selected sample from the population of a deck of cards. Each student, then, has drawn one random sample. Since every student has calculated the mean for his sample, there are 75 sample means (75 $\overline{X}$'s). In the particular hand illustrated above, the value of the sample mean is 6.8. Of course, there will be a variety of values for the sample means among the 75 samples.

In order to discover the distribution of values of sample means among the 75 hands, a frequency distribution may be constructed as in Table 7–1. This frequency distribution is constructed, of course, by counting the number of hands (samples) whose means fall into a particular interval. Thus if two samples have means of 8.6 and 9.3, their means will be placed in the interval 8.55 to 9.55. From this frequency distribution a histogram can be constructed as in Figure 7–9. It is important to note that the scores being plotted in Figure 7–9 are not values of individual cards, but are the values of 75 sample means, each sample being of size 10.

The mean of a deck of 52 cards (the population in this situation) is 7.0. We shall call the deck of cards the *original population*. Since it is a population, its mean is symbolized by μ, but in order to emphasize that it is the mean of the original population, the letter o will be subscribed to it. Thus, $\mu_o = 7.0$. Examining Figure 7–9 it can be seen that the values

Table 7–1
Frequency Distribution for the Sample Means of Hands Consisting of 10 Cards

Interval	Frequency
0.55 – 1.55	0
1.55 – 2.55	0
2.55 – 3.55	0
3.55 – 4.55	3
4.55 – 5.55	8
5.55 – 6.55	13
6.55 – 7.55	22
7.55 – 8.55	19
8.55 – 9.55	7
9.55 – 10.55	2
10.55 – 11.55	1
11.55 – 12.55	0

of the sample means are clustering in the center near the value of 7.0. An interesting point now emerges. It would, of course, be possible to draw more than 75 samples of 10 cards each from the original population (a deck of cards). In fact, we could draw so many samples, and plot the mean value of each sample in the distribution started in Figure 7–9, that

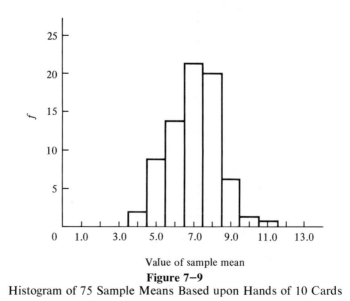

Value of sample mean
Figure 7–9
Histogram of 75 Sample Means Based upon Hands of 10 Cards

this distribution itself could be considered a population. The interesting point is: What would be the value of the mean for such a population? Since we are interested in the *mean of a population* that is made up of the values of *sample means,* we will symbolize it as $\mu_{\bar{x}}$. As might be anticipated, $\mu_{\bar{x}}$ turns out to always equal μ_o, which in this case is 7.0.

Since a distribution consisting of the values of sample means is obtained by drawing repeated random samples from the original population, computing the mean for each sample and plotting it, such a distribution is called a *sampling distribution of the mean.* More generally, a *sampling distribution* refers to any distribution that is obtained by drawing repeated random samples from a population, computing a descriptive statistic for each sample, and then plotting the value of this statistic for each sample. There are many types of sampling distributions in addition to the sampling distribution of the mean, some of which will be considered later.

Another interesting property of sampling distributions of the mean is their shape compared to the shape of the original population on which they are based. The distribution for a deck of cards is graphed in Figure 7 – 10. This distribution is rectangular and certainly nonnormal in shape. The histogram in Figure 7–9, however, bulges in the middle and tapers off at its ends and, although not normal, it more closely resembles a normal distribution than the original population. In fact, it can be shown that as the size of the random samples that are selected from the original population gets larger, the sampling distribution of the mean for those samples will approach a normal distribution in shape, *regardless of the shape of the original population!* Mathematically, the shape of the sampling distribution of the mean *approaches* a normal curve as the sample size gets larger. For practical purposes, however, the normal curve is a good approximation to the sampling distribution of the mean even with samples of only moderate size. It is this property that underlies part of the great utility of the normal curve. It means that even though the dis-

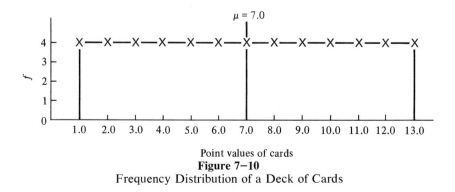

Point values of cards
Figure 7–10
Frequency Distribution of a Deck of Cards

tribution of some variables may be nonnormal, it may still be possible to apply normal-curve models because the sampling distribution of the mean can be usefully approximated by a normal curve if the sample size is moderately large.

A final property of sampling distributions of the mean is the relationship between the standard deviation of the original population and the standard deviation of the sampling distribution. If we represent the standard deviation of the original population as σ_o and the standard deviation of the sampling distribution of the mean as $\sigma_{\bar{x}}$, then $\sigma_{\bar{x}} = \sigma_o/\sqrt{N}$, where N is the size of the samples that were drawn in constructing the sampling distribution.

These properties of sampling distributions of the mean are derived from a mathematical law known as the *central limit theorem*. From it we are able to say that: If repeated random samples of size N are drawn from any original population (of whatever form) which has a mean of μ_o and a standard deviation of σ_o, then as N becomes large, the sampling distribution of the sample means approaches normality with a mean ($\mu_{\bar{x}}$) equal to μ_o and a standard deviation ($\sigma_{\bar{x}}$) equal to $\sigma_o/\sqrt{N}$.

THE STANDARD
ERROR OF THE MEAN

The standard deviation, as a measure of the dispersion in a distribution, indicates how far scores deviate from the mean. When the mean of a random sample is computed, the deviations of the observations from the population mean are averaged. For example, if the mean of a population of test scores is 70, one randomly selected score might be 60 and another 85. The first deviates from the mean by 10 units and the second by 15 units. But if these two scores are treated as a random sample of size 2 and their mean computed, $\overline{X} = (60 + 85)/2 = 72.5$, the mean of the sample deviates from the mean of the population by only 2.5 units. In other words, when a random sample is selected, the values of scores in the sample above the population mean are likely to be balanced by other scores in the sample which fall below the population mean. The result is that the mean of the sample tends to be close to the mean of the population. Of course, it is not necessary that this balancing has to occur. It is possible that through random selection all the scores in a sample could have extreme values at one end of the population distribution. In such a situation, the mean of the sample would remain very far from the mean of the population. But as the size of the sample becomes larger, it becomes more likely that balancing will occur and less likely that all the scores in the sample would have extreme values from the same side of the distribu-

The Standard Error of the Mean

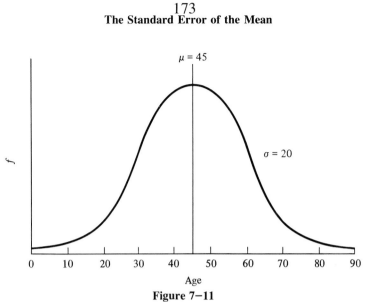

Figure 7–11
Distribution for the Population of the Ages of a City's Residents

tion. Consequently, as the size of the samples gets larger, the means of the samples in a sampling distribution cluster more and more closely about the value of the mean of the original population. Thus the degree of dispersion in a sampling distribution will become less as the size of the samples selected from the original population gets larger. The standard deviation of a sampling distribution ($\sigma_{\bar{x}}$) is a measure of the dispersion in that distribution. The formula $\sigma_{\bar{x}} = \sigma_o / \sqrt{N}$ indicates how much the dispersion in the sampling distribution will be reduced by samples of various sizes.

Suppose that we had a population of the ages of all the people living in a particular city. This population might have a mean of 45 and a standard deviation of 20. The distribution is represented in Figure 7–11. If repeated random samples of 50 city residents each were selected from this population, the distribution of the sample means would result in the sampling distribution in Figure 7–12a. This sampling distribution will have a mean equal to 45 (the value of the original population mean) and a standard deviation equal to 2.83. The value of the standard deviation of this sampling distribution was determined as follows:

$$\sigma_{\bar{x}} = \frac{\sigma_o}{\sqrt{N}}$$

$$= \frac{20}{\sqrt{50}}$$

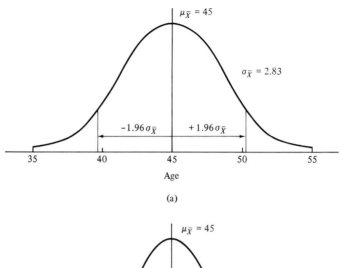

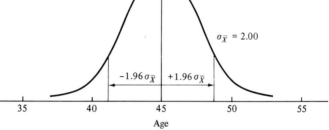

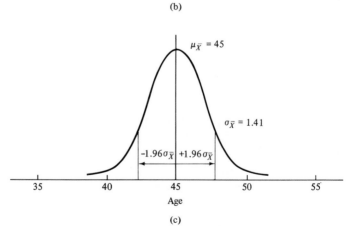

Figure 7–12

Sampling Distributions of the Mean for Three Sample Sizes Taken from the Population in Figure 7–11: (a) Size 50; (b) Size 100; (c) Size 200

$$= \frac{20}{7.07}$$

$$= 2.83$$

Notice that the dispersion in the sampling distribution is reduced by roughly a factor of 7. Obviously, the larger the sample size, the greater the reduction of dispersion in the sampling distribution. This reduction, however, is proportional not to the size of the samples selected but to the *square root* of the sample size. This means that as samples get larger, there is a decreasing amount of reduction in the dispersion of the sampling distribution. For samples of 50, the reduction is a factor of about 7 ($\sqrt{50} = 7.07$). For samples of 100, the reduction is a factor of 10 ($\sqrt{100} = 10$), and the reduction for samples of size 200 is a factor of about 14 ($\sqrt{200} = 14.14$). The sampling distributions of the mean for samples of size 100 and 200 selected from the population of city residents' ages are graphed in Figures 7–12b and c. From Figure 7–12 it can be seen how the sampling distribution clusters more tightly about the population mean as the sample size increases. But it should be noted that to reduce by half the amount of dispersion in the sampling distribution (from $\sigma_{\bar{x}} = 2.83$ to $\sigma_{\bar{x}} = 1.41$) is was necessary not to double but to quadruple the sample size (from 50 to 200). This stems from the fact that the dispersion in the sampling distribution of the mean is proportional to the square root of the sample size.

When a sample is selected and its mean computed we think of that mean as the result of computations made on a random selection of observations from the original population. It is also possible to think of that sample mean in another way, as one sample mean that is randomly selected from the population of the sampling distribution of the mean. In other words, the particular sample mean that is obtained is conceived of as one randomly selected mean from all possible sample means that could occur. Since the sampling distribution for the mean is approximately normal and its mean and standard deviation can be found from information about the original population, the normal curve can be used to represent the probabilities of selecting various values for the sample mean.

Referring to Figure 7–12a, the mean ($\mu_{\bar{x}}$) of the sampling distribution is 45 and its standard deviation ($\sigma_{\bar{x}}$) is 2.83. We know that in any normal distribution 68.26% of the observations lie within ±1 standard deviation of the mean. Therefore, 68.26% of the means *of all possible samples* of size 50 that could be drawn from the population of city residents will have values between 42.17 and 47.83. These values were found as follows:

upper limit of 68.26%
interval in sampling distribution
for samples of size 50 $\qquad = \mu_{\bar{x}} + 1\sigma_{\bar{x}}$

$$= 45 + 2.83$$

$$= 47.83$$

lower limit of 68.26%
interval in sampling distribution
for samples of size 50 $\qquad = \mu_{\bar{x}} - 1\sigma_{\bar{x}}$

$$= 45 - 2.83$$

$$= 42.17$$

We also know that in a normal distribution 95% of all the observations lie within ±1.96 standard deviations of the mean. Therefore, the range in which 95% of all the *means* of all possible samples of size 50 that could be selected can be found as follows:

upper limit of 95%
interval in sampling distribution
for samples of size 50 $\qquad = \mu_{\bar{x}} + 1.96\sigma_{\bar{x}}$

$$= 45 + (1.96)(2{:}83)$$

$$= 45 + 5.55$$

$$= 50.55$$

lower limit of 95%
interval in sampling distribution
for samples of size 50 $\qquad = \mu_{\bar{x}} - 1.96\sigma_{\bar{x}}$

$$= 45 - (1.96)(2.83)$$

$$= 45 - 5.55$$

$$= 39.45$$

What this range indicates is that if repeated random samples of size 50 were drawn from the population, in the long run 95% of all the sample means that would be obtained will fall in the range 39.45 to 50.55. This interval is plotted in each of the sampling distributions in Figure 7–12.

The standard deviation of a sampling distribution of the mean is also known as the *standard error of the mean*. A standard deviation measures how far scores deviate from the mean in a distribution. Since the mean of a sampling distribution is equal to the mean of the original population ($\mu_{\bar{x}} = \mu_o$), the standard deviation of a sampling distribution measures how far sample means deviate from the value of the true population mean. Consequently, the standard deviation of a sampling distribution is a numerical indicator of the deviation in the values of sample means, and thus their *error,* from the true population mean.

This point is made mathematically by calculating the interval in which, say, 95% of all possible sample means for a sampling distribution will fall. For the sampling distribution in Figure 7–12a, this interval has a range of 11.1 units. For sampling distributions based on larger sample sizes, the range of the interval in which 95% of all possible sample means fall becomes smaller as in Figure 7–12b and c. As the sample size gets larger, then, the sampling distributions cluster more tightly about the value of the mean for the original population. The standard deviation of the sampling distribution measures the degree of this clustering. Consequently, it is also a measure of the amount of error that can be expected to occur in sample means as a measure of the true mean of the original population. Hence the name "the standard error of the mean."

Question 7–5. If repeated samples of size 100 were drawn from a population of people's ages which has a mean of 40 and a standard deviation of 20:

a. What percentage of all possible sample means would be in "error" from the population mean by more than 3 years?
b. Within what range about the population mean would 95% of all possible sample means fall?
c. Within what range about the population mean would 99% of all possible sample means fall?

_____

CONFIDENCE INTERVALS

In practice, only rarely do we actually know the values of the true mean and standard deviation of the original. population. It is more usual to have selected a random sample from the population. In such situations our information is limited to knowledge about *sample* characteristics, such as the sample mean and sample standard deviation. And it is not possible to directly calculate characteristics of sampling distributions from information about one sample. It is possible, however, to use the estimate of the standard deviation from a sample to estimate the standard deviation (standard error) of the sampling distribution. Symbolizing the *estimate* of the standard error as $\hat{s}_{\bar{x}}$, then*

$$\hat{s}_{\bar{x}} = \frac{\hat{s}}{\sqrt{N}}$$

*In many books the estimate of the standard error of the mean ($\sigma_{\bar{x}}$) is symbolized as $\hat{\sigma}_{\bar{x}}$. $\hat{s}_{\bar{x}}$ is used in this book to emphasize that the value of $\hat{s}_{\bar{x}}$ is based upon sample information.

Suppose, for example, that a random sample of 50 families in a city is selected by a researcher and the number of children in each family is recorded. The researcher calculates $\overline{X}$ and $\hat{s}$ for this sample as follows:

Family Number	Number of Children (X)	X²	Family Number	Number of Children (X)	X²
1	2	4	26	3	9
2	1	1	27	4	16
3	4	16	28	2	4
4	0	0	29	1	1
5	4	16	30	1	1
6	2	4	31	3	9
7	3	9	32	1	1
8	6	36	33	2	4
9	7	49	34	0	0
10	0	0	35	7	49
11	7	49	36	6	36
12	1	1	37	4	16
13	0	0	38	1	1
14	3	9	39	5	25
15	4	16	40	3	9
16	3	9	41	2	4
17	2	4	42	4	16
18	5	25	43	3	9
19	1	1	44	5	25
20	5	25	45	2	4
21	4	16	46	6	36
22	8	64	47	7	49
23	5	25	48	0	0
24	0	0	49	4	16
25	2	4	50	5	25
				$\Sigma X = 160$	$\Sigma X^2 = 748$

$$\overline{X} = \frac{\Sigma X}{N} = \frac{160}{50} = 3.2$$

$$\hat{s} = \sqrt{\frac{\Sigma X^2 - [(\Sigma X)^2/N]}{N-1}} = \sqrt{\frac{748 - [(160)^2/50]}{49}}$$

$$= \sqrt{\frac{748 - (25,600/50)}{49}} = \sqrt{\frac{748 - 512}{49}} = \sqrt{\frac{236}{49}} = \sqrt{4.82}$$

$$= 2.20$$

An estimate of $\sigma_{\bar{x}}$ can now be calculated using $\hat{s}$. The sampling distribution for this standard error is the sampling distribution of the mean for samples where $N = 50$, since that is the size of the researcher's sample.

$$\hat{s}_{\bar{x}} = \frac{\hat{s}}{\sqrt{N}}$$

$$= \frac{2.20}{\sqrt{50}}$$

$$= \frac{2.20}{7.07}$$

$$= .31$$

Quite often the mean of a sample is to be used as an estimate of the mean of the population from which the sample was drawn. Although the sample mean might closely approximate the population mean, it is also possible that it may not. A way is needed to estimate the accuracy of a sample mean as a measure of the population mean. This can be done through the calculation of *confidence intervals* (abbreviated *C.I.*).

If the sample size is at least moderately large, the sampling distribution for the mean approximates a normal distribution and we know that for *any* normally shaped distribution, 95% of the scores lie within ±1.96 standard deviations of the mean. If repeated random samples are taken from a population, then 95% of the means of all possible samples must lie within a distance of $1.96\sigma_{\bar{x}}$ of the value of the true population mean. Now, if one sample were drawn and the mean of the sample were one that fell in the 95% range, then if a distance of $1.96\sigma_{\bar{x}}$ were marked off about the *sample* mean, the interval would include within its range the value of the true population mean. This process is illustrated in Figure 7–13. In the figure a distance equal to $1.96\sigma_{\bar{x}}$ is drawn about the value of the true population mean. The means of five samples are marked on the graph. Three of those sample means—$\overline{X}_2$, $\overline{X}_3$, and $\overline{X}_4$—are among the 95% of all possible sample means which fall into the range of ±$1.96\sigma_{\bar{x}}$ about the population mean. Two sample means, $\overline{X}_1$ and $\overline{X}_5$, fall outside this range (and thus belong to the 5% of all possible sample means which are not included in the range ±$1.96\sigma_{\bar{x}}$ about the population mean).

A distance equal to $1.96\sigma_{\bar{x}}$ has been marked off above and below each of the five sample means. As can be seen, these intervals about $\overline{X}_2$, $\overline{X}_3$, and $\overline{X}_4$ include the value of the true population mean ($\mu_{\bar{x}} = \mu_o$). The intervals about $\overline{X}_1$ and $\overline{X}_5$, however, miss the population mean. In other words, if a particular sample mean is *any* one of the 95% of those which lie within $1.96\sigma_{\bar{x}}$ of the population mean, then an interval of $1.96\sigma_{\bar{x}}$ drawn about the sample mean will include the value of the true popula-

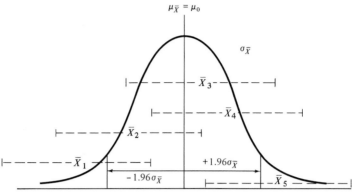

Figure 7–13
Sampling Distribution of the Mean with Confidence Intervals Drawn about Five
Sample Means

tion mean. Thus if we were to calculate this interval about the mean of a
sample, we could be 95% "confident" that the value of the true population mean lies somewhere in that interval. But even if the sample mean
resulting from a particular random sample is one of those that lies in the
middle 95% of the sampling distribution of the mean, we would not
know exactly where in the interval drawn about the sample mean the
true population mean falls, since the sample mean might be any of those
in the 95% area.

An interval drawn about a sample mean in this way, then, provides
an estimate of the value of the population mean. Such an interval is
called a *confidence interval for the mean*. If a distance equal to $1.96\sigma_{\bar{x}}$ is
marked off above and below the sample mean, the interval is called a
95% confidence interval. The upper and lower limits of confidence intervals are usually abbreviated UL and LL, with the confidence level as a
subscript.

Since $\sigma_{\bar{x}}$ is not available to calculate confidence intervals when we
possess information about a single sample, $\hat{s}_{\bar{x}}$ is used as an estimate in its
place. Using the information for the sample of 50 families, a 95% confidence interval is calculated as follows:

$$UL_{.95} = \bar{X} + 1.96\hat{s}_{\bar{x}}$$

Substituting from the information calculated on pages 178 and 179,

$$UL_{.95} = 3.2 + (1.96)(.31)$$

$$= 3.2 + .61$$

$$= 3.81$$

$$LL_{.95} = \overline{X} - 1.96\hat{s}_{\overline{x}}$$

$$= 3.2 - (1.96)(.31)$$

$$= 2.59$$

This confidence interval tells the researcher that, based upon the information in his sample, he can be 95% confident that the true population mean of the number of children per family lies somewhere in the range 2.59 to 3.81. We can say this because, in the long run, 95% of all possible random samples of this size that could be selected will lie within $1.96\sigma_{\overline{x}}$ of the true population mean.

It is possible to construct confidence intervals for any desired level of confidence. All that needs to be done is to find the area under the normal curve which corresponds to the desired level of confidence. Suppose that in our present example the researcher was not satisfied with a 95% level of confidence and he desired an interval in which he could be 99% confident that it contained the true population mean. In a normal distribution 99% of the area under the curve falls within ±2.58 standard deviations of the mean. Therefore,

$$UL_{.99} = \overline{X} + 2.58\hat{s}_{\overline{x}}$$

$$= 3.2 + (2.58)(.31)$$

$$= 3.2 + .80$$

$$= 4.0$$

$$LL_{.99} = \overline{X} - 2.58\hat{s}_{\overline{x}}$$

$$= 3.2 - (2.58)(.31)$$

$$= 3.2 - .80$$

$$= 2.4$$

The 99% confidence interval tells the researcher that based on the information contained in his sample, he can be 99% confident that the true mean number of children per family lies somewhere between 2.4 and 4.0.

Notice that the researcher pays a price to increase his confidence that the population mean is contained in a particular range. For the 95% level of confidence, the interval is 1.22 units wide (3.81 − 2.59), but for the 99% level of confidence, the interval is 1.6 units wide (4.0 − 2.4). If the researcher would wish to maintain a certain level of confidence and decrease the size of the confidence interval, he must draw a larger sample size. Suppose that the researcher drew a sample of 100 families instead of 50. To avoid some calculations we will assume that this sample of 100 also has an $\overline{X} = 3.2$ and an $\hat{s} = 2.20$. The value for the estimate of the standard error will now decrease,

$$\hat{s}_{\overline{X}} = \frac{\hat{s}}{\sqrt{N}} = \frac{2.20}{\sqrt{100}} = \frac{2.20}{10} = .22$$

and a 99% confidence interval will be

$$UL_{.99} = \overline{X} + 2.58\hat{s}_{\overline{X}}$$
$$= 3.2 + (2.58)(.22)$$
$$= 3.2 + .57$$
$$= 3.77$$
$$LL_{.99} = \overline{X} - 2.58\hat{s}_{\overline{X}}$$
$$= 3.2 - (2.58)(.22)$$
$$= 3.2 - .57$$
$$= 2.63$$

This 99% confidence interval based upon a sample of 100 has a smaller range, 1.14 (3.77 − 2.63), than the range of the 95% confidence interval based upon a sample of 50, 1.22.

Question 7−6. In a survey of 1000 adults in the United States, the respondents were asked to indicate their feelings toward the women's liberation movement on a scale from 0 to 100. If they felt strong dislike toward the movement, they were told to indicate a number near 0. If they had strong positive feelings, they were told to indicate a number near 100. If they had neutral feelings, they were told to indicate a number near 50. Strength of feelings between these were to be indicated by corresponding numbers. (This procedure is called a *thermometer scale* since it is usually presented to respondents as a thermometer on which they are to indicate the "temperature" of their feelings.) It turned out for this sample that on the scale, $\overline{X} = 45$ and $\hat{s} = 20$. Based on this information, answer the following questions:

a. Find an estimate of the standard deviation of the sampling distribution for all possible samples of size 1000 on this variable.
b. What is another name for the standard deviation specified in part a? Explain the use of this name.
c. Calculate a 95% confidence interval for the mean.
d. Calculate a 99% confidence interval for the mean.
e. Calculate a 90% confidence interval for the mean.

CONFIDENCE INTERVALS
FOR THE MEAN USING SMALL SAMPLES

In computing confidence intervals it is necessary to know the value of the standard deviation of the sampling distribution. This value is multiplied by the z score for a standard normal distribution which encloses the area under a normal curve that corresponds to the desired level of confidence. The value of the standard error of the mean can be found through the formula $\sigma_{\bar{x}} = \sigma_o/\sqrt{N}$. This requires that the standard deviation of the original population be known, which it usually is not when working with a sample. (If it were known, we would probably also know the value of the population mean and thus have no need at all for a confidence interval.) This problem was solved by using $\hat{s}_{\bar{x}}$, where $\hat{s}_{\bar{x}} = \hat{s}/\sqrt{N}$, in place of $\sigma_{\bar{x}}$. This substitution is adequate where the size of the sample is large. The reason is that where the sample size is large, $\hat{s}$ provides a close and stable estimate of σ, thus justifying the substitution of $\hat{s}_{\bar{x}}$ for $\sigma_{\bar{x}}$. When the sample size is small, however, the value of $\hat{s}$ fluctuates a good deal from sample to sample, which in turn causes $\hat{s}_{\bar{x}}$ to vary substantially. In such situations $\hat{s}_{\bar{x}}$ cannot justifiably be used as an estimate of $\sigma_{\bar{x}}$. Consequently, normal-curve methods cannot be used to calculate confidence intervals when small samples are involved.

The solution to this problem lay in developing a new type of probability distribution. This was done by the statistician W. S. Gosset, who published his work under the name "Student." The new distribution, known as Student's t distribution, gives the probabilities for the possible values of a "test statistic" called t. A *test statistic* is a statistic that is calculated from information about ordinary sample statistics such as $\bar{X}$ and $\hat{s}$. The formula for the t statistic is

$$t = \frac{\bar{X} - \mu_o}{\hat{s}_{\bar{X}}}$$

What a t distribution is about can best be understood by visualizing the following process. A random sample is selected from an original population whose mean (μ_o) is known. The mean of the sample ($\bar{X}$) and its standard deviation ($\hat{s}$) are calculated. Then $\hat{s}_{\bar{x}}$ is calculated for this sample. Substituting the values of μ_o, $\bar{X}$, and $\hat{s}_{\bar{x}}$ into the formula above yields the value of t for the sample. This value is then plotted on a graph. If this process is repeated indefinitely, drawing a random sample, making the calculations, and plotting the value of t for each sample on the graph, we will eventually wind up with a graph of the t distribution. [Notice that the value of $\hat{s}_{\bar{x}}$ is affected by the sample size ($\hat{s}_{\bar{x}} = \hat{s}/\sqrt{N}$), and thus the sample size also affects the value of t. The result is that a different distribution of t is obtained for different sample sizes. We shall return to this point later.] A graph of a t distribution for sample sizes of 10 is compared to the normal distribution in Figure 7–14.

When z scores are employed to construct confidence intervals, the sampling distribution of the mean is translated into a standard normal distribution. When so translated, the sampling distribution represents the difference between a sample mean $(\overline{X})$ and the true population mean (μ_o) divided by the standard deviation of the sampling distribution $(\sigma_{\overline{X}})$. The z-score formula is

$$z = \frac{\text{score} - \text{mean}}{\text{standard deviation}}$$

For the distribution called the sampling distribution of the mean this becomes

$$z = \frac{\overline{X} - \mu_{\overline{X}}}{\sigma_{\overline{X}}} \qquad (\text{recall that } \mu_{\overline{X}} = \mu_o)$$

Consequently a 95% confidence interval is found by multiplying the z score that encompasses 95% of the area in a standard normal distribution by the standard deviation of the sampling distribution.

A t distribution is analogous to a standardized sampling distribution except that it is shaped differently. Scores in it are composed of the difference between a sample mean $(\overline{X})$ and the value of the true population mean (μ_o) divided by $\hat{s}_{\overline{X}}$. Thus, if the t score that encompasses 95% of the area in a t distribution can be found, it can be multiplied by $\hat{s}_{\overline{X}}$ to construct a confidence interval in the same way that a z score is multiplied by $\sigma_{\overline{X}}$.

In Figure 7–14 it can be seen that the t distribution is flatter in the middle and thicker in the tails than the normal distribution. This means

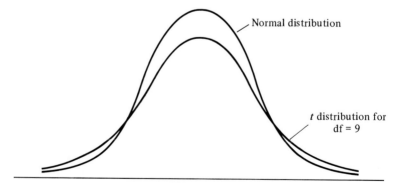

Normal distribution

t distribution for df = 9

Figure 7–14
Comparison of the Normal and t Distributions

that we have to go out farther from the mean to encompass 95% of the area in the t distribution than we do to encompass 95% of the area in a normal distribution. For 95% of the area in a normal distribution it is necessary to go out a distance of 1.96 units from the mean. For 95% of the area of the t distribution graphed in Figure 7–14, it is necessary to go a distance of 2.2622 units from the mean. The distance is greater because more of the area in a t distribution is contained in the tails of the curve than is the case for a normal distribution.

Statisticians have investigated t distributions and constructed tables that give the proportions of the area in a t distribution encompassed by various t scores just as the table for the standard normal distribution gives areas encompassed for various z scores. Thus, confidence intervals for the mean can be found on the basis of $\hat{s}_{\bar{x}}$ using t scores instead of z scores when the sample size is small. But before we can do this, one additional concept needs to be developed.

We saw that the t score will be affected by the sample size and that, as a result, in contrast to the normal distribution, there is a different t distribution for different size samples. Just as the normal curve is speci-fied by a mathematical formula, there is a formula for t distributions. The single parameter of this formula, though, is not the sample size but something called "degrees of freedom."

Suppose we were told that the mean for a sample of five observa-tions was 6. Since $\bar{X} = \Sigma X/N$, then $\Sigma X = N\bar{X}$, and for this sample $\Sigma X = (5)(6) = 30$. Now we are told to guess the scores for the five observations. We might guess that

$$X_1 = 8 \qquad X_3 = 0$$
$$X_2 = -2 \qquad X_4 = 10$$

After guessing the scores of four of the observations, we are not free to guess the fifth. Because the mean of the sample is specified, the last score must be a number that makes $\Sigma X = 30$. After we have guessed four of the observations, the score of the fifth is determined. Since

$$X_1 + X_2 + X_3 + X_4 = 8 + (-2) + 0 + 10$$
$$= 16$$

the score of the fifth observation must be 14, as this is the only number that will make $\Sigma X = 30$, given our guesses for the first four scores. The *degrees of freedom* (abbreviated *df*), then, is the number of scores in a sample whose values can be fixed arbitrarily. Every time a mean is calcu-lated, 1 degree of freedom will be "lost." In constructing confidence in-tervals, the sample mean is calculated and 1 degree of freedom is lost. Therefore, df $= N - 1$. Care should be taken to understand that nothing more is meant by the concept "degrees of freedom" than what has just been said. It is used because it, rather than the sample size, is the parameter that appears in the mathematical formula for the t distribution.

Tables for areas of t distributions are laid out differently than the table for areas under the normal curve. Since different t scores encompass the same areas in t distributions for different degrees of freedom, tables of t scores such as A–4 in the Appendix give only the more commonly used areas for many t distributions. In the body of the table are t scores. Each row of the table gives t scores for the distribution indicated by the degrees of freedom in the leftmost column. At the top of each column is the proportion of the area in *both* tails of the t distribution that is cut off by the t scores in that column. For the t distribution plotted in Figure 7–14, the degrees of freedom are 9. Entering the row for df $= 9$, under the column headed 0.05, the t score is 2.2622. This tells us that in this particular t distribution, 2.5% of the area in the distribution lies to the right of the t score of $+2.2622$ and 2.5% of the area in the distribution lies to the left of the t score of -2.2622, for a proportion of the total area of .05. Conversely, 95% of the area in the distribution lies *between* the t scores -2.2622 and $+2.2622$. If it were desired to find the t scores that encompass 99% of the area, we would move across the row of df $= 9$ to the column labeled 0.01.

We are now ready to construct confidence intervals using the t distribution. Suppose that a researcher is interested in the percentage of their budgets that cities over 50,000 devote to social services. He selects a random sample of 16 of these cities and finds their percentages to be as follows:

Percentage of Budget Devoted to Social Services (X)	X^2
5.7	32.49
6.0	26.00
4.2	17.64
8.0	64.00
5.5	30.25
3.9	15.21
4.3	18.49
5.7	32.49
9.0	81.00
3.0	9.00
4.2	17.64
5.8	33.64

$$
\begin{array}{cc}
7.7 & 59.29 \\
4.6 & 21.16 \\
5.4 & 29.16 \\
\underline{5.0} & \underline{25.00} \\
\Sigma X = 88.0 & \Sigma X^2 = 522.46
\end{array}
$$

$$\overline{X} = \frac{\Sigma X}{N} = \frac{88}{16} = 5.5$$

$$\hat{s} = \sqrt{\frac{\Sigma X^2 - [(\Sigma X)^2/N]}{N - 1}} = \sqrt{\frac{522.46 - [(88)^2/16]}{15}}$$

$$= \sqrt{\frac{522.46 - (7744/16)}{15}} = \sqrt{\frac{522.46 - 484}{15}}$$

$$= \sqrt{\frac{38.46}{15}} = \sqrt{2.546}$$

$$= 1.60$$

$$\hat{s}_{\overline{x}} = \frac{\hat{s}}{\sqrt{N}} = \frac{1.60}{\sqrt{16}} = \frac{1.60}{4}$$

$$= .4$$

The researcher knows that the mean percentage of the budget the cities in his sample allocated to social services is 5.5%. He now wishes to know within what range he can expect to find the mean percentage of their budgets that all cities devote to social services with a confidence of 99%. Because the sample size is small, it is necessary for the researcher to use t scores in constructing a confidence interval. Since a 99% confidence interval corresponds to a t score for the area of .01 in the table of t distributions, the formulas for the limits of the confidence interval are written

$$UL_{.99} = \overline{X} + t_{.01}(\hat{s}_{\overline{x}})$$
$$LL_{.99} = \overline{X} - t_{.01}(\hat{s}_{\overline{x}})$$

Using a table such as A–4, the researcher finds that $t_{.01}$ for 15 degrees of freedom is 2.9467. Substituting into the formulas,

$$UL_{.99} = \overline{X} + t_{.01}(\hat{s}_{\overline{x}})$$
$$= 5.5 + (2.9467)(.4)$$
$$= 5.5 + 1.18$$
$$= 6.68$$

$$LL_{.99} = \overline{X} - t_{.01}(\hat{s}_{\overline{X}})$$
$$= 5.5 - (2.9467)(.4)$$
$$= 5.5 - 1.18$$
$$= 4.32$$

Thus on the basis of his sample of 16 cities, the researcher can be 99% confident that the mean percentage of their budgets that cities over 50,000 devote to social services lies in the range 4.32 to 6.68.

When should t scores or z scores be used in constructing confidence intervals? An answer to this question depends upon both the shape of the original population from which the sample is drawn and the size of the sample. If the original population is *known* to be normally distributed, z scores may be used to construct confidence intervals with samples as small as 30 observations. If the sample size is smaller than 30, t scores should definitely be used.

Technically, both t and z scores are to be used for confidence intervals only for samples drawn from a normal population. The reason t scores were developed is for use on small samples, not because of non-normal populations. Fortunately, investigation has shown that if the sample size is fairly large and the original population is at least roughly uni-model and symmetric, t scores may be used safely. In such situations t scores should be used even if the sample size is larger than 30. After a sample size of 100 is reached, values for the t distribution are very closely approximated by the normal distribution. This is because as the degrees of freedom increase, the t distribution approaches the normal distribution as a limit. Consequently, it makes little difference whether t scores or z scores are used if the sample size is over 100.

Question 7–7. A freshman class at a university took a reading-comprehension test. A random sample of 25 freshmen was selected which had an $\overline{X} = 65$ and an $\hat{s}$ of 20 on the test. Assuming that the reading-comprehension scores for the whole freshman class are normally distributed:

a. Construct a 95% confidence interval for the mean based on this sample.
b. Construct a 99% confidence interval.
c. Construct a 90% confidence interval.

RELATIONSHIPS
AMONG TYPES OF DISTRIBUTIONS

In this chapter a number of types of distributions have been discussed. Perhaps the strangest distribution that has been encountered is the t distribution. The t distribution is an instance of the use of a test statistic. A

Table 7–2
Characteristics and Relationships among Various Types of Distributions

Name of Distribution	Population or Sample	Nature of Scores in Distribution	Symbol for a Score	Shape of Distribution	Mean	Standard Deviation	Relationship to Original Population
Original population	Population	Raw score	X	Anything, but helpful if normal	μ_o	σ_o	
Random sample	Sample	Raw score	X		$\overline{X}$	s	Selected from it by random sampling
Standard normal distribution	Population	Standard score	z	Normal	$\mu = 0$	$\sigma = 1$	$z = \dfrac{\overline{X} - \mu_o{}^*}{\sigma_o}$
Sampling distribution of the mean	Population	Sample means	$\overline{X}$	Normal if sample size is large	$\mu_{\overline{X}} = \mu_o$	$\sigma_{\overline{X}} = \dfrac{\sigma_o}{\sqrt{N}}$	Constructed from repeated random samples from original population
t distribution	Population	Test statistic	t	t-shaped for small samples; approaches normal as sample size increases	$\mu = 0$	Varies	Calculated from repeated random samples selected from original population through $t = \dfrac{\overline{X} - \mu_o}{\hat{s}_{\overline{X}}}$

*if original population is normal

distribution for a test statistic is developed through the same procedures as for sampling distributions. Once the distribution has been constructed, it is possible to use it as a probability model to estimate the probability of a particular value of the test statistic occurring in a randomly selected sample.

All the distributions that have been studied are in some way related to an original population. The original population may be about anything, and usually it is the characteristics of this distribution that are ultimately of concern. The distribution of a sample is related to the original distribution by virtue of being randomly selected from it. The standard normal distribution is related to the original population (if it is normally shaped) through the z-score transformation. The sampling distribution of the mean is composed of the means of repeated random samples drawn from the original population. And the t distribution is composed of a t statistic calculated on repeated random samples drawn from the original population.

Since these distributions are in some way constructed from the original population, their characteristics are related to the characteristics of the original population. It is useful to emphasize these relationships, and this is done in Table 7–2. The distributions, their characteristics, and their relationship to an original population are presented in this table.

EXERCISES FOR CHAPTER 7

1. Assuming that all of the following distributions are normally shaped, for each distribution find within what range (1) 68.26% of the scores in the distribution fall, and (2) 95% of the scores fall.

   a. The distribution of the weights of professional football players, where the mean is 220 and the standard deviation is 15.
   b. The distribution of the heights of professional basketball players in inches, where the mean is 78 and the standard deviation is 3.
   c. A distribution of test scores in which the mean is 82 and the standard deviation is 7.
   d. The age of members in the U.S. House of Representatives, where the mean is 50 and the standard deviation is 12.

2. Suppose that for a particular year on the law school admissions test (LSAT) the mean score for all people taking the test is 500, the standard deviation is 90, and the scores are normally distributed:

   a. What percentage of people had scores over 600?
   b. What percentage of people had scores less than 300?
   c. What percentage of people had scores between 700 and 750?
   d. If a person had a score of 630 on the test, what percentage of people had scores less than his? Greater than his?
   e. If 5000 people took the test that year, how many had a score between 300 and 400?

3. Suppose that a college entrance examination is given to all entering college students. It is found that the scores are normally distributed with a mean of 450 and a standard deviation of 75.

   a. What is the probability that if a student were selected at random, his score on the test would be greater than 550? The probability that it would be less than 550?
   b. What is the probability that if a student were selected at random, his score would lie between 350 and 400?
   c. If the z score of a student on this test were -1.5, what was his original score on the test?
   d. If random samples of size 200 were selected from this population, what would be the value of the standard error of the mean?
   e. What is the probability that if a random sample of 200 were selected from the population, the sample mean would deviate from the population mean by more than 10 points?

4. In a random sample of 500 adults in a city it was found that the average amount of federal income tax for the people in the sample was $1200 and the standard deviation $300. Based on this information:

   a. Calculate a 95% confidence interval for the mean.
   b. Calculate a 99% confidence interval for the mean.
   c. Calculate a 68.26% confidence interval for the mean.

5. A sample of 100 observations has a mean of 55 and a standard deviation of 15. If the true population mean is 58, which confidence-interval setting, .95 or .99, is necessary to include the population mean?

6. A random sample of 16 students at a college report the following number of hours spent studying each week:

25	18	30	9
37	8	22	27
15	35	25	20
42	2	25	28

Compute an interval that will contain the mean amount of study hours per week for all students at the college with a 95% level of confidence; for a 90% level of confidence.

ANSWERS TO
QUESTIONS IN CHAPTER 7

7–1: a. 130.4 pounds to 169.6 pounds. b. 124.2 pounds to 175.8 pounds.
   c. 15.87%, 84.13%. d. 99.5%, .5%.

7–2: a. .1915. b. .3085.
   c. .3085. d. .1359.
   e. About 13.

7–3: a. About 9. b. About 31.
   c. 11.8. d. 16.

7–4: a. $P = .2513.$ b. $P = .2039.$
   c. About 68.

7–5: a. 13.36%. b. 36.08 to 43.92.
   c. 34.84 to 45.16.

7–6: a. .63.
   b. Standard error of the mean. It is a measure of the amount of deviation and thus the error that can be expected in sample means from the true population mean.
   c. 43.76 to 46.23. d. 43.37 to 46.62.
   e. 43.94 to 46.04.

7–7: a. 56.74 to 73.26. b. 53.81 to 76.19.
   c. 58.16 to 71.84.

8

Hypothesis Testing: Differences Between Two Groups

In many problems a researcher conducts an investigation because he has an idea that certain variables are related. A statement of such an idea is called an *empirical hypothesis*. Some examples of empirical hypotheses might be:

The more education people have, the more supportive they will be of civil liberties.

The more competitive the party system in his district, the more a legislator will vote in support of social-welfare policies.

The greater the population density of an area, the higher its suicide rate.

Men are more politically active than women.

Men are more supportive of militaristic policies than women.

Increasing unemployment causes increases in crime.

Empirical hypotheses assert that certain variables are related in some way: that one variable "causes" or "affects" or is "associated" with another. Whether these relationships exist, whether the hypotheses are true, is what the researcher attempts to establish by examining evidence. Most hypotheses deal with two variables, and this is the case with the hypotheses listed above. It is possible, however, to state hypotheses that deal with many variables.

It is also possible to formulate empirical hypotheses which assert that there is no relationship between two variables. For example, it might be hypothesized that: "Smoking marijuana does not lead to heroin use." Although the hypothesis can be stated in this way, it can be reformulated to read that smoking marijuana *does* lead to heroin use, but it is thought that the evidence will not support it. That is, what is desired is to be able to eliminate a hypothesis that does assert a relationship. Often the elimination of hypotheses is as important as their establishment. The point is that empirical hypotheses assert relationships between variables, and we examine evidence to see if the assertion can or cannot be supported. In this chapter some of the problems that arise in using sample information to evaluate empirical hypotheses will be considered.

Hypothesis testing in statistics is not primarily concerned with establishing the truth or falsity of an empirical hypothesis. Rather, it is concerned with evaluating the evidence that is being brought to bear upon a hypothesis. If the evidence consists of information from a sample that has been randomly selected from a population, the possibility arises that the results in the sample are due to random sampling error. That is, because of the random sampling procedure used in selecting the sample, it is possible that the results in the sample happened by chance and are not due to the fact that the empirical hypothesis is really true. Before making any conclusions about the implications of the sample results upon an empirical hypothesis, it is necessary to decide whether or not the results are due to chance factors that come into play in the random sampling procedure. Hypothesis testing is a statistical procedure that provides a way to decide this question.

In hypothesis testing the researcher sets up two *statistical hypotheses*. Although these hypotheses are related to and may be derived from the empirical hypothesis, it is often the case that neither of them is the same as the empirical hypothesis. In most situations, one of the statistical hypotheses says in effect that any results in the sample are due only to chance and is called the *null hypothesis* (symbolized as H_0). The other statistical hypothesis says that the results in the sample are not due to chance and is called the *alternative hypothesis* (symbolized as H_1). Since the null and alternative hypotheses are opposite sides of the same coin, a decision to accept one means to reject the other. It is necessary then only to work with one of these hypotheses. In hypothesis testing it is usually the null hypothesis that is tested. If the testing shows that the null hypothesis can be rejected, this implies that the alternative hypothesis may be accepted: that whatever results appear in the sample are not due to chance. Notice that this does *not* say that the results in the sample support the *empirical hypothesis*. It says that the results in a sample may be used as evidence bearing upon the empirical hypothesis without much fear that those results, whatever they are, occurred only by chance.

If the testing shows that the null hypothesis cannot be rejected, then the possibility that the sample results are due only to chance must be seriously considered. If the empirical hypothesis asserted that something should happen beyond a chance level and hypothesis testing shows that there is a serious possibility that the results are due only to chance, then obviously the researcher is not justified in interpreting the sample results as support for the empirical hypothesis. In fact, in many situations a failure to reject the null hypothesis would be interpreted as evidence *against* the empirical hypothesis. In any case, we certainly would not want to proceed in using the sample results, regardless of how dramatic they might be, to support such an empirical hypothesis if the null hypothesis cannot be rejected.

THE LOGIC
OF HYPOTHESIS TESTING

The procedures involved in hypothesis testing may be illustrated through the coin-flipping game considered in Chapter 5. In this game suppose that we win when the coin comes up tails and our opponent wins when it comes up heads. On the first six flips the coin comes up heads each time. At this point we become suspicious and formulate an empirical hypothesis that the coin is dishonest—that it is heavily weighted on the heads side. But before we can accept the results of the first six flips as evidence in support of this empirical hypothesis, the possibility that the coin is really honest and came up heads six times in a row by chance alone must be evaluated. Since this possibility stipulates that the coin is honest and that the obtained results are due only to chance, this possibility can be stated as the null hypothesis:

H_0: $P_H = .5$ (coin is honest and results are due only to chance)

A number of alternative hypotheses could be formulated to this null hypothesis. It is possible, for example, to formulate an *exact* alternative hypothesis such as H_1: $P_H = .7$. An exact alternative hypothesis is one that specifies a precise alternative value for the parameter under consideration. In such situations hypothesis testing proceeds by determining under which hypothesis, the null or the alternative, it is more probable that the obtained results would occur. In most practical problems, however, the formulation of an exact alternative hypothesis is not possible. Even in our present example, we think that the coin is weighted in some way but we really have no idea how much it is weighted, so we cannot propose a specific alternative value for P_H. Consequently, we will be concerned only with *inexact* alternative hypotheses, ones which say that the parameter does not have the value stipulated in the null hypothesis but which do not offer a specific alternative value for the parameter.

There are still three inexact alternative hypotheses that can be formulated:

1. H_1: $P_H \neq .5$.
2. H_1: $P_H < .5$ (P_H is less than .5).
3. H_1: $P_H > .5$ (P_H is greater than .5).

The choice among various alternative hypotheses is guided by any additional assumptions that we might be willing to make. If the coin being used in the game belonged to our opponent, it would be reasonable to suppose that if the coin is weighted it is weighted on the heads side, so we will *assume* that alternative hypothesis number 2 can be eliminated. Once we make this assumption, number 1 can also be eliminated, since we have decided only to consider the idea that the coin is weighted on the heads side. Therefore, the alternative hypothesis that will be tested is

$$H_1: P_H > .5$$

Note that in this particular problem the alternative statistical hypothesis closely corresponds to our empirical hypothesis. This need not necessarily be true in all hypothesis testing.

The next step is to select a model of chance that is appropriate to the situation. Such a model will enable us to calculate the *probabilities* of various outcomes occurring by chance alone if the null hypothesis were true. The model will be used to calculate the probability that the *obtained outcome* in an experiment (six heads in this situation) could occur by chance if the null hypothesis is true. If this probability is fairly large, the model tells us that there is a good possibility that under the null hypothesis the obtained result could have occurred by chance. This would be consistent with the idea that the null hypothesis is true. Therefore, if this happens, we will decide to *accept* the null hypothesis (or fail to reject it). If, however, the model says that the probability is very small that the obtained outcome could have occurred by chance under the assumption that the null hypothesis is true, this would be inconsistent with the idea that the null hypothesis is true. Therefore, we would *reject* the null hypothesis and *accept* the alternative hypothesis.

The model that is appropriate for coin flipping is the binominal distribution. In this situation the choice of a model is very straightforward. In more practical situations the choice is more complicated and it is often necessary to make certain simplifying assumptions. Even in this situation it is necessary to assume that the coin-flipping procedure is a random process, since the binomial computes probabilities on the basis of what will happen by chance. If the null hypothesis is true, the probability that the obtained result of six heads would occur by chance can be calculated from the binomial as follows:

$$P_{6H} = C_6^6(.5)^6(.5)^{6-6}$$

$$= (1)(.016)(1)$$

$$= .016$$

The calculation shows that if the null hypothesis were true, the probability that the obtained result could occur by chance is only about 2 times in 100. This probability is so small that we can be pretty safe in concluding that the outcome of six heads in playing the game would not have happened just by chance if the null hypothesis ($P_H = .5$) were true. Therefore, we reject the null hypothesis and accept the alternative hypothesis that $P_H > .5$ and we should stop playing the game or get a new coin.

Question 8–1. A researcher studying the effects of the stress of modern life on people hypothesizes that urban areas should have a higher divorce rate than rural areas because of the greater stresses involved in urban living. He selects a random sample of 50 urban counties and a random sample of 60 rural counties and obtains the divorce rate for each county. The mean divorce rate for each sample is calculated and it is found that $\overline{X}_u = 28$ and $\overline{X}_r = 25$.

a. Verbally state what the null and alternative hypotheses are in this situation.
b. Restate the null and alternative hypotheses in terms of symbols.

THE ALPHA LEVEL

In hypothesis testing a decision is made between a null and an alternative hypothesis. This is done by calculating the probability of the obtained result in a sample occurring by chance if the null hypothesis were true. If this probability is small, the null hypothesis can be rejected. But what size probability is small enough to reject the null hypothesis? This problem is known as selecting the alpha (α) level. Alpha is the maximum value of probability that will be considered small enough to reject the null hypothesis. There are really no hard and fast rules about selecting an alpha level. But since rejection of the null hypothesis depends upon it, a value for alpha should be chosen before a test is conducted so that the results of the test itself are not allowed to influence us in its selection. In social science, alpha levels are usually set by convention at .05 or .01. But these levels are not sacrosanct. Later some principles that guide the choice of an alpha level will be considered, but for the moment we shall accept the conventional values.

Once an alpha level is selected, decisions about the null and alternative hypotheses must be made in accordance with it. If an alpha level is set at .05 and a probability of .06 is obtained in a particular test, the null hypothesis cannot be rejected. And if an alpha of .05 had been chosen and the probability works out to .03, the alpha level cannot be changed to .01, and again the null hypothesis cannot be rejected. Once the rules are set, the game must be played in accordance with them. This seems fair since we get to make the rules before the game starts.

Question 8–2. If the researcher selects an alpha level of .01 in the problem posed in Question 8–1, explain the conditions under which the null hypothesis would be rejected; would be accepted.

DIFFERENCES
BETWEEN TWO GROUPS

We are now ready to consider hypothesis testing in the context of empirical hypotheses that involve two variables. A common problem is to determine whether two groups do or do not differ on an interval or ratio level variable. Suppose that a researcher is interested in the problem of sexual discrimination among high school teachers. His interest leads him to ask if there is a difference in the salaries paid to male and female teachers. This hypothesis involves two variables, sex and salary. One variable, sex, is a simple dichotomous variable which divides the observations into two groups. The hypothesis then asks whether or not the two groups differ on a second variable, salary. To study this problem suppose that the researcher selects a random sample of 50 male teachers and a random sample of 50 female teachers and computes the mean salary for each sample. He finds that the mean salary for the sample of male teachers is $8300, or $\bar{X}_m = 8300$. And that the mean salary for the sample of female teachers is $8100, or $\bar{X}_w = 8100$. Now it is an undeniable fact that there is a difference of $200 between the means of the two samples. But before the researcher draws any conclusions about what this difference portends, he must consider the possibility that in the population of teachers there is no difference in the mean salaries of men and women and that the difference in the sample means appeared by chance alone because of the random sampling procedure.

The researcher's problem at this point is to decide whether the means of the two groups in the population are the same or whether they differ. We shall assume that no matter what the situation is in the population, the distributions in the population are normally shaped. This is often a realistic assumption and helps to simplify the discussion at this point. The consequences of nonnormal distributions in the population are not serious if the sample sizes are large and will be discussed later. The various possibilities that could exist in the population are presented graphically in Figure 8–1.

In Figure 8–1a the distributions for male and female teachers on

Differences Between Two Groups

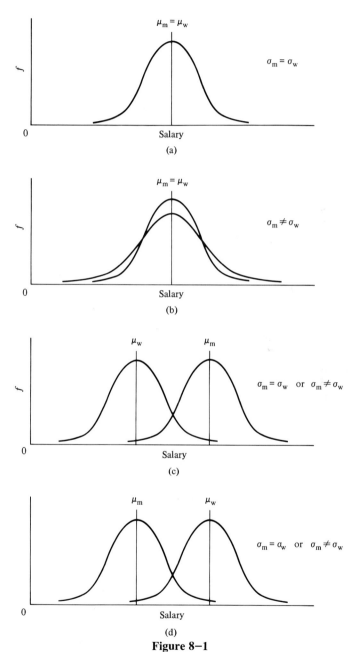

Figure 8–1
Possible Population Distributions of Salaries for Male and Female High School Teachers

salary are identical. This means not only that their means are equal, which is what the researcher is concerned about, but also their standard deviations. In Figure 8–1b the means of the two groups are equal, but their standard deviations differ. If either of the situations represented in these two figures is the true situation in the population, any difference in the sample means is due to chance alone.

In Figure 8–1c the mean salary for men is higher than that for women. Again it would be possible for the standard deviations of the two distributions to be either equal or unequal. If this figure describes the true situation in the population, the difference in the sample means is probably a reflection of this fact. Finally, it is possible that the mean salary for women is higher than the mean salary for men, as shown in Figure 8–1d. Before the researcher collected his data he felt that there might be a difference between the mean salaries for men and women because he thinks there is discrimination in the teaching profession. But he might have felt that the discrimination could be against *either* men or women and was not willing, therefore, to assert which group would have the higher mean salary. Consequently, the researcher must also consider the situation in Figure 8–1d as a possibility.

The researcher, of course, has no way of knowing what the true situation is in the population. Hypothesis testing provides a way to make a decision about the situation in the population, based upon the information contained in the samples. To do this the researcher sets up a null and an alternative hypothesis. Since the null hypothesis usually asserts that any results in the sample are due only to chance, the null hypothesis for this problem is

$$H_0: \mu_m = \mu_w$$

This hypothesis asserts that the means for men and women in the population are identical, and thus that any difference in the means of random samples of men and women are due only to chance, the chance processes being those that exist in the random-sampling procedure. The alternative hypothesis will be

$$H_1: \mu_m \neq \mu_w$$

This hypothesis says that the means for men and women in the population are not equal, and therefore the difference in the means of the random samples are a reflection of this fact. Notice that the alternative hypothesis does not specify the direction of the difference, that the mean of one group is larger than the mean of the other, but only that there is a difference. This is because the researcher did not have any reason to believe that one group made more money than the other *before* he gathered his data.

THE SAMPLING DISTRIBUTION OF THE DIFFERENCE BETWEEN TWO MEANS

In order to test the null hypothesis, an appropriate model is needed from which the probabilities of various results occurring by chance can be calculated. For this problem the appropriate model is a sampling distribution, called the *sampling distribution of the difference between two means*. This sampling distribution is constructed in a way similar to that of the sampling distribution of the mean. From two original populations that have the same mean, random samples are independently selected from each and the sample means calculated. Then the *difference* between the two sample means is taken:

$$D = \overline{X}_1 - \overline{X}_2$$

This difference *(D)* is now plotted on a graph. If this process is repeated indefinitely, so many *D* scores will be plotted in the graph that it will become a population itself. This population of *D* scores is called *the sampling distribution of the difference between two means,* which is a very good description of how it is created.

As might be anticipated, this sampling distribution is normally shaped if the two original populations from which the samples are drawn are normal. If the two original populations are not normal, the sampling distribution will approach normal as the size of the samples gets large. This is why it was stated earlier that nonnormal populations did not create problems if the sample sizes were large.

If the means of the two original populations are equal, then the mean of the sampling distribution of the difference between two means will be equal to zero. This point can be seen intuitively in the following way. Each sample mean is an estimate of its population mean. The two population means, though, are equal. Therefore, the two sample means are estimates of the same value. Any fluctuation in the values of the sample means is a random one due only to the random sampling procedure. Sometimes the value of the first sample mean will be larger than that of the second sample mean, which will make the *D* score positive. At other times the value of the second sample mean will be larger than that of the first sample mean, which will make the *D* score negative. Each of these situations in the long run should happen with equal frequency, like heads and tails in flipping an honest coin, and "balance" each other. Since the positive and negative scores will balance each other, the mean of the population of *D* scores, or the mean of the sampling distribution of the difference between two means, symbolized as $\mu_{(\overline{X}_1 - \overline{X}_2)}$, will equal zero.

If the means of the two original populations are *not* equal, the mean of the sampling distribution of the difference between two means will be equal to the *difference* between the two original population means. It is quite possible to work with such a sampling distribution. In most situations in social science, however, the question of interest is whether the means of two groups are equal, not whether the means of the two groups in the population differ by a specific amount. We shall consider only situations where the null hypothesis states that the means of the two original populations are equal and consequently only deal with sampling distributions of the difference between two means which have a mean equal to zero.

As is the case for the sampling distribution of the mean, there is a relationship between the standard deviation of the sampling distribution of the difference between two means [symbolized as $\sigma_{(\bar{X}_1 - \bar{X}_2)}$] and the standard deviations of the two original populations. This relationship is specified by the equation

$$\sigma_{(\bar{X}_1 - \bar{X}_2)} = \sqrt{\frac{\sigma_1^2}{N_1} + \frac{\sigma_2^2}{N_2}}$$

where N_1 and N_2 are the number of observations in the two samples being selected from the two original populations and σ_1^2 and σ_2^2 are the variances of the two original populations. This standard deviation is also called the *standard error of the difference between means*. Despite its fancy name and symbol, it is simply the standard deviation of a particular distribution. (It should be noted that there is no requirement that the size of the two samples must be equal. The characteristics of the sampling distribution of the difference between two means hold for equal or unequal sample sizes.)

The sampling distribution of the difference between two means can be used as a model to calculate the probability of obtaining a particular difference between two sample means by chance when the means of the two original populations are equal. Such probabilities are found by calculating areas under a normal curve in the way that was done in Chapter 7. Suppose that for a particular sampling distribution of the difference between two means, $\sigma_{(\bar{X}_1 - \bar{X}_2)} = 7$. It will be recalled that in *any* normal distribution 68.26% of all the scores lie in an interval of ±1 standard deviation about the mean. Therefore, we know that 68.26% of all possible differences that could occur by chance alone between the means of two samples (and based on the same sample sizes used to construct the sampling distribution) have values between −7 and +7. We also know that in a normal distribution, 95% of all the scores lie within ±1.96 standard

deviations of the mean. Therefore, 95% of all possible differences between two sample means will lie in the following interval:

$$\text{upper limit} = \mu_{(\bar{X}_1 - \bar{X}_2)} + 1.96\sigma_{(\bar{X}_1 - \bar{X}_2)}$$

$$= 0 + (1.96)(7)$$

$$= 13.72$$

$$\text{lower limit} = \mu_{(\bar{X}_1 - \bar{X}_2)} - 1.96\sigma_{(\bar{X}_1 - \bar{X}_2)}$$

$$= 0 - (1.96)(7)$$

$$= -13.72$$

Consequently, the probability is .95 that any one difference between the
means of two random samples will have a value that lies in the interval
from -13.72 to $+13.72$. Because the sampling distribution of the difference between two means is normally shaped, standard scores (z scores)
can be used to answer questions about the probability of a particular difference between two sample means occurring by chance.

ESTIMATING
THE STANDARD ERROR

In order to use the sampling distribution of the difference between
means as a probability model, it is necessary to know the value of the
standard error of the difference between the two means. This requires that
the standard deviations of the two original populations (σ_1 and σ_2) be
known, and in most real situations they are not. Fortunately, if both
sample sizes are at least moderately large, each greater than 30 observations, $\hat{s}^2$ is a good estimate of σ^2, and it is possible to use the unbiased
sample variance for each sample in place of the original population variances and calculate an estimate of the standard error of the difference
between means, $\hat{s}_{(\bar{X}_1 - \bar{X}_2)}$, as follows:

$$\hat{s}_{(\bar{X}_1 - \bar{X}_2)} = \sqrt{\frac{\hat{s}_1^2}{N_1} + \frac{\hat{s}_2^2}{N_2}}$$

Question 8–3. Suppose that for the problem in Question 8–1, the researcher found that the standard deviations for his samples were
$\hat{s}_u = 6$ and $\hat{s}_r = 5$.

a. Calculate an estimate of the standard error of the difference between
two means for this problem ($N_u = 50$ and $N_r = 60$).

b. If the null hypothesis formulated in Question 8 – 1 is true, within what range of values would 68.26% of differences between the means of all possible samples of urban and rural counties fall by chance? 95%?

CALCULATING THE TEST STATISTIC

Let us now return to the problem of the difference in salaries for male and female teachers. If the null hypothesis is true, there is no difference in the population means of the salaries for men and women, and any difference in the values of the sample means would be due only to chance. If the model calculates that the probability of the obtained difference happening by chance is small, the researcher will decide that the difference in the sample means did not occur by chance and will reject the null hypothesis. The researcher decides that any probability less than or equal to 5 times out of 100 will be small enough for him to reject the null hypothesis or $\alpha = .05$.

The sampling distribution of the difference between two means is the model from which the probability of the *obtained difference* in the sample means occurring by chance will be calculated. Since the sampling distribution is normally shaped, calculating this probability becomes a problem of finding areas under a normal curve. There is, however, not just one sampling distribution of the difference between two means, but many. Which one is of interest in a particular problem depends on the means and standard deviations of the original populations and the sizes of the samples taken from them. But since they are all normally shaped, the obtained difference between two sample means can be converted into a z score, and the single table for the standard normal curve can be used to compute areas. The formula for a z score is

$$z = \frac{\text{score} - \text{mean}}{\text{standard deviation}}$$

For a sampling distribution of the difference between two means, a score is the obtained difference between two sample means, which we will represent as D. The mean of the sampling distribution is $\mu_{(\bar{x}_1 - \bar{x}_2)}$, and its standard deviation is $\sigma_{(\bar{x}_1 - \bar{x}_2)}$. Therefore, the z-score formula becomes

$$z = \frac{D - \mu_{(\bar{x}_1 - \bar{x}_2)}}{\sigma_{(\bar{x}_1 - \bar{x}_2)}}$$

Since $D = \overline{X}_1 - \overline{X}_2$,

$$z = \frac{(\overline{X}_1 - \overline{X}_2) - \mu_{(\overline{X}_1 - \overline{X}_2)}}{\sigma_{(\overline{X}_1 - \overline{X}_2)}}$$

Because the null hypothesis specifies that the means of the two original populations are equal, $\mu_{(\overline{X}_1 - \overline{X}_2)}$ for the sampling distribution of interest will equal zero. And in real problems we have seen that it is necessary to estimate $\sigma_{(\overline{X}_1 - \overline{X}_2)}$ by $\hat{s}_{(\overline{X}_1 - \overline{X}_2)}$. Therefore,

$$z = \frac{(\overline{X}_1 - \overline{X}_2) - 0}{\hat{s}_{(\overline{X}_1 - \overline{X}_2)}}$$

leaving

$$z = \frac{\overline{X}_1 - \overline{X}_2}{\hat{s}_{(\overline{X}_1 - \overline{X}_2)}}$$

This is the test statistic that will be used to find the probability of the obtained difference between the two sample means occurring by chance if the null hypothesis is true.

Suppose that the researcher has found the standard deviations of his samples to be $\hat{s}_m = 360$ and $\hat{s}_w = 340$. Then

$$\hat{s}_m^2 = (360)^2 = 129{,}600$$

$$\hat{s}_w^2 = (340)^2 = 115{,}600$$

The estimate of the standard error is found as follows:

$$\begin{aligned}
\hat{s}_{(\overline{X}_1 - \overline{X}_2)} &= \sqrt{\frac{\hat{s}_m^2}{N_m} + \frac{\hat{s}_w^2}{N_w}} \\
&= \sqrt{\frac{129{,}600}{50} + \frac{115{,}600}{50}} \\
&= \sqrt{2592 + 2312} \\
&= \sqrt{4904} \\
&= 70.03
\end{aligned}$$

The sample means were $\overline{X}_m = 8300$ and $\overline{X}_w = 8100$. Thus,

$$z = \frac{\overline{X}_m - \overline{X}_w}{\hat{s}_{(\overline{X}_1 - \overline{X}_2)}}$$

$$= \frac{8300 - 8100}{70.03}$$

$$= \frac{200}{70.03}$$

$$= 2.86$$

Question 8–4. For two independent random samples from the same population, $\bar{X}_1 = 285$, $\bar{X}_2 = 275$, $\hat{s}_1 = 40$, $\hat{s}_2 = 42$, $N_1 = 80$, and $N_2 = 84$.

a. Calculate an estimate of the standard error for the sampling distribution of the difference between two means based on these samples.
b. Calculate a standardized score for the difference between the two sample means.

_____

LOCATING
THE CRITICAL REGION

In Figure 8–2 the standard normal distribution is presented. For the purposes of this problem, this distribution represents the *standardized* sampling distribution of the difference between two means. From it the probabilities of various values of the test statistic occurring by chance can be found. Obviously, values close to zero in the middle of the distribution are fairly likely to occur by chance. And as the value of the test

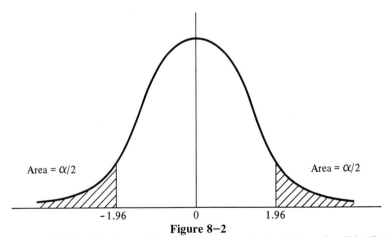

Area = $\alpha/2$ Area = $\alpha/2$

-1.96   0   1.96
Figure 8–2
Location of Critical Regions (Shaded) in the Standardized Sampling Distribution of the Difference between Two Means for Alpha Equal to .05

statistic moves away from zero in either direction, the probability of a value occurring by chance gets smaller. In other words, the farther a particular value of the test statistic is from the center of the distribution, the less likely it is that value will occur by chance under the null hypothesis. The researcher is interested in those values that are very unlikely to occur by chance if the null hypothesis is true. These values, then, are those which occur in the tails away from the middle of the distribution.

Since the researcher set the alpha level equal to .05, he wants to locate the values of the test statistic that have a probability of occurring by chance in the sampling distribution of .05 or less. Because the researcher did not assert anything about the direction of the difference between the means of the two groups, he must consider both positive and negative values of the test statistic. Therefore, it is necessary to determine where the most unlikely 2.5% of the positive values and where the most unlikely 2.5% of the negative values lie in the distribution. These values will constitute the 5% (the alpha level) of most unlikely values to occur by chance under the null hypothesis.

From our work on confidence intervals, we know that 95% of all the scores in a normal distribution lie within ±1.96 standard deviations of the mean and 5% fall outside this interval. This means that for the standardized sampling distribution in Figure 8–2, the probability is .05 that the test statistic would have, by chance, a value less than −1.96 or greater than +1.96. Thus, the areas in both tails beneath the curve to the left of −1.96 and to the right of +1.96 equal .05. These areas are called the *critical regions* of the curve, since any values of the test statistic that fall into these regions are among the most unlikely 5% to occur by chance in the distribution. What percentage of the most unlikely values of the test statistic are of concern is determined by the alpha level. (For this reason the critical regions are sometimes called the *alpha regions*.) The values of the test statistic that bound these areas are called the *critical values* because they mark the beginning of the critical regions.

If the *obtained value* of the test statistic, the value calculated from the samples, "exceeds" the critical value, less than −1.96 or greater than +1.96, the researcher knows that the probability of the obtained value of the test statistic occurring by chance under the null hypothesis is less than his selected alpha level, and consequently *the null hypothesis* can be rejected. This means that if the null hypothesis were true, the probability of the obtained value of the test statistic occurring by chance alone is very small (very small being defined by the alpha level). Since the probability is very small that the obtained value of the test statistic would occur by chance if the null hypothesis is true, the researcher can *reject* the idea that the null hypothesis is true.

In the present problem, the obtained value of the test statistic of

2.86 exceeds (lies beyond) the critical value of +1.96. This means that the probability is very small of this obtained value of the test statistic occurring by chance alone if the null hypothesis is true. The researcher, therefore, can reject the null hypothesis: He can reject the idea that there is no difference between the population means for male and female teachers and that the difference in his sample means occurred just by chance. In doing so, he accepts the alternative hypothesis that there is a difference in the population means. Since the sample mean for men was greater than the sample mean for women, the researcher concludes that there is a difference in the mean salaries for men and women, with men having a higher mean salary than women.

Question 8–5. What are the critical z-score values for an alpha level of .01? Is the probability smaller than this alpha level of the z-score test statistic obtained in Question 8–4b occurring by chance alone under the null hypothesis? _____

STATISTICAL SIGNIFICANCE

When the null hypothesis is rejected in a particular test, the results in the sample are said to be *statistically significant*. What this means, *and all that it means*, is that the sample results are not likely to be due to chance. If alpha equals .05 and the obtained value of the test statistic exceeds the critical value, it lies among the 5% of all possible values in the sampling distribution that are most unlikely to occur by chance if the null hypothesis is true. The researcher then decides that it is not probable that such an unlikely occurrence should happen the one time he draws samples if the null hypothesis were true, so he rejects the idea that the null hypothesis is true.

The phrase "statistical significance" does *not* mean that the results are interesting, or useful, or that they prove the empirical hypothesis. It is up to the researcher to interpret what the sample results mean. It is possible that in the case of high school teachers, there is discrimination against women and this results in their having lower salaries than men. But it also might be that women teachers as a group have not worked as long as men and that the difference in salaries is due to increases related to length of service. Or the difference might be for some other reason. Hypothesis testing per se does not prove the researcher's empirical hypothesis. The researcher must draw upon other knowledge that he has, additional facts and theories to guide him in interpreting the results. What hypothesis testing does is to show that there is a result in the sample worth looking into: that the sample result, whatever it is, is probably not just due to chance.

If the obtained value of the test statistic does not "exceed" the critical value, the results in the sample are said to be *not statistically significant*. This means that if the null hypothesis is true, the probability that the results in the sample occurred by chance is greater than the alpha level. Since the researcher selects the alpha level, he cannot reject the null hypothesis and must seriously consider the possibility that the results in the sample are only due to chance.

Notice that the probability of the precise value of the test statistic occurring by chance is not found in hypothesis testing. This is because it is only desired to know if the probability of the obtained value of the test statistic is greater or less than the alpha level. It is possible, however, to calculate a more precise probability for the obtained value of the test statistic. It will be recalled that in a continuous distribution, the probabilities of exact values cannot be found and it is necessary to find the probability of a value falling into a particular interval. In hypothesis testing, we are interested in the probability of unlikely values, so the probability for the interval of the test statistic having a value equal to or greater than the obtained value can be found. Examining Table A−3 in the Appendix, the area from the mean to a z score of 2.86 is found to be .4979. Thus, the area to the right of 2.86 is .0021 (.5000 − .4979). Doubling this value because the researcher did not know the direction of the difference in the means before the data were collected yields .0042. Thus, the probability that the test statistic would have the value of 2.86 or larger by chance is .0042. Sometimes these more precise values for test statistics are given. This is often done when results are published, since it allows the reader to use a different alpha level in evaluating the results than the one used by the author. But for the researcher's purposes, if he specified an alpha level before the test, his interest is in whether the probability of the test statistic occurring by chance under the null hypothesis is greater or less than the alpha level. This is done by seeing if the obtained value of the test statistic exceeds the critical value and the more precise calculation is not necessary.

Question 8−6. Use samples of the size described in Question 8−4 to calculate the probability that a difference between the sample means of the size reported there or larger could occur by chance alone.

TYPE I AND
TYPE II ERRORS

When an alpha level is selected in hypothesis testing, a decision will be made to reject the null hypothesis if the probability of the obtained value of the test statistic under the null hypothesis is less than or equal to the alpha level. But no matter what value is selected for alpha there is a chance that even though the test statistic exceeded the critical value (statistically significant), it still occurred by chance and the null hypothesis is really true. In this situation a decision will be made to reject the null hypothesis when in fact the null hypothesis is true. This decision is an error. In statistics this kind of an error is called a *Type I error: rejecting the null hypothesis when in fact it is true.* The probability of making a Type I error is equal to the alpha (α) level, since alpha is the probability that the test statistic could exceed the critical value by chance when the null hypothesis is true.

In the problem of the difference in means between male and female teachers' salaries, a Type I error would be made if in the population there were no difference in the means of the two groups and the value of the test statistic exceeded the critical value just by chance. Of course, the only way to know these things for certain is if the values of the population parameters (the population means) were known, and then it could be seen if a correct decision were made in the test. But if we had this knowledge, there would be no need for any test at all. A hypothesis test does not tell us anything for certain but only with a high degree of probability, and there is always a risk, equal to the alpha level, of falsely rejecting the null hypothesis. This does not negate the utility of the test. A good question to ask about these things is: If we had to bet money on a decision, how would we bet? If one alternative had a 95% chance of being correct and the other alternative had a 5% chance of being correct, we would obviously bet on the first alternative. This is what is done in hypothesis testing. Note, though, that if 100 such bets (each with a 95% and 5% alternative) were made, and every time we bet on the 95% alternative, we could expect to lose about 5 of the 100 bets. Similarly, if 100 hypothesis tests were conducted where alpha was .05 for each and each test turned out to be statistically significant, it can be expected that in about 5 of the 100 tests a Type I error will be made. This is an important reason to avoid searching through a large number of hypothesis tests for "statistically significant" results. If a few tests out of many are significant, there is a good chance that they represent Type I errors.

One way to avoid making a Type I error is to select a very small alpha level. Instead of .05 or .01, alpha could be set equal to .001 or even .00001. Although this reduces the probability of making a Type I error, it increases the probability of making another type of error, known as a Type II error. *A Type II error is accepting the null hypothesis when in fact it is false.* This would mean failing to detect a real difference (or some other result), and what we are usually interested in doing is to detect a real difference if there is one.

The probability of making a Type II error is called beta (β). Whereas the value of alpha can always be specified, the value of beta can be found only when "exact" alternative hypotheses are used. Since we are concerned only with "inexact" alternative hypotheses, we will not determine values for beta. But in general, for a given sample size, alpha and beta are inversely related; as one increases, the other decreases. There is not, however, a simple relationship between the two, so beta cannot be calculated just from a knowledge of alpha. The probability of making one type of error, then, cannot be reduced without at the same time increasing the probability of making the other type of error. The only way to reduce the probability of both types of error at the same time is to increase the sample size. This makes sense intuitively, since a larger sample size should be more "accurate," less open to error. The various possibilities in making a decision about null hypotheses can be diagrammed as follows:

H_0 is Really

		True	False
Researcher Decides to	Accept H_0	Correct decision	Type II error
	Reject H_0	Type I error	Correct decision

When an alpha level is selected, the probability of making a Type I error (α) is set and so is the probability of making a Type II error (β). The selection of an alpha level, then, should be guided by the consequences of making a Type I or Type II error.

Suppose that a potentially useful chemical food additive that prevents food spoilage is developed. It is thought, however, that it might also cause cancer. The additive is fed to a group of randomly selected white rats, and another group of randomly selected white rats is used as a control group. After a time the incidence of cancer in the two groups is noted. A hypothesis test is then conducted to see if there is a statistically significant difference in the incidence of cancer in the two groups. The null hypothesis in this situation is that the incidence of cancer in the two groups is the same. What alpha level should be selected in this experiment?

The alpha level in this experiment should be selected so as to minimize the probability of making the error that has the most serious consequences. If a Type I error is made, rejecting the null hypothesis when it is true, the researcher will mistakenly decide that there is a difference between the two groups and would conclude that the evidence supports the idea that the additive causes cancer. The consequences of this mistaken decision would be to withhold the additive from use and thus to have continued, unnecessary food spoilage. If a Type II error is made, accepting the null hypothesis when it is false, the researcher will mistakenly decide that there is no difference in the incidence of cancer in the two groups and would conclude that the evidence does not support the idea that the additive causes cancer. The consequences of this decision might be disastrous if it led to the use of the additive and people got cancer as a result. Obviously, the consequences of a Type II error in this situation are much more serious than those of a Type I error. Therefore, the probability of a Type II error (β) should be reduced even if it means increasing the probability of making a Type I error (α). Although it is not possible to directly calculate values of beta, the selection of a higher alpha level will reduce beta. Therefore, an alpha level of .20 or even .30 might be selected in this situation.

Although this example is highly contrived, the investigation of Type I and Type II errors is an essential part of the strategy of hypothesis testing. Indeed, in some problems the formulation of the null and alternative hypotheses depends upon an analysis of these errors. But in most situations in social science, because exact alternative hypotheses cannot usually be formulated and because it is difficult to evaluate what the consequences are of an error, it is not usually possible to make an extensive analysis of Type I and Type II errors. Under these conditions it makes some sense to select one of the conventional values for alpha of .05 or .01. Using these values may result in too many Type II errors (overlooking a real difference), but they ensure that among all the "statistically significant" results that are reported, there will be relatively few Type I errors. This means that at least not too many of the results that are published happened by chance.

THE POWER OF A TEST

If β is the probability of *accepting* the null hypothesis when it is false, then $1 - \beta$ is the probability of *rejecting* the null hypothesis when it is false. This probability is called the *power of a test:*

$$\text{power of a test} = \begin{array}{c}\text{probability of rejecting}\\ \text{the null hypothesis when it}\\ \text{is false}\end{array} = 1 - \beta$$

In the case of the difference between two means, the null hypothesis states that there is no difference. If there really is a difference, we want to detect it—to reject the null hypothesis. The power of a test is the probability that if there is a difference, the test will detect it by rejecting the (false) null hypothesis.

Statistical tests differ in their power. For example, nonparametric tests, ones that do not assume interval or ratio level data, are generally less powerful than corresponding parametric tests such as the difference-of-means test that has just been discussed. The power of any test can be increased by increasing the sample size. Although we will not discuss it here, it is possible, if the researcher can specify what size difference he expects to find between two groups, to calculate the probability that given a sample of a certain size, the test will reject the null hypothesis. Consequently, it is possible to calculate what size of sample should be used to have a high probability of rejecting the null hypothesis.

ONE-TAILED TESTS

In the problem of the salaries of high school teachers, the researcher had no idea before he collected the data whether male or female teachers would have higher salaries. Consequently, in conducting his hypothesis test, he had to allow for the possibility that the difference in the mean salaries could be in either direction. In calculating the test statistic through the formula

$$z = \frac{\overline{X}_m - \overline{X}_w}{\hat{S}_{(\overline{X}_1 - \overline{X}_2)}}$$

z would have a positive value if $\overline{X}_m$ were larger than $\overline{X}_w$ and z would have a negative value if $\overline{X}_w$ were larger than $\overline{X}_m$. Because the researcher had no idea which of these situations would occur, he had to locate the critical region in both tails of the curve so that the test statistic could become significant in either a positive or negative direction. Such a procedure is called a *nondirectional* or *two-tailed test.*

In some situations, however, the researcher is willing to stipulate the direction of the difference as part of his empirical hypothesis. In these cases it would be wasteful to use a two-tailed test, since the empirical hypothesis is saying that the test statistic can take on values only in one direction, and there is no use in locating half of the critical region in the other direction. This waste can be avoided and the test itself made more powerful by locating all of the critical region in the expected direction. Such a test is called a *directional* or *one-tailed test.* A one-tailed test should be used when the researcher is willing to incorporate the direction of the expected difference into his empirical hypothesis or when the empirical hypothesis for some reason implies a direction. Of course, the specification of a one-tailed test must be done before the test statistic is calculated so that the results of the test itself do not influence the choice between a one- and two-tailed test.

Suppose a researcher believed that female students as a group do better in college than males. He bases this empirical hypothesis, perhaps, on the idea that there is not as much social and economic pressure on females to attend college as there is on males, and therefore there is a tendency for only the more highly motivated and able females to attend college compared to males. The researcher then selects random samples of male and female college students and calculates the mean cumulative grade-point average for each group. His null hypothesis is

$$H_0: \mu_w = \mu_m$$

The alternative hypothesis, now, is *not* that the mean for women is not equal to the mean for men, because this does not take account of the direction of the difference implied by the empirical hypothesis. Rather, the alternative hypothesis will be

$$H_1: \mu_w > \mu_m$$

and a one-tailed test should be used.

If an alpha level of .05 is selected, the researcher wants to locate all the critical region in the positive tail of the curve, since this is the expected direction of any difference. The critical value will be the z score that cuts off the farthest 5% of the area in the right-hand tail. This area is illustrated in Figure 8–3. Examination of Table A–3 shows that .4500 of the area under the curve lies between the mean and a z score of about 1.65. Consequently, .05 of the area under the curve (.5000 − .4500) lies to the right of 1.65, and this is the critical value for a one-tailed test with an alpha level of .05.

The critical value for a one-tailed test is smaller than the critical

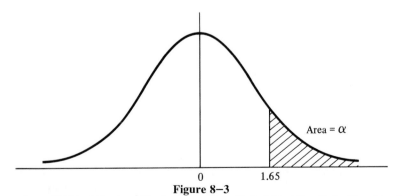

Area = α

0   1.65

Figure 8–3
Location of Critical Region (Shaded) for a One-Tailed Test for Alpha Equal to .05

value for a two-tailed test (1.96). Consequently, a one-tailed test will re-ject the null hypothesis with a smaller difference between the means than will a two-tailed test. Thus, a one-tailed test is more powerful than a two-tailed test in the same situation. This makes sense intuitively when we consider that in a one-tailed test, some additional information is stip-ulated, and there should be some advantage for doing so.

A one-tailed test should be used when the empirical hypothesis clearly implies a difference or change in a specific direction. This direc-tion should be specified in the alternative hypothesis. We do not just guess a direction for the sample result in order to gain the advantage of the greater power of a one-tailed test. If there are any doubts about the direction of the difference implied by the empirical hypothesis, a two-tailed test should be used.

The question of direction should be part of the empirical hypothesis and not just what the researcher thinks might be the situation. This point is illustrated in the following example. Suppose that it is desired to see if a manufacturer is putting less cereal in his boxes than is advertised on the label. The researcher may have no notion at all whether the boxes are underfilled, overfilled, or exactly right. But the empirical hypothesis specifies a direction, and a one-tailed test should be used. If the manu-facturer really overfills his boxes, that is very nice of him, but it is not of concern to the empirical hypothesis. The empirical hypothesis is con-cerned only with whether or not the boxes are underfilled. This is not unfair to the manufacturer, for any amount of overfilling will result in acceptance of the null hypothesis and rejection of the idea that the boxes are underfilled. The manufacturer will just not be acknowledged for doing it. A one-tailed test should be used since it will be more likely to un-cover any underfilling than a two-tailed test, and this is what the empiri-cal hypothesis is concerned about.

Question 8–7. For the problem discussed in Question 8–1, $\bar{X}_u = 28$ and $\bar{X}_r = 25$. Use the estimate of the standard error you calculated in Ques-tion 8–3 to determine whether the difference in the mean divorce rates for urban and rural counties is statistically significant for an alpha of .01 by a one-tailed test. What is the critical value of the test statistic in this problem?

_____

DIFFERENCES BETWEEN TWO
GROUPS USING SMALL SAMPLES

In order to use a z score as a test statistic to evaluate a difference be-tween two group means, it is necessary to estimate $\sigma_{(\bar{x}_1 - \bar{x}_2)}$ by $\hat{s}_{(\bar{x}_1 - \bar{x}_2)}$. This estimation is adequate when the two samples each contain at least

30 observations. When the samples fall below this size, though, the estimation is no longer a good one and a z score cannot be used as the test statistic. At this point the t distribution comes to our aid.

Statisticians have shown that when the sample sizes are small, the formulas on page 205 do not result in a standardized sampling distribution that is normal in shape but in one that is t-shaped. Thus, the test statistic becomes

$$t = \frac{(\overline{X}_1 - \overline{X}_2) - \mu_{(\overline{x}_1 - \overline{x}_2)}}{\hat{s}_{(\overline{x}_1 - \overline{x}_2)}}$$

It is emphasized that although this formula is the same as for the z-score test statistic, it generates a nonnormal t distribution when the sample sizes are small because of the instability of $\hat{s}_{(\overline{x}_1 - \overline{x}_2)}$ when small samples are used. As the sample sizes get larger, the t distribution approximates the normal distribution as a limit and, as can be expected from the fact that the formulas would then be the same, the t scores come to equal z scores. The t score then represents a standardized sampling distribution of the difference between two means just as the z score did, except that it must be referred to a different distribution. Since we are only concerned with hypotheses that specify no difference between the means of the two original populations, $\mu_{(\overline{x}_1 - \overline{x}_2)}$ will equal zero and the t-score formula will reduce to

$$t = \frac{\overline{X}_1 - \overline{X}_2}{\hat{s}_{(\overline{x}_1 - \overline{x}_2)}}$$

It will be recalled that the use of the t distribution requires the calculation of a parameter called *degrees of freedom*. Degrees of freedom refer to the number of scores that can be arbitrarily determined, and 1 degree of freedom is lost each time a mean is calculated. In problems involving the difference between two means, the number of scores is equal to $N_1 + N_2$. And since two means are calculated in the t score formula ($\overline{X}_1$ and $\overline{X}_2$), 2 degrees of freedom are lost. Therefore,

$$df = N_1 + N_2 - 2$$

ASSUMPTIONS IN THE USE OF THE t DISTRIBUTION

Two assumptions must be made, at least formally, to use the t distribution that were not necessary when a z score was used to test hypotheses about group means. The first assumption is that the two original populations are normally distributed. For the distribution of the t statistic to be precisely t-shaped, in theory the two original populations from which t scores are calculated must be normal in shape. In practice, however, even large departures from normality make little difference if the samples are at least moderate in size, and this assumption can usually be relaxed. A more serious assumption formally required by the t distribution is that the variances of the two original populations be equal. This property is called *homogeneity of variance*. Again if the distribution of the t statistic is to be precisely t-shaped, there must be homogeneity of variance in the two original populations from which t scores are calculated. It is possible to relax this assumption also under certain conditions and if some "corrections" are employed. Here, however, we shall restrict ourselves to discussing situations in which the researcher is willing to make the assumption that the variances of the two original populations are equal.

THE POOLED
ESTIMATE OF THE VARIANCE

If the variances of the two original populations are assumed to be equal, the variance of each sample is an estimate of the same value. In such situations a "pooled" value for the sample variance will yield a more accurate estimate of the population variance than either sample variance. A pooled estimate for the variance can be used to obtain a more accurate estimate of the standard error. Essentially, pooling combines two independent estimates of a value to obtain a single more accurate estimate of that value. The formula for a pooled estimate of the original population variance ($\hat{s}_p^2$) is

$$\hat{s}_p^2 = \frac{(N_1 - 1)\hat{s}_1^2 + (N_2 - 1)\hat{s}_2^2}{N_1 + N_2 - 2}$$

The formula for the standard error is

$$\hat{s}_{(\bar{x}_1 - \bar{x}_2)} = \sqrt{\frac{\hat{s}_1^2}{N_1} + \frac{\hat{s}_2^2}{N_2}}$$

and the value for the standard error using the pooled estimate is obtained by substituting $\hat{s}_p^2$ for $\hat{s}_1^2$ and $\hat{s}_2^2$:

$$\hat{s}_{(\bar{x}_1 - \bar{x}_2)} = \sqrt{\frac{\hat{s}_p^2}{N_1} + \frac{\hat{s}_p^2}{N_2}}$$

Question 8–8. A researcher investigating the effects of handgun regulations upon murder rates believes that strict regulation of handguns results in lower murder rates. He selects a random sample of cities which strictly regulate handguns and a random sample of cities in which handguns are relatively unregulated. The murder rates per 100,000 population for the cities in the two samples are:

Cities with Strict Regulation	Cities with Little Regulation
29	34
24	30
32	37
31	24
20	25
18	26
21	28
	36

a. Find the mean murder rates for the two samples.
b. Find the unbiased estimate of the variance for each sample.
c. Calculate a pooled estimate of the variance.
d. Calculate an estimate of the standard error of the difference between two means using the pooled estimate of the variance.

EXAMPLE OF A HYPOTHESIS TEST OF THE DIFFERENCE BETWEEN MEANS USING THE t DISTRIBUTION

Suppose that a political scientist is interested in determining if there is a difference between urban and rural counties in the turnout rates among registered voters in local elections. He selects a random sample of rural counties and a random sample of urban counties and obtains the percentage of registered voters who voted in the last local election for each county as follows:

Urban Counties: Percentage Turnout of Registered Voters in Each County		Rural Counties: Percentage Turnout of Registered Voters in Each County	
X_u	X_u^2	X_r	X_r^2
37	1,369	38	1,444
44	1,936	30	900

Hypothesis Test of Difference between Means Using t Distribution

38	1,444	41	1,681
31	961	37	1,369
45	2,025	27	729
32	1,024	38	1,444
37	1,369	28	784
49	2,401	43	1,849
33	1,089	26	676
48	2,304	35	1,225
39	1,521	42	1,764
47	2,209		
$\Sigma X_u = 480$	$\Sigma X_u^2 = 19,652$	$\Sigma X_r = 385$	$\Sigma X_r^2 = 13,865$

$$\bar{X}_u = \frac{\Sigma X_u}{N_u} = \frac{480}{12} = 40 \qquad \bar{X}_r = \frac{\Sigma X_r}{N_r} = \frac{385}{11} = 35$$

$$\hat{s}_u^2 = \frac{\Sigma X_u^2 - [(\Sigma X_u)^2/N_u]}{N_u - 1} \qquad \hat{s}_r^2 = \frac{\Sigma X_r^2 - [(\Sigma X_r)^2/N_r]}{N_r - 1}$$

$$= \frac{19,652 - [(480)^2/12]}{11} \qquad = \frac{13,865 - [(385)^2/11]}{10}$$

$$= \frac{19,652 - (230,400/12)}{11} \qquad = \frac{13,865 - (148,225/11)}{10}$$

$$= \frac{19,652 - 19,200}{11} \qquad = \frac{13,865 - 13,475}{10}$$

$$= \frac{452}{11} \qquad = \frac{390}{10}$$

$$= 41.09 \qquad = 39.00$$

Because the empirical hypothesis does not specify whether rural or urban counties will have the higher turnout rate, it is decided to conduct a two-tailed hypothesis test of the difference between the means. Therefore, the statistical hypotheses are

$$H_0: \mu_u = \mu_r$$
$$H_1: \mu_u \neq \mu_r$$

The researcher decides he can assume that the two original populations have equal variances and that they do not depart seriously from normality. He can therefore use the t distribution to test the null hypothesis and selects an alpha level of .05.

The first step in calculating the value of the test statistic for the samples, referred to as the obtained value of the test statistic, is to pool the variances. This requires the calculation of the unbiased sample variances, which has been done above. The formula for the pooled variance in this problem will be

$$\hat{s}_p^2 = \frac{(N_u - 1)\hat{s}_u^2 + (N_r - 1)\hat{s}_r^2}{N_u + N_r - 2}$$

The values for $\hat{s}_u^2$ and $\hat{s}_r^2$ were found to be 41.09 and 39.00. Therefore,

$$\hat{s}_p^2 = \frac{(11)(41.09) + (10)(39.00)}{12 + 11 - 2}$$

$$= \frac{452 + 390}{21}$$

$$= \frac{842}{21}$$

$$= 40.10$$

The estimate of the standard error then will be

$$\hat{s}_{(\bar{x}_1 - \bar{x}_2)} = \sqrt{\frac{\hat{s}_p^2}{N_u} + \frac{\hat{s}_p^2}{N_r}}$$

$$= \sqrt{\frac{40.10}{12} + \frac{40.10}{11}}$$

$$= \sqrt{3.34 + 3.64}$$

$$= \sqrt{6.98}$$

$$= 2.64$$

Since $\bar{X}_u$ and $\bar{X}_r$ were found above to be 40 and 35, the obtained t score for the samples is

$$t = \frac{\bar{X}_1 - \bar{X}_2}{\hat{s}_{(\bar{x}_1 - \bar{x}_2)}}$$

$$= \frac{40 - 35}{2.64}$$

$$= \frac{5}{2.64}$$

$$= 1.89$$

The researcher is now ready to compare this obtained value to the critical value of t for $\alpha = .05$ to see if the probability is .05 or less that the obtained t score could have occurred by chance under the null hypothesis.

The critical value of *t* may be found in Table A−4 by looking in the row for the proper degrees of freedom and under the column labeled .05. For this problem the degrees of freedom are

$$\text{df} = N_u + N_r - 2$$
$$= 12 + 11 - 2$$
$$= 21$$

The critical value in the table is 2.0796. The obtained value of *t* is less than the critical value, which means that the probability of the obtained value occurring by chance under the null hypothesis is greater than the alpha level of .05. Therefore, the null hypothesis *cannot* be rejected. The researcher must accept the null hypothesis that the difference between his sample means was likely to have occurred by chance when the means of the two original populations are equal. In terms of his empirical hypothesis, the researcher must conclude that the data do *not* support the idea that there is a difference between urban and rural counties in the turnout rates of registered voters in local elections.

To illustrate a one-tailed test using the *t* distribution, let us change the empirical hypothesis about voter turnout to read: Urban counties will have a higher turnout rate among registered voters than rural counties in local elections. It is not legitimate to alter the empirical hypothesis to make a one-tailed test after calculating the test statistic and then do the test on the same data. It would be proper for the researcher to alter the empirical hypothesis on the basis of the above results and then conduct a one-tailed test on a *new* set of data. It is emphasized that only for purposes of illustration and convenience will a hypothesis test be repeated on the same data using a one-tailed test.

The statistical hypotheses for a one-tailed test will be

$$H_0: \mu_u = \mu_r$$
$$H_1: \mu_u > \mu_r$$

Again, an alpha level of .05 will be employed. This time, however, all of the critical region will be located in the tail which corresponds to the direction of the difference when the mean for urban counties is greater than the mean for rural counties. Because of the way the test statistic was calculated, this will be the positive or right-hand tail of the curve. The critical value of the test statistic, then, will be the *t* score which cuts off the uppermost 5% of the area under the *t* distribution for 21 degrees of freedom.

Since the *t* distribution is symmetric, Table A−4 gives the critical value of *t* whose positive value cuts off half the critical region in the

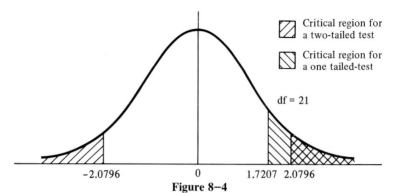

Critical region for
a two-tailed test

Critical region for
a one tailed-test

df = 21

-2.0796   0   1.7207 2.0796

Figure 8–4
Location of Critical Regions and Critical Values in a *t* Distribution for One- and
Two-Tailed Tests for Alpha Equal to .05

right-hand tail *and* whose negative value cuts half the critical region in
the left-hand tail. For the column headed 0.05, the critical values in this
column cut off 2.5% of the area in each tail. Thus if we wish to locate
5% of the area in the right-hand tail, the positive critical values in the
column headed 0.10 will do this since half of .10 equals the alpha level
of .05. These areas are illustrated in Figure 8–4.

For 21 degrees of freedom, the critical value for a one-tailed test
where alpha equals .05 is 1.7207. The obtained value of *t* from the samples was 1.89. Therefore, in this one-tailed test the researcher can reject
the null hypothesis. The results in the sample are said to be statistically
significant for alpha equal to .05 by a one-tailed test.

Question 8–9. For the problem in Question 8–8, state symbolically a null
and an alternative hypothesis. Use the estimate of the standard error
calculated in Question 8–8 to determine if the difference between the
mean murder rates of the two samples is statistically significant. Use a
one-tailed test and an alpha level of .05.

STEPS IN
HYPOTHESIS TESTING

It is useful to make explicit the various steps involved in testing statistical
hypotheses.

   1. *Formulating the empirical hypothesis.* Although the empirical
      hypothesis is really not part of statistics, decisions about what
      statistics to use and what is to be tested are based upon it. The
      most important thing about the empirical hypothesis for statistics

is that it be as explicit as possible. Vague hypotheses are not going to provide useful guides in selecting statistical procedures. The empirical hypothesis should specify what it is that we wish to know from the data. Research is often conducted with only some loose generalizations in mind. But at the time a hypothesis test is to be conducted, something specific must be said for the purposes of the test.

2. *Formulating the statistical hypotheses.* From the empirical hypothesis a null and an alternative hypothesis are formulated. This involves deciding what characteristics of the data shall be tested. So far only the testing of means has been discussed, but means are not the appropriate characteristic to test in all situations. Once it is decided what characteristics will be involved in the test, the null hypothesis is stated. Generally, the null hypothesis is formulated in such a way as to say that there is no difference (or other result) present in the population and that the sample results just occurred by chance. Then the alternative hypothesis is stated. It says, generally, that there really is a difference (or other result) present in the population and that the sample results are not just due to chance under the null hypothesis. If the empirical hypothesis implied a direction to the result, it should be incorporated into the alternative hypothesis so that a one-tailed test can be conducted.

3. *Selecting a model.* Hypothesis tests are conducted by calculating from an appropriate model of chance the probability that the result could occur by chance under the null hypothesis. The justification of the use of a model to make this calculation is that it closely approximates the empirical operation of chance in the actual situation. It should always be kept in mind that the model itself is a purely mathematical entity. Usually, the models are sampling distributions which have been constructed through probability theory. Although we have not employed mathematics, we have verbally discussed the construction of these models.

The researcher must decide which of the many models available in statistics is appropriate to his problem. This decision is determined by the characteristics specified in the null hypothesis and other characteristics of the data, such as the sample size and the level of measurement of the data. When a model is selected as appropriate, its use in calculating probabilities is justified when various assumptions can be made about the empirical situation that are necessary to the development of the model. The most basic assumption is random sampling. All the tests in this and most statistics books assume that samples of data are obtained through simple random sampling. We have seen that random sampling is not a simple problem. A researcher, then, needs to take steps to ensure that his data represent a simple random sample, or at least closely approximate one, before using these tests.

A second basic assumption concerns the level of measurement of the data. Means, standard deviations, and other statistics based on them assume that the numbers in the data represent interval or ratio level measurements. Thus, parametric tests, such as the difference of means tests using either a z score or a t score as the test statistic, require at least interval level measurement. Nothing prevents anyone from using these parametric tests on nominal or ordinal data. Statistics deals only with numbers and is not concerned with what the numbers may represent. But a researcher who does this is faced with the problem of interpreting his results. Since the arithmetic operations of addition, subtraction, multiplication, and division, operations essential to the calculation of parametric statistics, do not make sense for nominal and ordinal measurements, what interpretation can be put on the results of a parametric test using nominal or ordinal data?

Sometimes researchers treat data that are ordinal and might be roughly interval as interval data. This is often done in the case of various kinds of attitude measurements. When a researcher does this and then proceeds to use a parametric test, he should be quite aware that he is making a questionable assumption and is running a risk that the test could produce a misleading result. The point here is that in parametric tests, the data must be at least interval or *assumed to be interval* for the results of the test to have any meaning. If such an assumption cannot be made, the researcher will be forced to pick a different model, one that does not require the assumption of interval data. Some of these non-parametric models will be discussed later.

In addition to random sampling and level of measurement, other assumptions may be required by a particular test. In the case of the difference of means test using the t distribution, the model formally requires that the two original populations are normally distributed and have equal variances. Technically, then, to use this model, the researcher must know (have evidence) that these things are true or be willing to assume that they are true. If a researcher makes assumptions like these and the data do not really meet them, serious distortions can occur in the results of some tests.

If these additional assumptions for a test can be violated without much effect, the test is said to be *robust*. This, we saw, is true of the t distribution in testing the difference between two means. Not all tests, however, are as robust as this, and serious distortions may occur if their assumptions are violated. This is one reason why even a person who is only a "user" of statistics needs to have an idea of the theory and assumptions involved in using a particular test.

4. *Selecting an alpha level and locating the critical region.* Ideally the selection of an alpha level would be based upon an evaluation of the consequences of making Type I and Type II errors. In most practical situations, it will be very difficult, if not impossible, to do this, and it is likely that the researcher will resort to one of the conventional alpha levels of .05 or .01. But there should not be a mindless acceptance of these alpha levels. The researcher should ask himself if he can evaluate the consequences of a Type I or Type II error in his problem. If there are serious consequences attached to one or both types of error, it might be worth the effort to consult an expert to see if an evaluation of the errors can be made. If serious practical consequences are not involved, the researcher should still ask himself if he is able to do anything about reducing the probability of the errors in his situation. One obvious way is to increase the size of the sample if this is feasible and desirable. At the least, a researcher should ask if his particular problem actually requires the conventional alpha levels. If the research, for example, is exploratory and whatever results are uncovered in it will be investigated further in a larger study, the researcher might be willing to entertain a greater risk of a Type I error in order to decrease the probability of missing possible results (decreasing the probability of a Type II error). This can be done by increasing alpha to perhaps .10 or .15.

After an alpha level has been selected, the location of the critical region must be determined. If the empirical hypothesis implies nothing about the direction of the result, a two-tailed test is appropriate and half of the critical region will be located in each tail of the sampling distribution. If a direction is implied, a one-tailed test may be used and all the critical region will be located in the appropriate tail. The location of the critical region, of course, determines the critical value of the test statistic.

5. *Computing the test statistic.* The test statistic will be determined by whatever model is chosen to evaluate the null hypothesis. The computation of the test statistic may be fairly simple or rather complicated, but the purpose is always the same: to find the probability of the obtained value of the test statistic occurring by chance from the model if the null hypothesis is true.

6. *Making a decision about the null hypothesis.* Once the test statistic for the data is calculated, it is compared to the critical value. If it "exceeds" the critical value, it means that the probability of the obtained value occurring by chance if the null hypothesis were true is less than the selected alpha level. When this happens, the result in the sample is said to be *statistically significant,* which means that the null hypothesis for the problem was rejected. If the obtained value does not exceed the critical value, the

probability that the obtained result could occur under the null hypothesis is greater than the alpha level, and the null hypothesis must be accepted. The results in the sample are then said to be *not statistically significant.*

At this point, a third nontechnical possibility arises. *The researcher may opt not to make any decision about the null hypothesis.* Where the conventional alpha levels are used, a single decision-making rule is being applied across the board to many different types of situations. But there is no single decision rule that is correct for all situations, even if it employs such terms as "critical value," "null hypothesis," "alpha," and "beta." If the researcher has, for example, other reasons for believing the null hypothesis to be false but the results in his data did not turn out to be statistically significant, but came close, he may want to decide *to defer making a judgment* about the null hypothesis until he examines some more data.

What, then, is the purpose of a hypothesis test if it is not necessary to make a decision? Regardless of what the researcher is going to do with the information, he needs to know what the probability is that the sample results could have occurred by chance under the null hypothesis. Basically a hypothesis test gives us this information, and that is all. It tells us whether the sample result was likely to have occurred by chance under the null hypothesis where likely and unlikely is defined by alpha. In evaluating the sample results, the researcher should then incorporate this piece of information along with everything else he knows about the problem he is working on.

7. *Interpreting the results of the test in terms of the empirical hypothesis.* This step, as was the first one, is not part of statistics. As researchers, we are not interested just in the statistical question of whether or not the null hypothesis can be rejected. We want to know what the implications are of our data for the empirical hypothesis. It is up to the researcher to take the information from the hypothesis test and incorporate it with the sample results and say something about how the data bear upon the empirical hypothesis. This is not a mathematical procedure. It requires thought, insight, and knowledge of the substantive area with which the problem deals. A statistical procedure is not a substitute for thinking.

THE USE
OF HYPOTHESIS TESTS

In this chapter hypothesis testing about the difference between two means has been covered. But there are many other characteristics besides means that can be tested. It will be useful to briefly mention some of these to indicate the range of strategies that statistics provides. The value of a single sample mean against a particular value for the population mean can be tested. Proportions, such as the proportion of people who rate the president as doing a good job, can be tested. It is possible to test standard deviations and variances. And it is possible to test statistics that measure the degree of association between two variables, such as the correlation coefficient discussed in Chapter 11. There are many other tests in addition to these. The selection of which test a researcher uses is guided by the empirical hypothesis, which is why its explicit formulation is so important. But all tests follow the same general procedures outlined above.

All hypothesis tests, or significance tests as they are also called, have one general purpose: to determine if a "random mechanism" can account for the results in the data. Such a random mechanism can take place in three distinct contexts, two of which are of formal concern to statistics.

1. *Random samples obtained from a population.* This is the context that has been of concern here. The formal random mechanism is the random sampling procedure used to select the sample from the population. The purpose of a significance test here is to see if this random sampling procedure was likely to produce the results in the sample if the null hypothesis is true.

2. *Random assignments in experiments.* Experiments in the social sciences most often appear in psychology, although some experimentation is done in sociology and, to a much lesser extent, in political science. Although these experiments can become quite complicated (there is, in fact, a field of study devoted to the design of experiments), a simple illustration will explain the role of a random mechanism in experiments. Suppose that a psychologist wished to see if a drug quickens the reactions of people. The problem for the psychologist is to try to control out extraneous influences so that if he observes changes in reaction time, he can attribute the change to the experimental treatment, the drug. To test the drug, the psychologist injects one group of people with the drug (called the *experimental group*) and another group with a sugar solution (called the *control group*) and of course does not let people know what they are injected with. The purpose of this procedure is that the process of being injected might itself in some way affect reaction time. The idea is to treat the two groups in exactly the same way except for the experimental variable.

   Now if the reaction time of the two groups is measured, the psychologist still has a problem. It might be that the people who were administered the drug had faster reaction times to begin with than the people who did not get the drug. One way to solve this problem is for the psychologist to randomly assign people to the two groups. Then any difference in the *average* reaction times for the experimental and control groups before the treatment is due to chance. After the experiment, the experimenter can conduct a hypothesis test on the difference between the means of the two groups. If the difference is larger than what could have been expected to occur by chance, the experimenter can say that the drug had an effect on reaction time. Thus, in experiments, hypothesis tests are used to evaluate whether the results of the experiment were likely to occur because of the random assignment procedures of subjects to experimental and control groups.

Note that there is no question here of selecting a sample from a population. Psychologists rarely take a random sample of a population for their experiments. Thus, they cannot make a *statistical* inference from their experimental results to a larger population. This does not mean that inferences from an experiment to a larger population cannot or are not made. But it does mean that the type of statistical inference from sample to population that has been discussed here cannot be done.

3. *Randomness as an empirical hypothesis about results in a population.* In political science it is not infrequent to find that data have been gathered upon an entire population. Thus information about all congressmen, or all state governors, or all counties in the United States, or even all cities over 50,000 population can be gathered. There is no sample involved in these situations, and thus there is no statistical inference to be made from a sample to a population. Since the entire population is measured, whatever the results in the data are, they are true, period. Because a sample is not involved and because there is no experimenter making random assignments, there is no formal random mechanism present to be evaluated by a hypothesis test. But there may be a hidden or implicit random process at work to cause the results in the population.

Suppose that a political scientist wished to see if there is a difference in the average age of Democratic and Republican senators. Whatever difference does occur is a fact; it is a parameter of the population and its value is not in question. The researcher wishes to account for this fact. One possibility is that there is something systematic at work which results, let us say, in Democratic senators being younger than Republican senators. Another possibility, however, is that age is pretty much a random variable (within certain limits) as far as election to the Senate is concerned, and the difference in the mean age of Democrats and Republicans results just from this random variable.

In this case, randomness is being put forward as an *empirical* hypothesis to explain a fact in the population. Consequently, a hypothesis test can be conducted to indicate if the fact is compatible with what one would expect in a random situation. If the null hypothesis can be rejected, the empirical hypothesis of randomness can be eliminated, and the political scientist can start to look for a systematic reason to account for the difference in age.

Perhaps the most important thing to keep in mind in interpreting hypothesis tests is that the finding of a result to be statistically significant does not necessarily mean that it is interesting or useful. It will be recalled that the power of a test is its probability of rejecting the null hypothesis when it is really false. In Figure 8–5 are the power curves for the *t* test using different sample sizes. The curves show, for example, that for samples of size 10, the probability is .80 that the null hypothesis will be rejected if the difference between the two original population means is equal in size to 1 standard deviation of the sampling distribution. Examining the curve for samples of size 100 shows that even very small differences are likely to be found statistically significant by the *t* test. In the case of the difference in the mean salaries of male and female teachers, for example, a difference in the sample means of about $40 would be found to be statistically significant if samples of 500 were taken. Is a difference of $40 of substantive interest even if it is statistically significant? Researchers who have great control over their data are sometimes advised not to use large sample sizes, so that they will not be bothered with detecting small differences that are of little interest. This problem is of

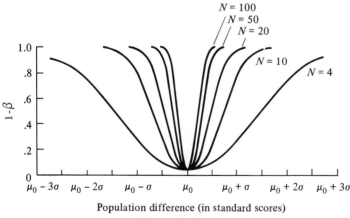

Figure 8–5

Power Curves for the *t* test; $\alpha = .05$; Two-Tailed test. (Reprinted from W. J. Dixon and F. J. Massey, *Introduction to Statistics.* New York: McGraw-Hill Book Company, 1957.)

special concern to sociologists and political scientists, who often deal with sample surveys whose size runs into the thousands. At this point very small differences are likely to be statistically significant. Perhaps the point can best be made by noting that any difference, no matter how small, can be made statistically significant if a large enough sample is used. Statistical significance says something about what could happen by chance, not that something important has happened. Hypothesis testing does not relieve the researcher of the responsibility of demonstrating the importance of his findings.

EXERCISES FOR CHAPTER 8

1. A researcher wishes to use the t distribution to see if there is a statistically significant difference between the means of two samples. What assumptions must be made to use the t distribution in this test?

2. A researcher interviews random samples of Protestants and Catholics and calculates the average number of days of church attendance per month for each sample. He plans to conduct a significance test of the difference between the mean church attendance for the two groups. State a null and an alternative hypothesis for this problem. In the context of this problem, what would it mean if a Type I error is made? A Type II error? If the researcher selects an alpha level of .10, what is the probability that a Type I error will be made? Explain.

3. Given two independent random samples of 100 observations each from the same population with $\hat{s}_1 = 11$ and $\hat{s}_2 = 12$, what is the probability that the two sample means will differ by 2 or more units? If the process of drawing two independent random samples from the population and calculating the difference between the sample means were repeated indefinitely, within what range of values would 95% of all possible differences fall? 99%?

4. A researcher hypothesizes that voting-turnout rates in local elections are lower in southern counties than in northern counties. He draws a random sample of 60 southern counties and a random sample of 70 northern counties and finds the mean turnout for the samples to be 40% for the southern sample and 47% for the northern sample. He finds the standard deviations of the samples to be: $\hat{s}_s = 23$ and $\hat{s}_n = 25$. State a null and an alternative hypothesis for this situation. Is the difference between the sample means statistically significant at an alpha level of .05?

5. A researcher wishes to determine if serving on the law review in college has an effect upon lawyers' incomes. In a city he obtains a random sample of 160 lawyers who served on their law review while in college and a random sample of 150 lawyers who did not serve on their law review. He finds that the average incomes for his samples are $36,000 for lawyers who were members of their law review and $34,500 for those who were not. The standard deviations $(\hat{s})$ for the samples were $4000 for lawyers who were members of their law review and $3000 for the lawyers who were not members. Is this difference statistically significant at an alpha of .005? To what population can a statistical inference be made?

6. A political scientist hypothesizes that congressmen from rural districts are more likely to be reelected to office than congressmen from

urban districts. He selects random samples of congressmen from urban and rural districts and finds out how many terms each congressman has served in the House of Representatives as follows:

Number of Terms Served

Urban Districts	Rural Districts
5	7
4	3
6	6
3	7
10	12
3	4
5	8
7	5
5	9
1	7
6	9

State a null and an alternative hypothesis for this problem. Is the difference between the means statistically significant at an alpha level of .05?

7. A researcher wishes to see if there is a difference in the suicide rates in cities over 100,000 population versus cities with a population less than 100,000. He selects random samples of the two types of cities and finds the suicide rates per 10,000 people for each city as follows:

Suicide Rates per 10,000 People

Cities under 100,000	Cities over 100,000
.5	.6
1.5	2.8
.3	.4
.7	2.3
.8	.5
.7	.4
1.5	1.2
2.0	.4
1.6	1.2
.3	1.8
.9	2.3
1.2	2.1

Is there a statistically significant difference in the mean suicide rates of the two samples at an alpha level of .05?

ANSWERS TO
QUESTIONS IN CHAPTER 8

8–1: a. H_0 : there is no difference in the mean divorce rates for urban and rural areas.

H_1 : the mean divorce rate for urban areas is greater than the mean divorce rate for rural areas.

b. $H_0 : \mu_u = \mu_r$, $H_1 : \mu_u > \mu_r$.

8–2: The null hypothesis will be rejected if the probability that the mean divorce rates for the two samples could have occurred by chance is equal to or less than 1 out of 100. The null hypothesis will be accepted if the probability that the difference beteen the sample means could have occurred by chance is greater than 1 out of 100.

8–3: a. $\hat{s}_{(\bar{X}_1 - \bar{X}_2)} = \sqrt{\dfrac{36}{50} + \dfrac{25}{60}} = 1.07.$

b. Upper limit for 68.26% = $+1\hat{s}_{(\bar{X}_1 - \bar{X}_2)} = +1.07,$
Lower limit for 68.26% = $-1\hat{s}_{(\bar{X}_1 - \bar{X}_2)} = -1.07,$
Upper limit for 95% = $+1.96\hat{s}_{(\bar{X}_1 - \bar{X}_2)} = +2.10,$
Lower limit for 95% = $-1.96\hat{s}_{(\bar{X}_1 - \bar{X}_2)} = -2.10.$

8–4: a. $\hat{s}_{(\bar{X}_1 - \bar{X}_2)} = \sqrt{\dfrac{1600}{80} + \dfrac{1764}{84}} = 6.40.$

b. $z = \dfrac{D - \mu}{\hat{s}_{(\bar{X}_1 - \bar{X}_2)}} = \dfrac{(285 - 275) - 0}{6.40} = 1.56.$

8–5: 2.58, −2.58, no.

8–6: .119.

8–7: Critical values are +2.58 and −2.58:

$z = \dfrac{28 - 25}{1.07} = 2.80$

The difference is statistically significant.

8–8: a. $\bar{X}_1 = 25$, $\bar{X}_2 = 30$.   b. $\hat{s}_1^2 = 32$, $\hat{s}_2^2 = 26$.

c. $\hat{s}_p^2 = \dfrac{(6)(32) + (7)(26)}{7 + 8 - 2} = 28.77.$ d. $\hat{s}_{(\bar{X}_1 - \bar{X}_2)} = \sqrt{\dfrac{28.77}{7} + \dfrac{28.77}{8}} = 2.78.$

8–9: $H_0 : \mu_1 = \mu_2$, $H_1 : \mu_1 < \mu_2$, $t = (25 - 30)/2.78 = -1.80$. The critical value of t for a one-tailed test where alpha is .05 and for 13 degrees of freedom is 1.77. Therefore, the difference in the mean murder rates of the two samples is statistically significant.

9
Chi-Square

Chi-square is the name of a sampling distribution of a test statistic just as the *t* distribution is a sampling distribution of the *t* test statistic. The chi-square distribution is another mathematical model of probability and can be used to calculate probabilities in the same way as was done with distributions such as the *t* distribution or the sampling distribution of the difference between two means. The chi-square distribution, though, is an appropriate model of probability for different types of empirical situations than those that have been discussed so far. Although the chi-square distribution has a number of uses, its most frequent use in social science occurs in testing whether the association between two nominal or ordinal level variables is statistically significant.

COUNTING AND MEASUREMENT

From our earlier discussion of the levels of measurement, it will be recalled that it is not meaningful to perform the operations of addition, subtraction, multiplication, and division on numbers that represent nominal or ordinal level measurements. Consequently, statistics that employ these operations, such as the difference of means tests of the last chapter, cannot be used on such data.* Nominal and ordinal data occur very frequently in social science. Fortunately, various techniques, called *nonparametric statistics,* have been developed to deal with this kind of data.

One way to deal with the problems of lower-level measurements is by working with the frequencies of occurrence of the values of a variable rather than with the values themselves. An illustration will best demonstrate this point. Suppose that in a survey of 500 people in a city the

---

*See "The Problems of Measurement" in Chapter 2 for a discussion of the levels of measurement.

respondents' religious preference is obtained. The responses are then coded according to the following scheme:

1. Protestant
2. Catholic
3. Jewish
4. Other
5. None

Thus, a person who is a Protestant is assigned a 1 on this variable, a Catholic a 2, and so on. This procedure yields nominal level measurements. The numbers 1, 2, 3, 4, and 5 identify categories into which people are placed. There is no underlying dimension here: A person coded as a 4 is not necessarily more or less religious, or more or less of anything, than a person coded as a 1, 2, or 3. This point is made obvious by the fact that some other coding scheme, such as the category of Protestant being a 3 and the category of Jewish being a 1, would have served just as well as the one above.

Suppose that the survey produced the results shown in Table 9–1. While the values of the variable (1, 2, 3, 4, and 5) cannot be manipulated arithmetically since they are nominal level measurements, it is possible to say that three times as many people in the survey are Protestant than are Catholic. Similarly, there are twice as many Catholics in the survey as Jews. These statements involve ratios which employ arithmetic operations: $\frac{300}{100} = 3$ and $\frac{100}{50} = 2$. In other words, although it is not possible to manipulate the values of the variable arithmetically, it is possible to manipulate the frequencies of occurrence of the values, the *counts*, arithmetically. Counting the number of objects in each category yields exact numbers that can be ordered along a dimension of frequency of occurrence and are ratio level measurements. Consequently, these numbers, the counts for each category, may be legitimately manipulated by any

Table 9–1

	Category					
	1 Protestant	2 Catholic	3 Jewish	4 Other	5 None	Total
Number of observations	300	100	50	15	35	500
Percentage of total	60	20	10	3	7	100

arithmetic operation. One of the most common manipulations is to determine what percentage of people in a sample fall into each category. This is done by simply dividing the number of people in a category (its count) by the total number of people in the survey. The resulting percentages are descriptive statistics. These percentages, calculated in Table 9–1, more understandably describe the distribution of the respondents across the categories than do the raw counts.

Question 9–1. According to the information in the survey whose results are presented in Table 9–1, what is the estimated probability that if one person were selected at random in the city, he would be Catholic? The probability that he would have no religious preference?

TABLES

When data are sorted into various categories, the information is presented in the form of tables. Table 9–1 is called a *one-way table* because it involves only one variable. When interest focuses on the relationship between two variables, a *two-way table* is used. The process of producing a two-way table is called *cross-tabulation*. Such tables are also known as *contingency tables,* the idea being to see if the occurrence of values on one variable is contingent upon the values of another variable.

Suppose that a researcher selected a random sample of 200 students at a college and asked for whom they voted for president in 1972. The researcher's hypothesis is that sex affects voting (that the way a person votes is contingent upon his sex) so he is going to cross-tabulate sex with voting choice in 1972 to see if there is an association between these two variables in the sample.

It will be recalled that an independent variable is one which is thought to cause or affect something else and that the dependent variable is the variable that is thought to be caused or affected. What is the independent variable and what is the dependent variable is specified by the researcher's empirical hypothesis. In this case it is obvious that the way a person votes cannot affect his sex, and therefore sex must be the independent variable and voting the dependent variable. In many situations, however, the determination of the independent and dependent variable is not so simple. The researcher must take care to formulate the empirical hypothesis clearly so as to specify this information, for it is often the only guide to go by. The specification of independent and dependent variables is not a permanent thing and changes according to the problem. For a geneticist or biologist interested in trying to account for the sex of children, for example, sex would be the dependent variable.

Table 9–2

Sex

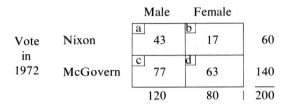

		Male	Female	
Vote in 1972	Nixon	43	17	60
	McGovern	77	63	140
		120	80	200

Suppose that the cross-tabulation by the researcher of voting by sex produces the results shown in Table 9–2. In such a two-way table, it is conventional to place the independent variable across the top of the table and the dependent variable along the side. Each square in the table is called a *cell*. The number of cells is determined by the number of values there are for each variable. The variable "Sex" has two values (male and female), as does the variable "Vote in 1972" (Nixon and McGovern). The number of cells in a table will be equal to the product of the number of values for each variable. Thus, for Table 9–2, $(2)(2) = 4$. Tables that have two values for each variable are often called "2×2" (read "2 by 2") tables.

For each value of a variable there is a row or a column in the table. Again it is conventional to speak of rows as running horizontally through the table and of columns as running vertically through the table. All the observations appearing in a particular column of the table have the same value on the independent variable, and all the observations appearing in a particular row have the same value on the dependent variable. Thus, in Table 9–2 all the people appearing in cells a and c are male and all the people in cells b and d are female. Similarly, all the people in cells a and b voted for Nixon in 1972 and all the people in cells c and d voted for McGovern in 1972.

At the end of each row and column is a number, which is the sum of all the observations appearing in that row or column. These sums are the *marginals*. A marginal at the end of a row is called a *row marginal*, and a marginal at the bottom of a column is called a *column marginal*. In the lower right-hand corner outside the table is a number called the *total*, which is the total number of observations appearing in the table. The column marginals, when summed, should equal this total, as should the sum of row marginals. These terms are illustrated on the following 2×3 table:

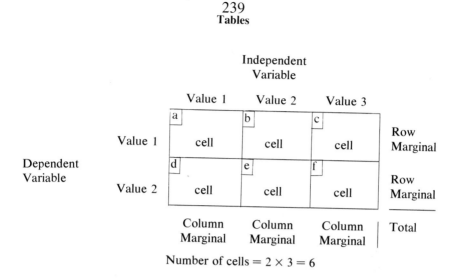

Independent
Variable

Number of cells = 2 × 3 = 6

When it is necessary for some reason to indicate individual cells in a table, they are usually labeled with lowercase letters, starting with the upper-leftmost cell and moving sequentially across the rows.

Question 9–2. A researcher wishes to see if union membership is related to the holding of interventionist or isolationist attitudes on foreign policy. He obtains a random sample of 300 people, and cross-tabulation produces the following table.

Union Membership

Foreign Policy Attitude		Union	Nonunion	
	Interventionist	20	100	120
	Isolationist	55	125	180
		75	225	300

a. What is the independent variable in this table? What is the dependent variable? Why?

b. What is the value for the row marginal of the first row? What is the value of the column marginal of the first column?

c. How many people in the survey are nonunion members? How many people in the survey have isolationist foreign policy attitudes?

d. According to the information in the survey, what is the proportion of union members in the population? What is the proportion of people holding interventionist foreign policy attitudes in the population?

PERCENTAGES

Usually, the purpose of constructing a two-way table is to see if there is some sort of association between two variables. The actual number of observations in the data which appear in each cell is called the *observed value* for the cell. But it is often difficult, simply by inspecting the observed values in a table, to determine if a relationship exists and, if it does, what kind of relationship it is. This is the case in Table 9–2. The problem is that there are unequal numbers of men and women in the table (120 men and 80 women). Consequently, we cannot directly compare the number of men who voted for Nixon (McGovern) with the number of women who voted for Nixon (McGovern). The solution to this difficulty is to compute percentages for the cells. By constructing percentages, the unequal "bases" involved will be taken into account and allow comparisons to be made.

This leads to a very confusing point: In any table it is possible to compute a percentage for each cell in three different ways.

1. *Percentages of the total.* One way to construct percentages would be to divide the observed value in each cell by the total. This is done in Table 9–3 for the data appearing in Table 9–2. The numbers in parentheses are the observed values for each cell. When percentages are computed using the total as the base, they tell us what percentage of all the people in the table fall into each cell. Thus, 21.5% of all the people in the survey (200) were males who voted for Nixon. Similarly, 31.5% of all the people in the table were females who voted for McGovern. When percent-

Table 9–3
Percentaging by Total

Sex

		Male	Female	
Vote in 1972	Nixon	21.5% (43)	8.5% (17)	60
	McGovern	38.5% (77)	31.5% (63)	140
		120	80	200 100%

Table 9-4
1956–1960 Vote Change in Percentages within the Active Core of the Electorate*

1960 Vote	1956 Vote		Total
	Stevenson	Eisenhower	
Kennedy	33	17	50
Nixon	6	44	50
	39	61	100

*From Philip Converse *et al.*, "Stability and Change in 1960: A Reinstating Election," *American Political Science Review*, 55 (June, 1961), p. 272.

ages are based on the total, the sum of the percentages in *all* the cells will add up to 100.

Percentages based on the total are used when it is desired to know the percentage of all the observations that fall into a particular cell. Table 9-4 shows how a national sample of about 1100 people voted in 1956 and 1960. In this table the researchers were not interested in seeing if a person's vote in 1956 affected his vote in 1960 but in determining what percentage of the sample voted consistently Democratic or Republican in both elections and what percentage voted for different parties in the two elections. As can be seen, 33% of the sample voted for Stevenson in 1956 and for Kennedy in 1960, and 44% voted for Eisenhower in 1956 and for Nixon in 1960. Six percent of the sample, though, crossed from Stevenson to Nixon and 17% crossed from Eisenhower to Kennedy. Percentages of the total in this table, then, identified the proportions of consistent party voters and switchers in the sample for the 1956 and 1960 elections.

Although percentages of the total often provide useful and interesting information, they cannot be used to analyze the relationship between variables since they communicate information based on all the observations in the table. To examine the relationship between the variables, percentages must be based upon information about one of the variables.

2. *Percentages of the column.* If the independent variable has been placed across the top of the table and the dependent variable on the side, then the proper way to construct percentages to examine the relationship between the two variables is by column. This is done by dividing the observed value in each cell by its column marginal. Thus, for cell a in Table 9-2, the observed value is 43 and its column marginal is 120, so the percentage by column for cell a is 35.8. Table 9-5 presents column percentages for the data in Table 9-2. When percentaging has been done by column, the percentages for all the cells *in the same column* will add up to

Table 9–5
Percentaging by Column

Sex

		Male	Female	
Vote in 1972	Nixon	35.8% (43)	21.3% (17)	60
	McGovern	64.2% (77)	78.7% (63)	140
		120 100%	80 100%	200

100. In Table 9–5, for example, 35.8 + 64.2 = 100 and 21.3 + 78.7 = 100.

Percentages by column tell us what percentage of all the observations which have the same value on the independent variable have a particular value on the dependent variable. In Table 9–5, we see that of the 120 men, 35.8% (43) voted for Nixon and 64.2% (77) voted for McGovern. Of all 80 women, 21.3% (17) voted for Nixon and 78.7% (63) voted for McGovern. Comparing these percentages, we see that a greater proportion of men voted for Nixon than did women (35.8 versus 21.3) and a greater proportion of women voted for McGovern than did men (78.7 versus 64.2). These percentages indicate that in this sample there is a relationship between sex and voting in 1972: Men were more likely to vote for Nixon than were women and, conversely, women were more likely to vote for McGovern than were men. Note that although a majority of both men and women in the survey voted for McGovern, there still is a relationship between sex and voting because women voted for McGovern at a higher rate than men did. It is the *relative difference* between the percentages that counts, not the absolute size of the percentages themselves. Percentaging by column, then, enables us to examine the effect of the values of the independent variable upon how observations fall among the values of the dependent variable. This allows us to determine the degree and nature of the association between the two variables.

3. *Percentages of the row.* Finally, it is possible to construct percentages by row. There is no reason to do this, though, if the independent variable is across the top of the table, but it is done in Table 9–6 to illustrate the procedure. When percentages are

Table 9–6
Percentaging by Row

Sex

		Male	Female	
Vote in 1972	Nixon	71.7% (43)	28.3% (17)	100% 60
	McGovern	55% (77)	45% (63)	100% 140
		120	80	ǀ 200

computed by row, the percentages for all the cells *in the same row* will add up to 100. Note that the percentages in Table 9–6 are quite different from those in Table 9–5. This is because the base on which the percentages are computed, the row marginals, are different from the base for the column percentages, the column marginals.

The row percentages tell us of all the people who voted for Nixon (or McGovern) what percentage were male and what percentage were female. This might be interesting information, but it is not useful in determining if sex affects voting. If we are interested in seeing if sex affects voting, we want to know, say, what proportion of all men voted for Nixon (64.2%), not what percentage of Nixon voters were men (55%).

Percentaging by row should be done when for some reason the independent variable has been placed on the side of the table and the dependent variable across the top. *The general rule is that percentaging should be done on the values of the independent variable.* If the values of the independent variable are at the tops of columns, percentaging by column should be done. If the values of the independent variable are at the sides of rows, percentaging by row should be done.

It is very common to cross-tabulate data using computers. Usually the computer programs generate tables in which the observed values and all three types of percentages appear in the cells. In these circumstances care must be taken in identifying the types of percentages. This can be done by determining how the percentages add up to 100, as has been described above.

Question 9–3. For the table in Question 9–2:

a. What is the percentage of people in the survey who are union members and who hold interventionist foreign policy attitudes? What is the percentage of people in the survey who are nonunion members and who hold isolationist foreign policy attitudes?

b. Compute percentages for the table in Question 9–2 in such a way as to see if union membership affects foreign policy attitudes.

Union Membership

		Union	Nonunion
Foreign Policy Attitude	Interventionist		
	Isolationist		

c. State the type of relationship you find.

d. According to the table in Question 9–2, estimate the probability that if one person were randomly selected from the population, he would be a union member. Estimate the probability that a randomly selected person would have an isolationist foreign policy attitude.

STATISTICAL HYPOTHESES
FOR RELATIONSHIPS IN TABLES

Percentages are one way to describe the association that occurs between two variables in a table. In Chapter 10 other descriptive methods of association for nominal and ordinal data will be discussed. But regardless of how the association between two variables in a sample is described, the problem arises of determining whether the association between the variables exists in the population. If an association is found between two variables for the observations in a random sample, that association in the sample is a fact. The question is whether that association also exists in the population from which the sample was drawn. It is possible that in the population there is no association — that when the sample was drawn, the random-selection process happened to select observations in such a way as to make an association between the variables appear in the sample. If this is the case, the association in the sample is something that happened by chance alone, and the variables are in no way associated or related to each other. Obviously, no matter how great or strong an association might be in the sample, if there is a substantial probability that it occurred by chance we cannot use the sample results as evidence to support the idea that there is a relationship between the variables.

Recalling our work on hypothesis testing, it can be seen that a test of significance is needed for association between two variables in a table.

The chi-square distribution provides a way to make such a test. In addition, the chi-square test makes no assumptions about the level of measurement of the data. In fact, the chi-square test was designed for use on the lowest-level data, nominal data, but it is also commonly used on ordinal data, and it may legitimately be used on even higher-level data. (It may be mentioned that any statistic designed for use on a certain type of data may always be used on data at any higher level of measurement. It is not legitimate, however, to use a statistic on data at a lower level of measurement than the test was designed for.)

The null and alternative hypotheses for associations in a table that are to be tested using the chi-square distribution are:

H_0: There is no association between the two variables in the population (and thus any association between the variables is due only to chance).

H_1: There is some association between the two variables in the population (and the association between the variables in the sample reflects this).

Note that these statistical hypotheses say nothing about the type of association. It is possible for variables to be associated in complicated and sometimes perverse ways. It is up to the researcher to specify the type of association in his empirical hypothesis and to use descriptive statistics, such as percentages, to interpret the type of association in a table. Since the above statistical hypotheses are usually the ones employed in all significance tests of association using the chi-square distribution, they will not be repeated for specific problems.

THE CHI-SQUARE STATISTIC

The logic of the chi-square test statistic for the association between two variables in a table is very interesting and straightforward. Let us start with the marginals for the survey on sex and voting presented in Table 9–7. These marginals tell us that of the 200 people in the sample, there were 120 men, 80 women, 60 Nixon voters, and 140 McGovern voters. Now let us ask the following question: Based upon the information in the survey, if one student at the college were selected at random, what is the estimated probability that the student would be a male? Since 200 students were selected at random from the population and 120 of these randomly selected students were male, the information in the survey estimates the probability as $\frac{120}{200}$, or .60, that one randomly selected student from the population would be male. The information in the survey also estimates the probability that a single randomly selected student

Table 9–7

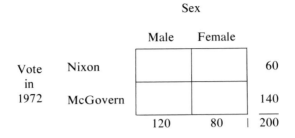

Sex

		Male	Female	
Vote in 1972	Nixon			60
	McGovern			140
		120	80	200

would be female as $\frac{80}{200}$, or .40. Similarly, the estimated probability that one randomly selected student would have voted for Nixon is $\frac{60}{200}$, or .30, and that a randomly selected student would have voted for Mc-Govern is $\frac{140}{200}$, or .70. These probabilities may be listed as follows:

$$P_m = .60 \qquad P_{\text{Nix}} = .30$$
$$P_f = .40 \qquad P_{\text{Mc}} = .70$$

Before proceeding, it should be emphasized that the "probabilities" just computed from the information in the sample are really *estimates* of probabilities. For example, the actual probability that a single randomly selected person from the population would be a male can only be determined by knowing the number of males in the population and the total number of people in the population. These numbers are unknown to us because we only have information on a random sample of the population. Consequently, when it is said that because 120 of the 200 people in this random sample were male and thus that the probability that one randomly selected person from the population would be male is $\frac{120}{200} =$ .60, we are really calculating an estimate of the probability. The same is also true of the probabilities of the various "joint occurrences" and "expected values" that will be computed below. The fact that estimates rather than actual probabilities are used, though, does not alter the basic logic of the argument. And the test of the final chi-square statistic takes into account that it is based upon estimates of probabilities. In discussing the development of the chi-square statistic, however, it will be convenient to talk about probabilities rather than estimates of probabilities. Thus, it should be kept in mind that whenever "probabilities" are calculated from information about a sample, these are really estimates of probabilities.

Returning to our problem, we can ask what is the probability that a student selected at random from the population would be male and have voted for Nixon *under the assumption that there is no relationship be-*

tween the two variables of sex and voting. When it is stated that there is no relationship between the variables, we are asserting that sex and voting are independent events – that the outcome of one event, sex, has no effect upon the probabilities of the outcomes of the other event, voting choice. In Chapter 6 it was found that the probability of the joint occurrence (two separate events occurring together) of two independent events, A and B, was

$$P_{(A \text{ and } B)} = (P_A)(P_B)$$

To ask what the probability is that one randomly selected student would be male and have voted for Nixon under the assumption of no relationship between sex and voting is to ask for the probability of the joint occurrence of two independent events. Therefore,

$$P_{(m \text{ and Nix})} = (P_m)(P_{\text{Nix}})$$
$$= (.60)(.30)$$
$$= .18$$

The probabilities of the other possible joint outcomes of the two variables under the assumption of no relationship are found in the same way:

$$P_{(m \text{ and Mc})} = (P_m)(P_{\text{Mc}})$$
$$= (.60)(.70)$$
$$= .42$$

$$P_{(f \text{ and Nix})} = (P_f)(P_{\text{Nix}})$$
$$= (.40)(.30)$$
$$= .12$$

$$P_{(f \text{ and Mc})} = (P_f)(P_{\text{Mc}})$$
$$= (.40)(.70)$$
$$= .28$$

These values are the probabilities of the various joint outcomes occurring by chance alone under the assumption that there is no relationship between the variables of sex and voting.

Since 200 students, not just 1, were randomly selected into the sample, the probability of a joint outcome can be multiplied by 200 to find the *number* of observations in the sample which could be "expected" to have a particular joint outcome by chance alone if the two variables are not related. These numbers are called *expected values*.

$$\text{expected value for the joint outcome of male and Nixon} = (P_{(m \text{ and Nix})})(200)$$

$$= (.18)(200)$$

$$= 36$$

$$\text{expected value for the joint outcome of male and McGovern} = (P_{(m \text{ and Mc})})(200)$$

$$= (.42)(200)$$

$$= 84$$

$$\text{expected value for the joint outcome of female and Nixon} = (P_{(f \text{ and Nix})})(200)$$

$$= (.12)(200)$$

$$= 24$$

$$\text{expected value for the joint outcome of female and McGovern} = (P_{(f \text{ and Mc})})(200)$$

$$= (.28)(200)$$

$$= 56$$

In Table 9–8 the expected values for the various joint outcomes are placed into the upper right-hand corner of each cell representing a particular joint outcome. Each cell of Table 9–8 contains two numbers. One number is the observed value, the actual number of people in the survey who were observed to fall into the cell. The second number is the expected value, the number of people who could be expected to fall into the cell if there were no relationship between the variables of sex and voting. The strategy should now become clear: If there really is no relationship between the two variables, then the observed values should

Table 9–8

Sex

		Male	Female	
Vote in 1972	Nixon	43 [36]	17 [24]	60
	McGovern	77 [84]	63 [56]	140
		120	80	200

closely approximate the expected values. Under the null hypothesis, any deviation of the observed values from the expected values will be due to chance alone. The chi-square distribution provides a model from which we can calculate the probability of the observed values deviating from the expected values by a particular amount. If this probability is large, greater than a selected alpha level, we say that there is a good possibility that the deviation of observed from expected values occurred by chance alone, and we must accept the null hypothesis of no relationship between the variables. If the model tells us that the probability is very small, less than a selected alpha level — that the deviation of observed from expected values could occur by chance under the the null hypothesis — the null hypothesis can be rejected and we can accept the alternative hypothesis that there is a relationship between the variables. Of course, the more that the observed values deviate from the expected values, the less likely it is that the null hypothesis is true.

Before we can compute probabilities using the chi-square distribution, it is necessary to measure the amount of deviation of the observed values from the expected values. This is done through the chi-square statistic. The formula for the chi-square statistic is

$$\chi^2 = \sum \frac{(O - E)^2}{E}$$

This formula tells us to subtract the expected value for a cell from the observed value for the cell, square the difference, and divide the result by the expected value of the cell. This operation is repeated for each cell of the table, and then the results for all the cells are summed. This sum is the obtained value of chi-square for the table.

The reason for squaring the difference between observed and expected values is to get rid of minus signs. If this were not done the differences would sum to zero, just as the deviations of observations about the mean of a distribution sum to zero. The squared differences are divided by the expected value for the cell to control for differences in the numbers of observations in the cells. A difference of 10 between an observed and expected value, for example, is a smaller deviation if the number of observations in the cell is 100 than if the number of observations is 50.

The obtained value of chi-square for Table 9–8 is found as follows:

$$\chi^2 = \sum \frac{(O - E)^2}{E}$$

$$= \frac{(43 - 36)^2}{36} + \frac{(17 - 24)^2}{24} + \frac{(77 - 84)^2}{84} + \frac{(63 - 56)^2}{56}$$

$$= \frac{7^2}{36} + \frac{-7^2}{24} + \frac{-7^2}{84} + \frac{7^2}{56}$$

$$= \frac{49}{36} + \frac{49}{24} + \frac{49}{84} + \frac{49}{56}$$

$$= 1.36 + 2.04 + .58 + .88$$

$$= 4.86$$

DEGREES OF FREEDOM

As in the case of the t distribution, the shape of the chi-square distribution changes according to the parameter "degrees of freedom." In the t distribution, degrees of freedom are the number of observations whose scores can be determined arbitrarily. In tables, though, we deal with the scores for the cells (the counts) rather than with the scores of individual observations. Degrees of freedom for a table refer to the number of cells in the table whose values can be arbitrarily determined given the marginals of the table. Suppose that the marginals of a 2 × 2 table were as follows:

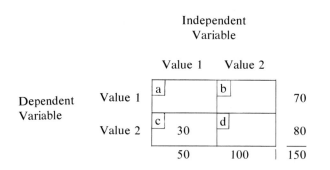

Independent
Variable

Now we arbitrarily assign the value of 30 to cell c. This action immediately fixes the values of all the other cells of the table because the values of the cells in a particular row or column must add up to the marginal for that row or column. For cells a and c the column marginal is 50. If cell c is set equal to 30, then cell a must be a number which when added to 30 will equal 50. Cell a then must be 20. Similarly, cell d must be a number which when added to 30 yields 80 (the row marginal for cells a and d). Therefore, cell d must equal 50. Finally, cell b must equal 50 because this is the only number which will yield 70 as its row marginal and 100 as its column marginal given the values already determined for cells a and d.

Independent
Variable

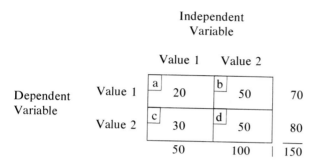

In any table, the values for the cells must sum to the values of the row and column marginals. The number of cells to which it is necessary to assign arbitrary values in order to fix the values of all the cells in the table is the degrees of freedom in the table. In a 2 × 2 table, if an arbitrary value is assigned to any one cell of the table, the values of the other cells are fixed. Thus, a 2 × 2 table has 1 degree of freedom.

In general, the degrees of freedom in any table is equal to the number of rows in the table minus 1, times the number of columns in the table minus 1. If R equals the number of rows in a table and C the number of columns,

$$df = (R - 1)(C - 1)$$

For a 2 × 2 table, $R = 2$ and $C = 2$:

$$df = (R - 1)(C - 1)$$
$$= (2 - 1)(2 - 1)$$
$$= (1)(1)$$
$$= 1$$

For a 3 × 4 table, $R = 3$ and $C = 4$:

$$df = (3 - 1)(4 - 1)$$
$$= (2)(3)$$
$$= 6$$

The 2 × 2 table examining the relationship between sex and voting, then, has 1 degree of freedom.

The obtained value of chi-square for Table 9–8 is 4.86. We can now consult a table of the distribution of values for chi-square for 1 degree of freedom to see if the probability of this obtained value is less than a selected alpha level of, say, .05. Table A–5 in the Appendix presents probabilities for the chi-square distribution. Across the top of this table are probabilities for chi-square values. Each row of the table presents values of the chi-square distribution for the degree of freedom given in the leftmost column. Thus, the first row of the table gives chi-square values for 1 degree of freedom. Looking under the column labeled .05 (the alpha level selected in this problem), it can be seen that the value of chi-square for 1 degree of freedom is 3.84. This means that the probability of a chi-square value of 3.84 *or larger* occurring by chance alone is .05. This, then, is the *critical value* of chi-square for this problem. Since the obtained value of chi-square in our problem exceeds this critical value (4.86 is larger than 3.84), the probability that the obtained value of chi-square occurred by chance under the null hypothesis is less than .05. Thus, we can reject the null hypothesis – that the association between sex and voting in Table 9–8 occurred by chance – and accept the alternative hypothesis – that there is a relationship between sex and voting in the population and that the results in the table reflect this fact.

It cannot be emphasized too strongly that the chi-square statistic does not describe the nature of the association between two variables in a table. The fact that a table is found to contain a statistically significant relationship does not mean that the results in the table necessarily support the researcher's empirical hypothesis. It only means that the results, whatever they might be, were unlikely to have occurred by chance. It is possible for very strange relationships to appear in tables that are statistically significant but which are substantively incompatible with a researcher's empirical hypothesis. The researcher must interpret any relationship in a table using a descriptive statistic such as percentages to find out the nature of the association between the variables.

Question 9–4. The table in Question 9–2 was:

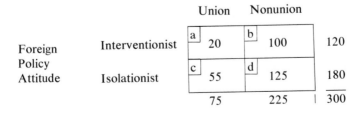

Union Membership

		Union	Nonunion	
Foreign Policy Attitude	Interventionist	a 20	b 100	120
	Isolationist	c 55	d 125	180
		75	225	300

a. According to the information in this table, find estimates of the following probabilities: P_{union}, $P_{nonunion}$, P_{inter}, and P_{iso}.

b. According to the information in the table, find the following joint probabilities under the assumption that there is no relationship between union membership and foreign policy attitude: $P_{(union\ and\ inter)}$, $P_{(union\ and\ iso)}$, $P_{(nonunion\ and\ inter)}$, and $P_{(nonunion\ and\ iso)}$.

c. Calculate the expected values for cells a, b, c, and d in the table.

d. Compute the value of the chi-square statistic for the table.

e. What is the critical value of chi-square for this problem for an alpha level of .01? Is the obtained value of the chi-square statistic significant at an alpha level of .01?

_____

LOCATION OF THE CRITICAL
REGION IN A CHI-SQUARE DISTRIBUTION

The chi-square distribution has a shape that is different from that of either the normal or the t distribution. As the number of degrees of freedom increases, though, chi-square approaches a normal distribution in shape. But the chi-square distribution, unlike a standard normal distribution or a t distribution, can never take on values less than zero (chi-square cannot have negative values). Chi-square values, no matter what the degrees of freedom of the distribution, begin at zero and increase in a positive direction. Figure 9–1 presents chi-square distributions for 1, 8, and 11 degrees of freedom.

Chi-square equals zero when observed values precisely equal the expected values in a table. For 1 degree of freedom, it can be seen from Figure 9–1 that it is fairly likely that, by chance, observed values will closely approximate expected values and result in a chi-square value close to zero. But as the degrees of freedom increase, the probability that the observed values will precisely equal the expected values by chance becomes small. That is, when there are many cells in a table, it is to be expected that observed values will deviate somewhat from the expected values *by chance*. In such situations, an association is detected when the deviation of observed from expected values is greater than the deviation that could be expected by chance. These facts indicate that the location of the critical region in a chi-square distribution differs from that in normal and t distributions.

In locating the critical region for an alpha level of .05, say, what we are interested in is locating the most unlikely 5% of the values in the chi-square distribution which correspond to a great deal of deviation of observed values from expected values, since this indicates association in a table. The point is that the left-hand tail of any chi-square distribution

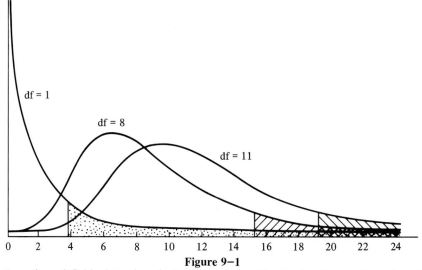

Figure 9–1
Location of Critical Regions in the Chi-Square Distribution for an Alpha Level of .05

represents the probability of observed values closely approximating expected values. This corresponds to a situation of no relationship between the two variables in a table. It is only the right-hand tail of a chi-square distribution which represents the probability of observed values deviating substantially from the expected values and which thus corresponds to a situation of association. Since we want to reject the null hypothesis only when there is an association between the variables, *all* of the critical region is located in the right-hand tail in a chi-square distribution, as indicated in Figure 9–1. It does not make sense to locate any part of the critical region in the left-hand tail (even when the chi-square distribution begins to be normal in shape), because this tail corresponds to a situation of no association and we do not want to reject the null hypothesis when there is no relationship between the variables.

Since all of the critical region in a chi-square distribution is already located in the right-hand tail, it is not possible to make a more powerful one-tail test in the way that was done in the *t* distribution by combining the halves of the critical region in a two-tailed test into one of the tails for a one-tailed test. In the *t* distribution, the difference between one- and two-tailed tests involved taking direction into account. Associations between variables can differ by direction. Thus, in the relationship between sex and voting, an association where men are more likely to vote for Nixon than women is opposite in direction to an association where women are more likely to vote for Nixon than men. But the chi-square statistic, by squaring the differences between observed and expected values, ignores any information about direction. The chi-square distribution is a one-tailed test for the degree (but not the direction) of deviation of the observed from the expected values.

If, however, the direction of the association between the variables is posited before the data are seen, it is possible to make a more powerful chi-square test that is conceptually equivalent to a directional one-tailed test. If an alpha level of .05 has been selected and the direction of the association has been posited, the chi-square value for a probability of .10 provides the critical value for this directional test. If an alpha of .01 has been selected, the chi-square value for a probability of .02 will provide the critical value for a directional test.

A word of warning, however, is in order about such directional tests. First, to use a directional test, the direction of the association must be posited before the data are seen. To infer the direction of the association from the same data that are going to be tested would violate the premises of the test. Second, in 2 × 2 tables it is clear what is meant by the direction of an association. In larger tables, though, the idea of direction of association may not be at all clear. In such tables, an association might take many different forms. In social science it is usually the case that empirical hypotheses are not sophisticated enough to assert complex forms of association. Unless the researcher's empirical hypothesis specifies the nature of the direction of association quite precisely, it is probably best to avoid a directional test in larger tables.

COMPUTATIONAL AIDS IN CALCULATING EXPECTED VALUES

Before the value of chi-square can be found for a table, the expected values for each cell must be calculated. The calculation of expected values can be simplified. The expected value of any cell in a table may be

Table 9–9

Sex

		Male	Female	
Vote in 1972	Nixon	a ⌐36 43	b ⌐24 17	60
	McGovern	c ⌐84 77	d ⌐56 63	140
		120	80	200

found by multiplying together the row and column marginals for that cell and dividing the product by the total number of observations in the table.

For the problem concerning the relationship between sex and voting, the observed values are given again in Table 9–9. For cell a in Table 9–9 its row marginal is 60 and its column marginal is 120. Therefore,

$$\text{expected value for cell a} = \frac{(\text{column marginal})(\text{row marginal})}{\text{total}}$$

$$= \frac{(120)(60)}{200}$$

$$= \frac{7200}{200}$$

$$= 36$$

This is the same expected value that was found earlier for cell a by use of the longer method. The expected values for the other cells in the table may be found in the same way. For cell d, for example, its row marginal is 140 and its column marginal is 80. Therefore,

$$\text{expected value for cell d} = \frac{(\text{column marginal})(\text{row marginal})}{\text{total}}$$

$$= \frac{(80)(140)}{200}$$

$$= \frac{11,200}{200}$$

$$= 56$$

This method of finding expected values is really exactly the same as the longer method used earlier. Consider cell a again. To find its expected value, the probability of the joint occurrence of being male and voting for Nixon under the assumption that the two events are independent is multiplied by the total number of observations in the table. The probability of being male is $\frac{120}{200}$, or the column marginal divided by the total. The probability of having voted for Nixon is $\frac{60}{200}$, or the row marginal divided by the total. Therefore, the probability of the joint occurrence of these two events, assuming they are independent, is

$$P_{(m \text{ and Nix})} = (P_m)(P_{\text{Nix}})$$

$$= \left(\frac{120}{200}\right)\left(\frac{60}{200}\right)$$

or

$$P_{(m \text{ and Nix})} = \left(\frac{\text{column marginal}}{\text{total}}\right)\left(\frac{\text{row marginal}}{\text{total}}\right)$$

As was seen before, this probability is the probability that one student selected at random from the population would be both male and have voted for Nixon. To find out how many people in the sample could be expected to be male and have voted for Nixon, the expected value for cell a, this probability is multiplied by the total number of observations in the table:

$$\text{expected value for cell a} = \left(\frac{\text{column marginal}}{\text{total}}\right)\left(\frac{\text{row marginal}}{\text{total}}\right)(\text{total})$$

This reduces to the short method for expected values:

$$\text{expected value for cell a} = \left(\frac{\text{column marginal}}{\text{total}}\right)\left(\frac{\text{row marginal}}{\text{total}}\right)(\text{total})$$

$$= \frac{(\text{column marginal})(\text{row marginal})}{\text{total}}$$

Thus, the expected value for any cell may be found by multiplying together its row and column marginals and dividing the result by the total number of observations in the table.

The fact that the numbers in the cells of a row or column must sum to the value of the marginal for the row or column provides another short-cut: By calculating expected values for a number of cells equal to the number of degrees of freedom in a table, the expected values for the other cells may be found by subtraction from the marginals. Thus, in Table 9–9, the expected value for cell a is computed to be 36; then the expected value for cell b can be found by subtracting 36 from the row marginal. Thus, the expected value for cell b is $60 - 36 = 24$, which is the same expected value found through the longer methods above. For cell c, the expected value of cell a is subtracted from the column marginal for cells a and c; $120 - 36 = 84$. The expected value for cell d may be found by subtracting the expected value for cell c from the row marginal for cells c and d: $140 - 84 = 56$.

Question 9–5. Calculate the expected values for cells a, b, c, and d of the table in Question 9–4 by multiplying the row and column marginals for each cell and dividing by the total. Are these expected values the same as were obtained in Question 9–4?

As was indicated earlier, the chi-square statistic may be used to test the significance of associations in tables larger than 2×2. It also fre-

Table 9-10

Region

		North	South	East	West	
	Democratic	20.25 20	13.5 20	36 35	20.25 15	90
Party Identi- fication	Independent	11.25 10	7.5 5	20 15	11.25 20	50
	Republican	13.5 15	9 5	24 30	13.5 10	60
		45	30	80	45	200

quently happens that the expected values in a table have fractional values which make the computation of chi-square more tedious. Both these points are illustrated in Table 9-10, which examines the association between region of the United States and party identification in a sample of 200 people.

The calculation of chi-square for this table is done as follows:

$$\chi^2 = \frac{(20-20.25)^2}{20.25} + \frac{(20-13.5)^2}{13.5} + \frac{(35-36)^2}{36} + \frac{(15-20.25)^2}{20.25}$$

$$+ \frac{(10-11.25)^2}{11.25} + \frac{(5-7.5)^2}{7.5} + \frac{(15-20)^2}{20} + \frac{(20-11.25)^2}{11.25}$$

$$+ \frac{(15-13.5)^2}{13.5} + \frac{(5-9)^2}{9} + \frac{(30-24)^2}{24} + \frac{(10-13.5)^2}{13.5}$$

$$= \frac{(-.25)^2}{20.25} + \frac{(6.5)^2}{13.5} + \frac{(-1)^2}{36} + \frac{(-5.25)^2}{20.25}$$

$$+ \frac{(-1.25)^2}{11.25} + \frac{(-2.5)^2}{7.5} + \frac{(-5)^2}{20} + \frac{(8.75)^2}{11.25}$$

$$+ \frac{(1.5)^2}{13.5} + \frac{(-4)^2}{9} + \frac{(6)^2}{24} + \frac{(-3.5)^2}{13.5}$$

$$= \frac{.06}{20.25} + \frac{42.25}{13.5} + \frac{1}{36} + \frac{27.56}{20.25} + \frac{1.56}{11.25} + \frac{6.25}{7.5}$$

$$= \frac{25}{20} + \frac{76.56}{11.25} + \frac{2.25}{13.5} + \frac{16}{9} + \frac{36}{24} + \frac{12.25}{13.5}$$

$$= .003 + 3.13 + .03 + 1.36 + .14 + .83$$
$$+ 1.25 + 6.80 + .17 + 1.78 + 1.5 + .91$$
$$= 17.9$$

Since Table 9–10 is a 4 × 3 table, the degrees of freedom in the table are

$$\text{df} = (4 - 1)(3 - 1)$$
$$= (3)(2)$$
$$= 6$$

For six degrees of freedom, the critical value of chi-square for an alpha level of .01 is 16.81. Since the obtained value of chi-square for the table exceeds this value, we can reject the null hypothesis and conclude that there is an association between region and party identification.

Having established that there is a statistically significant association between region and party identification, the observed and expected values in the table can be compared to reveal some things about the nature of the association between the variables. Looking at the column for the North, it can be seen that the observed values are fairly close to the expected values. In the column for the South, however, the observed value for Democrats is much larger than the expected value (20 versus 13.5) and the observed value for Republicans is smaller than the expected value (5 versus 9). Thus, the South has a larger number of Democrats and a smaller number of Republicans than would be expected. A similar examination of the columns for the East and the West will reveal how these regions differ in their distribution of party identification from what would be expected if party identification were randomly distributed across the country. In this way, comparison of observed and expected values can be used to determine how and where the association occurs between variables in a table.

THE COMPUTATIONAL FORMULA FOR CHI-SQUARE IN 2 × 2 TABLES

Fourfold, or 2 × 2 tables, occur quite frequently in social science. A computational formula for chi-square has been derived for this special case which eliminates the need to calculate expected values. It is

$$\chi^2 = \frac{N(ad - bc)^2}{(a + b)(c + d)(a + c)(b + d)}$$

where *a, b, c,* and *d* are the *observed* values for cells a, b, c, and d in the table and *N* is the total number of observations in the table. The chi-square value for Table 9–9 using this computational formula is

$$\chi^2 = \frac{(200)[(43)(63) - (17)(77)]^2}{(43 + 17)(77 + 63)(43 + 77)(17 + 63)}$$

$$= \frac{(200)(2709 - 1309)^2}{(60)(140)(120)(80)}$$

$$= \frac{(200)(1400)^2}{80,640,000}$$

$$= \frac{(200)(1,960,000)}{80,640,000}$$

$$= \frac{392,000,000}{80,640,000}$$

$$= 4.86$$

This is the same value for chi-square as was found earlier with the method employing expected values.

Question 9–6. Here, again, is the table relating union membership to foreign policy attitudes.

Union Membership

		Union	Nonunion	
Foreign Policy Attitude	Interventionist	20	100	120
	Isolationist	55	125	180
		75	225	300

Use the computational formula to calculate the value of chi-square for the table. Is this the same value of chi-square you obtained in Question 9–4? _____

CONTINUITY AND SAMPLE SIZE IN THE CHI-SQUARE STATISTIC

There are some technical problems in the use of chi-square which arise because the chi-square *statistic* as calculated from a table only approximates an actual chi-square distribution. Theoretically, an exact fit is obtained only when the sample size is very large. This means that error will be present if a chi-square statistic is evaluated by a chi-square distribution if the sample size is not adequate. The error depends not just on the size of the sample, though, but also on the number of cells in the table. Therefore, it is necessary to talk about the number of observations necessary for each cell in a table.

For 2 × 2 tables a conservative rule is that the *expected* value for each cell must be at least 10. In larger tables, the conservative rule requires an *expected* value of 5 for each cell. If the number of cells in the table is large, however, a few of the cells may have expected values less than 5, but in no case may the expected value for a cell be less than 1.0. It is emphasized that these rules about size are in terms of the *expected* values, not the observed values. Thus, the observed value of a cell may be zero as long as its expected value meets the rules just stated.

Another problem arises in the case of 2 × 2 tables. The chi-square distribution is a continuous distribution, which means that the statistic should be able to take on all possible numerical values in an interval. Cell frequencies, however, are always in terms of integer numbers, and this disrupts the continuity of values that are possible for the chi-square statistic. This disruption is serious in 2 × 2 tables but not in larger tables. It has been recommended, therefore, that in the case of 2 × 2 tables, a correction in the value of the chi-square statistic should be made for continuity. This is known as *Yates' correction for continuity*.

Yates' correction involves reducing the size of the difference between the observed and expected values by .5. The formula for the chi-square statistic with Yates' correction is

$$\chi^2 = \sum \frac{(|O - E| - .5)^2}{E}$$

For Table 9–9 the value of chi-square using Yates' correction is

$$\chi^2 = \sum \frac{(|O - E| - .5)^2}{E}$$

$$= \frac{(|43 - 36| - .5)^2}{36} + \frac{(|17 - 24| - .5)^2}{24} + \frac{(|77 - 84| - .5)^2}{84}$$

$$+ \frac{(|63 - 56| - .5)^2}{56}$$

$$= \frac{(7 - .5)^2}{36} + \frac{(7 - .5)^2}{24} + \frac{(7 - .5)^2}{84} + \frac{(7 - .5)^2}{56}$$

$$= \frac{(6.5)^2}{36} + \frac{(6.5)^2}{24} + \frac{(6.5)^2}{84} + \frac{(6.5)^2}{56}$$

$$= \frac{42.25}{36} + \frac{42.25}{24} + \frac{42.25}{84} + \frac{42.25}{56}$$

$$= 1.17 + 1.76 + .50 + .75$$

$$= 4.18$$

The value for chi-square for this problem without Yates' correction was 4.86. As can be seen, Yates' correction for continuity will reduce the size of the chi-square statistic. The computational formula for the chi-square statistic with Yates' correction is

$$\chi^2 = \frac{N[|(ad - bc)| - (N/2)]^2}{(a + b)(c + d)(a + c)(b + d)}$$

There appears to be some disagreement among statisticians as to whether the use of Yates' correction is really warranted. However, it does appear frequently in social science and we should be familiar with it and the reason for its use. Since Yates' correction reduces the value of the chi-square statistic, its use results in a more conservative test. If it is desired to be very safe in deciding to reject the null hypothesis in a 2 × 2 table, Yates' correction may be used.

Question 9–7. Calculate the value of chi-square for the table in Question 9–4 using Yates' correction for continuity. Is the chi-square value still statistically significant at an alpha level of .01?

THE USE OF CHI-SQUARE
ON HIGHER-LEVEL DATA

Although the chi-square statistic was designed for use in examining the relationship between nominal level variables, it may also be used on ordinal, interval, or ratio level data. Suppose in a survey of 100 people, the respondents were asked: "How much of the time do you think you can trust the government in Washington to do what is right—just about always, most of the time, or only some of the time?" The researcher then wishes to see if people 40 years or older are more or less trusting of the government than people younger than 40. Age is a ratio level variable whereas the question on trust yields ordinal measurements. By "collapsing" people's ages into two categories of young (less than 40) and old (40 or over) the researcher can construct a table, such as Table 9–11, and use chi-square to see if there is a statistically significant association between the collapsed age variable and the question on trust.

Although there is nothing incorrect about this procedure, it ignores

Table 9–11

Age

		Young	Old	
How Often Trust Govern- ment To Do Right Thing	Some of the Time	15	30	45
	Most of the Time	15	10	25
	Just About Always	20	10	30
		50	50	100

a good deal of information present in the data. By collapsing age into two categories, people who differ a great deal in age are treated as being exactly the same. Thus, a person 41 years old is placed in the same category as a 65-year-old person and the difference in their ages is ignored.

The chi-square statistic also ignores information about the ordering of categories. Thus, on the trust variable the fact that category 3 indicates more trust than categories 1 and 2, and that category 2 indicates more trust than category 1, is not taken into consideration by the chi-square statistic.

Finally, it is possible that by collapsing variables, relationships are concealed. Using the variables in Table 9–11, suppose that 21- to 30-year-old people are very distrustful of government, that 31- to 50-year-old people are very trusting, and that people over 50 are very distrustful. Thus, the association between the two variables is an inverted U: Trust is low for young people, increases to a high level for middle-aged people, and decreases to a low level for old people. If age is now collapsed into two categories divided at the 40-year-old mark, the collapsing will conceal the relationship: The high-trust middle-aged people (31 to 50) in each category will balance out the low-trust young or old people in the category, so the two categories will have about the same level of trust and cause the relationship between age and trust to disappear. Whenever variables are collapsed, there is some risk that relationships may be concealed.

Thus, while the chi-square statistic may be used on ordinal, interval, or ratio level data, it results in a conservative test of the association between variables for such data and there is a danger that it may overlook real relationships. There are other significant tests available to test associations among ordinal, interval, and ratio level variables which are more likely to detect an association in these situations than is the chi-square statistic. But it is not incorrect to use chi-square in these situations as long as the researcher is aware of the problems involved.

EXERCISES FOR CHAPTER 9

1. A cross-tabulation of data from a survey on occupation and religion produces the following table:

		Religion			
		Catholic	Protestant	Jewish	
Occu-pation	Business	5	20	5	30
	Professional	10	15	15	40
	Labor	15	15	0	30
		30	50	20	100

Is there a statistically significant association between religion and occupation at an alpha level of .01? If the researcher is interested in seeing if religion affects occupation, percentage the table and use the percentages to describe the nature of the association.

2. In a survey, respondents were classified as party identifiers or as independents. The researcher was interested in seeing if having a party identification affected whether people voted in elections. He therefore constructed the following contingency table:

		Party Identification		
		Party Identifiers	Independents	
Voting Record	Voted	27	6	33
	Did Not Vote	13	14	27
		40	20	60

Use the method of expected values to calculate the value of chi-square for this table. Is there a statistically significant association between the variables at an alpha of .01? What is the independent variable and the dependent variable in this table? Why? Percentage the table to describe the association between the variables.

3. Calculate the value of chi-square for the table in Exercise 2 using the computational formula for 2×2 tables.

4. Calculate the value of chi-square for the table in Exercise 2 using Yates' correction for continuity. Is this value statistically significant at an alpha of .01?
5. In a survey it was determined whether the respondents' level of political information was high or low. Level of political information was then cross-tabulated by sex to produce the following table:

		Sex		
		Male	Female	
Level of Political Information	High	60	30	90
	Low	40	70	110
		100	100	200

Calculate the value of chi-square for this table. Is it significant at an alpha level of .05? If the researcher thinks sex affects the level of political information, percentage the table and describe the association. It was felt that level of education might be important to this problem, so the respondents in the preceding table were divided into a less-educated group and a highly educated group. A table examining the relationship between sex and level of political information was then constructed for each educational group, as follows:

Less-Educated Group

		Sex		
		Male	Female	
Level of Political Information	High	2	5	7
	Low	28	65	93
		30	70	100

Highly Educated Group

		Sex		
		Male	Female	
Level of Political Information	High	58	25	83
	Low	12	5	17
		70	30	100

Calculate the value of chi-square in both these tables. Is there a significant association between sex and level of political information in either of these tables? Percentage both tables as was done for the first table. Verbally describe what is happening in all three tables. Construct a theory that can account for the results in the three tables.

ANSWERS TO
QUESTIONS IN CHAPTER 9

9–1: $P_{\text{Catholic}} = \frac{100}{500} = .20$, $P_{\text{none}} = \frac{35}{500} = .07$.

9–2: a. Union membership; foreign policy attitude. The empirical hypothesis specifies that it is union membership which affects foreign policy attitude.
b. 120, 75.
c. 225, 180.
d. Proportion of union members $= \frac{75}{300} = .25$; proportion of people holding interventionist attitudes $= \frac{120}{300} = .40$.

9–3: a. Percentage who are union members and hold interventionist attitudes $= \frac{20}{300}(100) = 6.7\%$; percentage who are nonunion and hold isolationist attitudes $= \frac{125}{300}(100) = 41.7\%$.

b.

		Union Membership	
		Union	Nonunion
Foreign Policy Attitude	Interventionist	26.7%	44.4%
	Isolationist	73.3%	55.6%

c. Union members tend to hold isolationist foreign policy attitudes more than nonunion members do.
d. $P_{\text{union}} = \frac{75}{300} = .25$,
$P_{\text{iso}} = \frac{180}{300} = .60$.

9–4: a. $P_{\text{union}} = \frac{75}{300} = .25$,
$P_{\text{nonunion}} = \frac{225}{300} = .75$,
$P_{\text{inter}} = \frac{120}{300} = .40$,
$P_{\text{iso}} = \frac{180}{300} = .60$.
b. $P_{\text{(union and inter)}} = (.25)(.40) = .10$,
$P_{\text{(union and iso)}} = (.25)(.60) = .15$,
$P_{\text{(nonunion and inter)}} = (.75)(.40) = .30$,
$P_{\text{(nonunion and iso)}} = (.75)(.60) = .45$.
c. cell a $= (.10)(300) = 30$,
cell b $= (.30)(300) = 90$,
cell c $= (.15)(300) = 45$,
cell d $= (.45)(300) = 135$.

d.
$$\chi^2 = \sum \frac{(O - E)^2}{E}$$

$$= \frac{(20 - 30)^2}{30} + \frac{(100 - 90)^2}{90} + \frac{(55 - 45)^2}{45} + \frac{(125 - 135)^2}{135}$$

$$= \frac{-10^2}{30} + \frac{10^2}{90} + \frac{10^2}{45} + \frac{-10^2}{135}$$

$$= \frac{100}{30} + \frac{100}{90} + \frac{100}{45} + \frac{100}{135}$$

$$= 3.33 + 1.11 + 2.22 + .74$$

$$= 7.40.$$

e. The critical value of chi-square for df $= 1$ at an alpha level of .01 is 6.64; yes, it is statistically significant.

9–5: cell a $= \dfrac{(75)(120)}{300} = \dfrac{9000}{300} = 30,$

cell b $= \dfrac{(225)(120)}{300} = \dfrac{27,000}{300} = 90,$

cell c $= \dfrac{(75)(180)}{300} = \dfrac{13,500}{300} = 45,$

cell d $= \dfrac{(225)(180)}{300} = \dfrac{40,500}{300} = 135.$

9–6: $\chi^2 = \dfrac{(300)[(20)(125) - (100)(55)]^2}{(120)(180)(75)(225)}$

$$= \frac{(300)(2500 - 5500)^2}{364,500,000} = \frac{(300)(9,000,000)}{364,500,000}$$

$$= \frac{2,700,000,000}{364,500,000}$$

$$= 7.4.$$

9–7: $\chi^2 = \sum \dfrac{(|O - E| - .5)^2}{E}$

$$= \frac{(|20 - 30| - .5)^2}{30} + \frac{(|100 - 90| - .5)^2}{90} + \frac{(|55 - 45| - .5)^2}{45}$$

$$+ \frac{(|125 - 135| - .5)}{135}$$

$$= \frac{(|-10| - .5)^2}{30} + \frac{(|10| - .5)^2}{90} + \frac{(|10| - .5)^2}{45} + \frac{(|-10| - .5)^2}{135}$$

$$= \frac{(10 - .5)^2}{30} + \frac{(10 - .5)^2}{90} + \frac{(10 - .5)^2}{45} + \frac{(10 - .5)^2}{135}$$

$$= \frac{9.5^2}{30} + \frac{9.5^2}{90} + \frac{9.5^2}{45} + \frac{9.5^2}{135}$$

$$= \frac{90.25}{30} + \frac{90.25}{90} + \frac{90.25}{45} + \frac{90.25}{135}$$

$$= 3.01 + 1.00 + 2.00 + .67$$

$$= 6.68; \text{ yes.}$$

10
Nominal and Ordinal Measures of Association

The two preceding chapters were concerned with tests of statistical significance. In this chapter we begin consideration of the problem of measuring association between variables. The present chapter deals with association among nominal and ordinal variables and Chapter 11 focuses on association between interval and ratio level variables.

It will be recalled that tests of significance are concerned with determining whether or not a particular observed result was likely to have occurred by chance. In most situations, the observed result occurs in a randomly selected sample and the question is whether or not the random-selection process used in drawing the sample could have generated the observed result. If a test of significance shows that the probability of the observed result occurring by chance is very small, we reject the idea that the observed result occurred because of the random-selection process (reject the null hypothesis) and accept the idea that the observed result was not due to chance (accept the alternative hypothesis), and consequently that some mechanism must be at work to bring about the observed result. In such a situation we speak of the observed result as being *statistically significant*.

When the observed result involves a relationship between two variables, a significance test does not immediately indicate the extent to which the two variables are related. This is where measures of association come in. A *measure of association* is a statistic whose purpose is to indicate the degree to which two variables are related in a given set of data. We know, for example, that people's heights are related to their weights; that, generally speaking, as height increases so does weight. But to what extent are the two variables related? Suppose that in one society it is socially desirable to be short and fat. In this society there will be many short, fat people, which would result in there being only a weak, but still real, relationship between height and weight. In another society, any fatness might be socially undesirable, and consequently in this society there would be a strong relationship between height and weight. In both societies there would be real relationships between height and weight; but in one society the relationship would be weak and in the other strong. Here now is the important point: If we were to draw substantial random samples of people in both societies and conduct a test of significance on the relationship between height and weight in each sample, in both samples the relationship would turn out to be statistically significant. A significance test asks only if there is a relationship that is not likely to have occurred by chance. In each sample the answer is "yes," since in both societies such a relationship actually exists. What the significance test does not tell us is the extent to which the two variables are related. It is this question which measures of association are designed to answer.

An important point about measures of association is that they are descriptive statistics. What they do is to describe the extent of association between two variables in a particular set of data. It does not matter what these data are or how they were obtained. Thus, the data might be a random sample, a nonrandom sample, or a population. Regardless of what the data represent, a measure of association is used to indicate the degree of strength of relationship between two variables *as it exists in that data.*

When dealing with a random sample, a measure of association is not a substitute for a test of significance. Regardless of what the value of the measure of association may be, a significance test must still be used to determine if the value was likely to have occurred by chance. It is possible that in one sample a high value for a measure of association occurred by chance whereas in a different sample a low value may not be the result of chance. When dealing with a population, of course, a test of significance is not necessary. If we could measure the height and weight of everyone in a society (the population), the degree of association found between height and weight would be a fact for that society, and not open to doubt.

As might be suspected, though, when dealing with a randomly selected sample, there is a relationship between statistical significance and the strength of association between two variables. In general, the stronger the association between two variables in a population, the more likely it is that the relationship between the variables in the sample will be found to be statistically significant, other things being equal. One implication of this is that smaller sample sizes are needed to detect relationships where there is a strong association than where there is a weak association. Suppose that 90% of the male voters in a city favor the Republican candidate in an election while 90% of the female voters favor the Democratic candidate. It is likely that a test of significance for whether or not there is a relationship between sex and candidate preference would turn out to be significant for a relatively small sample. If the percentages were 60 instead of 90, however, a larger sample would probably be needed for the test to be significant. Thus, the stronger the association is in a population, the smaller the size of the sample needed to detect the fact that there is a relationship between the variables.

An interesting point here is that any association, no matter how small, will become statistically significant if the sample size is made large enough. For example, assume now that among our city voters 53% of the men favor the Republican candidate while "only" 50% of the women favor the Republican candidate. If a random sample of 5000 voters is selected and the same 3-percentage-point difference between the preferences of men and women is found in the sample, this result will be statistically significant. But is a 3-percentage-point difference between the preferences of men and women of any *substantive* interest? The significance test tells us that a difference between men and women probably exists in the population. But the difference is so small that it is unlikely to be of much substantive interest. The lesson is that just because an observed result is statistically significant does not necessarily mean that the result is important.

These points indicate that a test of significance will probably not provide us with a satisfactory amount of information about the relationship between two variables. A test of significance, if the null hypothesis is rejected, tells us that there is a relationship between the variables. But additional information about the size and nature of the association is needed to evaluate its importance. Measures of association are designed to help provide this information.

ASSOCIATION

Technically, the question of association between variables involves the relationship between the probability distributions of the variables. Let us briefly investigate what this means. It will be recalled that two events, A and B, are said to be independent if:

$$P_{(A|B)} = P_A \quad \text{and} \quad P_{(B|A)} = P_B$$

These two statements say that the probability of event A occurring is not changed by the fact that event B occurs first and that the probability of event B occurring is not changed if event A occurs first. In other words, the occurrence of one event has no effect upon the probability of occurrence of the other event. An example of two independent events is the flipping of two coins. Whether the first coin comes up heads or tails does not affect the probability of getting a head or tail on the second coin.

If either of the probability statements above is *not* true, then the two events are *not* independent. Suppose that among a group of school children, 10% have red hair. If one child were selected at random from this group, the probability that the child would have red hair is

$$P_{(red\ hair)} = P_A = .10$$

In this group of school children 20% have freckles. Therefore, the probability that if one child were selected at random he would have freckles is

$$P_{(freckles)} = P_B = .20$$

Now suppose we are told that 90% of the redheaded children in the group have freckles. If we now select one child at random and it turns out this child is a redhead, then the probability that a randomly selected child who has red hair also has freckles is

$$P_{(freckles\ given\ red\ hair)} = P_{(B|A)} = .90$$

Thus, $P_{(B|A)} \neq P_B$: the occurrence of the first event changes the probability of getting the second event and the two events are not independent.

Let us assume now that we are dealing with two variables, X and Y. Each variable has a probability distribution which indicates the probabilities of occurrence of the various values of the variable. If all the observations in a population are considered, the probability distribution for each variable can be determined from their frequency distribution. Suppose that X and Y have the following frequency and probability distributions in some population:

	Variable X			Variable Y	
Value	Frequency Distribution	Probability Distribution	Value	Frequency Distribution	Probability Distribution
X_1	100	.20	Y_1	200	.40
X_2	300	.60	Y_2	200	.40
X_3	100	.20	Y_3	100	.20

To determine whether or not there is an association between X and Y, it is useful to think of *each value* of X as being paired with some *distribution* of Y values. These *conditional distributions* can be found by cross-tabulating the observations; classifying each observation according to its scores on X and Y as in Table 10–1. Note that in Table 10–1 the frequency distribution for the X variable is given in the column marginals (the sums at the bottom of the columns in the table) and the frequency

Table 10-1

		Variable X			
		X_1	X_2	X_3	
Variable Y	Y_1	40	120	40	200
	Y_2	40	120	40	200
	Y_3	20	60	20	100
		100	300	100	500

distribution for the Y variable is given in the row marginals (the sums at the end of the rows in the table). It is for this reason that in a table the overall frequency distribution for a single variable is spoken of as the *marginal distribution*.

Now each value of X is paired with a particular conditional distribution of Y values. For value X_1 the conditional Y distribution is found in the first column of the table. The Y values are 40, 40, and 20. If we divide these values by the marginal for this column (100), we obtain the (conditional) probability distribution of Y values for observations that scored X_1 on the X variable. These probabilities are .40, .40, and .20, which are the same values as in the marginal probability distribution for variable Y. Moving to value X_2, the conditional Y distribution is 120, 120, and 60. Dividing these values by the marginal for the second column (300), we again obtain .40, .40, and .20. Finally, for value X_3, the conditional Y distribution is 40, 40, and 20, which when divided by the third column marginal (100) again yields .40, .40, and .20. When these conditional probability distributions are laid out as in Table 10–2, it is easy to see that different values of X do not affect the probability distri-

Table 10-2

		Variable X			
		X_1	X_2	X_3	
Variable Y	Y_1	.40	.40	.40	.40
	Y_2	.40	.40	.40	.40
	Y_3	.20	.20	.20	.20

Table 10–3

		Variable X			
		X_1	X_2	X_3	
	Y_1	60	120	20	200
Variable Y	Y_2	30	150	20	200
	Y_3	10	30	60	100
		100	300	100	500

bution of Y values. This situation is just like the outcome of one coin not affecting the probability of obtaining a head or a tail on a second coin, except that we are now dealing with distributions. Since each value of X is paired with exactly the same conditional Y distribution, the variables X and Y are independent of each other, and there is no association between them.

In Table 10–3 the variables X and Y have the same marginal distributions as in Table 10–1. Let us determine the conditional Y probability distributions for each value of X as was done before. For value X_1, the frequencies for the Y values are 60, 30, and 10. By dividing these values by the marginal for this column (100), the conditional probability distribution of Y values for X_1 is found: .60, .30, and .10. A similar process finds the conditional Y probability distribution for X_2 to be .40, .50, and .10; and for X_3 to be .20, .20, and .60. These conditional probability distributions are shown in Table 10–4. In this table the conditional Y probability distributions are different for each value of X. This means that a particular outcome on variable X changes the probability of occurrence of the Y values. Thus, each X value is paired with a different conditional

Table 10–4

		Variable X			
		X_1	X_2	X_3	
	Y_1	.60	.40	.20	.40
Variable Y	Y_2	.30	.50	.20	.40
	Y_3	.10	.10	.60	.20

Y distribution and the variables are *not* independent but are associated.

Another way of thinking about association between variables is the extent to which the scores of observations on one variable may be predicted from a knowledge of the scores of the observations on the other variable. In Table 10–2, the conditional Y probability distributions are identical to the marginal probability distribution of Y scores. This means that knowing the X scores of the observations in no way helps us to predict scores on the Y variable: By knowing X scores we can predict the Y scores of observations no better than if we did not know the X scores. Thus, the two variables are perfectly unrelated.

Let us now consider the situation in Table 10–4. If the X scores were unknown to us, we would use the marginal probability distribution of variable Y to make predictions about Y scores. This marginal distribution tells us that the Y scores Y_1 and Y_2 are equally likely to occur each with a probability of .40 and Y_3 has a .20 probability of occurrence. Thus to predict an observation's Y score (without knowledge of its X score) we should guess either Y_1 or Y_2, since they are the most likely ones to occur. If, though, the X score for an observation were known, we could improve our predictions. For X_1, there is a .60 probability of occurrence for Y_1, .30 for Y_2, and .10 for Y_3. Obviously, if an observation scored X_1 on the X variable, we should guess Y_1 for its score on the Y variable, since it is the most likely Y value to occur given X_1. For X_2, there is a .40 probability of occurrence of Y_1, .50 for Y_2, and .10 for Y_3. For observations that scored X_2, our best guess for them on the Y variable would be Y_2. For X_3, there is a .20 probability of occurrence of Y_1, .20 for Y_2, and .60 for Y_3. For observations that scored X_3 our prediction should be Y_3. Thus, because the two variables are associated, we can use the conditional distributions for each X value to make predictions and we shall greatly improve the correctness of our predictions. The more strongly variables are associated, the more accurate our predictions will be in guessing Y scores from a knowledge of X scores. If these predictions turn out to be perfectly correct for all observations, the two variables are "perfectly" related or associated in the set of data.

Variables can be associated in various ways. If, as scores on one variable increase, the scores on the second variable also increase, there exists a *positive* type or pattern of association between the variables. Height and weight are positively associated because as the scores on one increase, so do the scores on the other. An *inverse* or *negative* association exists between two variables if as the scores on one variable increase, the scores on the second variable decrease. Among adults, for example, it is generally thought that as age increases, IQ decreases. Various kinds of *curvilinear* associations are also possible. Physical strength among men, for example, increases with age to peak in the late thirties or early forties and then declines thereafter. The association in this situation is neither positive or negative but would resemble an inverted U.

Finally, association does not mean causation. If two variables are associated, it only means that they vary together in some way, so it is possible to predict the scores of one from a knowlege of scores of the other beyond a chance level. Why this prediction is possible is not specified by the fact of association. It *may* be that one variable does indeed "cause" the other. But it may also be the case that neither variable "causes" the other but that the two are associated for some other reason. Red hair and freckles, for example, usually go together (are associated), but neither causes the other. Inferring causation from statistical analysis is hazardous. It is up to the researcher to decide on the basis of his substantive knowledge of the problem whether to infer causation into a statistical association. *Association itself does not establish causation.* Problems of inferring causation from statistical analysis are taken up in Chapter 13.

PROBLEMS IN
THE MEASUREMENT OF ASSOCIATION

Unfortunately, there is no simple way to measure association that is satisfactory in all situations. There are a great many problems involved in measuring association between variables. As a result, there are different kinds of measures of association, each of which is designed for a particular type of situation.

Ideally, a measure of association should provide a description of the degree or strength of association between two variables and specify the nature of the relationship, such as positive or negative, between the variables. Generally speaking, measures of association are constructed so that if no relationship exists between the variables (they are independent of each other), the numerical value of the measure of association will be 0. If a perfect relationship exists between the variables — that scores on one variable can be predicted perfectly by knowing scores on the other variable — the numerical value of the measure of association will be 1.0. Some measures of association only try to indicate the strength or degree of association and ignore the question of the nature of association. These measures vary in value, theoretically, from 0 to 1.0. Other measures indicate whether the association is positive or negative in nature, in addition to indicating the strength of the association. This is usually done by having a positive relationship indicated by a positive numerical value for the measure of association and a negative relationship indicated by a negative numerical value. For such measures a perfect negative relationship would yield a numerical value of -1.0, a situation of no relationship a numerical value of 0, and a perfect positive relationship a numerical value of +1.0.

Another important property that a measure of association should have is that there be a meaningful interpretation for all possible numerical values of the measure. If 0 means no relationship and 1.0 means a perfect relationship, these are meaningful and easily understandable interpretations of the numbers. But what happens when the measure of association yields a value of .5? What does this number mean? Maybe it should mean that there is a "half-perfect" relationship. But what "half-perfect" means is not very clear. Perhaps it might mean that 50%, or half, of the predictions made in guessing scores on one variable from a knowledge of the scores on the other variable were correct and, by implication, half of these predictions were wrong. But if the dependent variable had, to begin with, two possible outcomes that were equally likely to occur, such as in flipping a coin, we would make correct predictions 50% of the time just by guessing randomly. And if an independent variable cannot make our predictions on the dependent variable better than what we could expect to do just by chance, we certainly do not want to say that the variables are associated. This illustrates only one of the difficulties that arise in trying to devise measures of association that have meaningful interpretations for values between 0.0 and 1.0. Because of such difficulties, there is no single "ideal" measure of association for nominal or ordinal data which is always easily interpretable. In fact, for some measures of association no easy interpretation exists at all for values between 0 and 1.0! And for measures where an interpretation does exist for the intermediate values, the interpretation is not usually simple or straightforward. Finally, to make matters worse, all measures of association behave peculiarly under certain conditions so that the same numerical value for a particular measure of association in different situations may not mean the same thing.

The basic determinant of what measure of association is to be used in a particular situation is the level of measurement of the variables involved. When nominal variables are involved, the variables do not possess an underlying dimension along which observations can be ordered. Observations are only sorted into categories on each variable, and the question of association is whether observations that fall into a particular category on one variable tend to fall into a particular category on the other variable. When ordinal variables are involved, there is an underlying dimension for each variable along which the observations can be ordered, but the distances between points or scores are not known. We only know that a higher score indicates more of a variable than a lower score but not how much more. The question of association here is the extent to which knowledge of the *ordering of observations* on one variable enables us to predict the ordering of observations on the other variable. Because ordinal variables do not give information about the distances between scores, the question of association involves predicting the ordering among observations rather than predicting the numerical values of the scores. With interval or ratio level variables, each variable has an underlying dimension along which observations can be ordered, and information is present about the distances between scores. At this level, the question of association becomes one of predicting the actual numerical scores of observations on one variable from a knowledge of their numerical scores on the other variable.

Because of the different information present in variables at the various levels of measurement, the question of association is different at each level of measurement. To use a measure of association that is constructed on the basis of predicting precise numerical values on variables at the nominal or ordinal levels of measurement is nonsensical, since variables at either of these levels do not meet the measurement assumptions necessary to make this type of prediction. (This statement needs some qualification. In practice, interval or ratio measures of association are sometimes used on ordinal data. But when this is done, the researcher must *assume* that the data meet interval or ratio level requirements and recognize the risks that such a procedure entails.) Similarly, it would be nonsensical to use on nominal variables a measure of association constructed on the basis of predicting the ordering of observations (ordinal measurement), because nominal variables have no underlying dimension along which observations can be conceptually ordered.

The differences in the question of association at the various levels of measurement is the basic reason for needing various measures of association. In addition, different measures of association are found at each level of measurement. Perhaps the main reason for this is that no "ideal" or "perfect" measure of association has yet been devised for each level of measurement. Consequently, there are different measures, each with its own advantages and disadvantages to be considered at each level of measurement.

PERCENTAGES AS
A MEASURE OF ASSOCIATION

Before turning our attention to specific measures of association, it will be useful to consider how percentages may be used to measure association. In Chapter 9 it was said that it is possible to percentage a table in three ways: by total, by column, or by row. Each of these ways involves asking what percentage the observed value in a cell is of a particular base. In percentaging by total, the base is the total number of observations in the table. In this type of percentaging, a particular cell percentage is the percentage of all the observations in the table that fall into that cell. In percentaging by column, the appropriate column marginal for each cell is used as the base. The cell percentages in this type of percentaging indicate what percentage of all the observations in a column fall into each cell of that column. Finally, row percentaging uses the appropriate row marginal for each cell as a base and indicates what percentage of all the observations in a row of cells fall into each cell of that row.*

In Chapter 9 it was said that the proper way to percentage a table to evaluate the association between two variables is to percentage on the values of the independent variable. If the independent variable is placed across the top of the table and the dependent variable along the side, as is customary, then the proper way to percentage the table is by column. The reason for this should be evident from the discussion at the beginning of this chapter: By percentaging on the values of the independent variable, the conditional distributions of scores on the dependent variable for each value of the independent variable can be compared. If there is an association between the variables, there will be differences between the conditional distributions of scores on the dependent variable. If the variables are unrelated, the conditional distributions will be very similar to each other.

Let us begin consideration of the use of percentages to measure association with 2×2 tables. In these tables we shall not work with the numbers of actual observations which fall into the various cells, but with percentages. That is, we shall work with tables that have already been percentaged on the values of the independent variables. For illustration, the independent variable will be called X and the dependent variable Y. To determine if there is an association between X and Y, it is necessary, as was stated earlier, to compare the conditional distributions of Y scores for each value of X. In Table 10–5 it can be seen that the condi-

Table 10–5

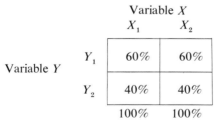

| | | Variable X | |
		X_1	X_2
Variable Y	Y_1	60%	60%
	Y_2	40%	40%
		100%	100%

*See "Percentages" in Chapter 9 for a more complete explanation of percentaging tables.

tional Y distributions are identical, and therefore there is no association between the variables.

Another way to note this fact is to determine the difference between pairs of percentages *across columns*. For the Y_1 row, the percentage of observations that score X_1 on the X variable and Y_1 on the Y variable is 60%. And the percentage of observations that score X_2 on the X variable and Y_1 on the Y variable is also 60%. Therefore. the difference between these percentages is zero, indicating no association. The same analysis could be made on the percentages falling in the Y_2 row. But since the percentages in the Y_2 row must always be equal to 100 minus the percentage in the Y_1 row, the results will be the same. Thus, in 2×2 tables it does not matter across which row of the table percentages are compared. This means that only one pair of percentages need be compared and allows us to arrive at a single difference between a pair of percentages to evaluate the association. If this difference is equal to zero, there is no association between the variables since this zero difference (in 2×2 tables) arises only when the conditional distributions of Y scores are identical.

In Table 10–6 there is a difference between the percentages of 10%. (Notice that this difference is the same for the pairs of percentages in either row Y_1 or row Y_2.) The fact that there is a nonzero difference between a pair of percentages indicates that there is some association between the variables. In Table 10–7 there is a difference between the percentages of 60%. In Table 10–7 it can be seen that if an observation has scored X_1 on variable X, it is very likely that it will have a score of Y_1 on variable Y, while an observation that scored X_2 on variable X is likely to have a score of Y_2 on variable Y. Thus, knowing the scores of observations on the X variable would enable us to predict fairly well their scores on the Y variable. This means that there is a strong association between variables X and Y. The difference in the percentages of 60% indicates this strong association.

Finally, in Table 10–8 there is a perfect association between X and

Table 10–6

| | | Variable X | |
		X_1	X_2
Variable Y	Y_1	65%	55%
	Y_2	35%	45%
		100%	100%

Table 10–7

		Variable X	
		X_1	X_2
Variable Y	Y_1	85%	25%
	Y_2	15%	75%
		100%	100%

Y. If observations have a score of X_1 on the X variable, then they always have a score of Y_2 on the Y variable, and observations that have a score of X_2 on the X variable always score Y_1 on the Y variable. This perfect association is indicated by a difference of 100% between a pair of percentages in a row. Thus, in 2×2 tables, the difference between a pair of percentages may vary from 0 percent, indicating no association or perfect independence, to 100 percent, indicating perfect association. And the larger the percentage difference, the stronger the association.

Percentages, then, allow us to derive a single numerical value to measure association between variables in 2×2 tables. This value is easy to calculate and interpret and is comparable across tables. In Table 10–7 the association between the variables is stronger than is the association in Table 10–6. The percentage difference as a measure of association reflects this in the fact that the percentage difference in Table 10–7 is 60%, whereas in Table 10–6 it is only 10%.

Percentages as a measure of association also have the advantage of being applicable to data at any level of measurement. Thus, they are appropriate for use with either nominal or ordinal variables or for combinations of variables at both levels. One disadvantage of percentages is that the single numerical value of the difference between a pair of percentages does not indicate the "direction" of association between ordi-

Table 10–8

		Variable X	
		X_1	X_2
Variable Y	Y_1	0%	100%
	Y_2	100%	0%
		100%	100%

nal variables. This is not a serious disadvantage, however, since the pattern of association can be determined from inspecting all the percentages in the table. In Table 10–7, for example, the direction of association is that a "change" from X_1 to X_2 on the X variable is associated with a change from Y_1 to Y_2 on the Y variable. In Table 10–8 the pattern is reversed: A change from X_1 to X_2 is associated with a change from Y_2 to Y_1.

For 2 × 2 tables, then, percentages provide a useful measure of association in the form of the difference between a pair of percentages. This difference has a number of advantages as a measure of association: It is easy to calculate, it is comparable across other 2 × 2 tables, its values can be interpreted in a relatively straightforward manner, and it can be used on data at any level of measurement.

Unfortunately, in tables larger than 2 × 2, percentages do not function nearly as well as a measure of association as they do in 2 × 2 tables. The major drawback is that there is no longer any way to calculate a single numerical value to indicate the strength of association between the variables. Percentages, though, still retain their ability to allow comparisons of the conditional distributions in large tables.

Consider Table 10–9. In this 3 × 3 table, the political position of people is cross-tabulated with their attitude toward the power of the federal government. Examination of the percentages shows that the proportion of conservative respondents who feel the federal government is too powerful (70%) is greater than the proportion of liberal respondents who feel this way (10%). In this table there is a steady progression in feeling that the federal government is too powerful as respondents become more conservative. The percentages as a whole clearly reveal

Table 10–9

		Political Position		
		Liberal	Moderate	Conservative
Power of Federal Government	Not Powerful Enough	50%	20%	10%
	About Right	40%	50%	20%
	Too Powerful	10%	30%	70%
		100%	100%	100%

this pattern. The problem, though, is that there is no *single* value that can be constructed from the percentages that can indicate the nature and strength of the association. If, for example, we tried to take a percentage difference as a measure of the strength of association, which pair of percentages should be used? For the row labeled "Too Powerful," the percentage for liberals is 10 and for conservatives 70, yielding a difference of 60. But for the row labeled "Not Powerful Enough," the percentage for liberals is 50 and for conservatives 10, yielding a difference of 40. And for the row labeled "About Right," the corresponding difference is only 20. These three differences would each give quite a different picture of the strength of association if we tried to use one of them as a measure of association. Thus, in this table it is not possible to use percentages to derive a single value to measure the strength of association as was possible in the 2 × 2 tables considered earlier. But what the percentages can show is that the conditional distributions of scores on the dependent variable do differ according to the values of the independent variable. Consequently, the pattern of association between the variables is revealed by the percentages.

The utility of using percentages to detail the pattern of association between variables is illustrated in Table 10–10. In this table the social class of respondents is cross-tabulated with their attitude toward the police. Examination of the table shows that there is no steady increase in favorableness (or unfavorableness) in attitude toward the police similar to the pattern in Table 10–9. Instead, there is a rather interesting U-shaped pattern present. Lower-class respondents on the whole have unfavorable attitudes toward the police, middle-class respondents have predominantly favorable attitudes, and upper-class respondents generally have unfavorable attitudes. Again, what the percentages do is to allow comparison of the conditional distributions, which in turn allows detection of the pattern of association.

As tables become larger, more and more patterns of association are

Table 10–10

		Social Class		
		Lower	Middle	Upper
	Favorable	25%	70%	20%
Attitude Toward Police	Neutral	20%	20%	15%
	Unfavorable	55%	10%	65%
		100%	100%	100%

possible. In such tables, percentages are a most useful tool in distinguishing among patterns of association. This is especially important, since many measures of association that do yield a single value to measure association will often result in a low value even when strong U-shaped patterns (or other "strange" patterns) are present. Thus, it is often desirable to use percentages to compare conditional distributions even when some other measure of association is being employed.

The major difficulty with percentages in tables larger than 2 × 2 is that a single numerical value cannot be derived to measure association. This, in turn, means that percentages cannot be used to easily compare association across different tables. Finally, as tables become very large, there are so many percentages present that it even becomes difficult to use them to detect patterns except in the most clear-cut situations.

We shall turn our attention now to specific measures of association. Some of these measures possess little advantage over the "percent difference" but are quite common in social science work. The development of these measures will also aid in our understanding of the difficulties involved in measuring association and will serve as a basis for introducing more sophisticated measures.

Question 10–1. Percentage the following table and verbally explain the nature of the association between age and party identification.

		Age			
		Young	Medium	Old	
	Democrat	50	80	70	200
Party Identification	Independent	70	70	20	160
	Republican	30	50	60	140
		150	200	150	500

NOMINAL
MEASURES OF ASSOCIATION

Phi

A common problem in the social sciences is the measuring of association between two nominal variables, each of which consists of two categories. This type of situation forms a 2 × 2 table. A statistic widely employed to measure the strength of association between two dichotomous variables of this sort is *phi* (ϕ). Using the labeling in Table 10–11, phi is defined as

Table 10–11

a	b	a + b
c	d	c + d
a + c	b + d	N

$$\text{phi} = \phi = \frac{ad - bc}{\sqrt{(a + c)(b + d)(a + b)(c + d)}}$$

In verbal terms, the numerator consists of the difference of the products of the frequencies in the cells on each of the diagonals of the table, and the denominator consists of the square root of the product of all the marginals of the table.

To illustrate the computation of phi, let us use the data for sex and voting that was used in Chapter 9. These data are reproduced in Table 10–12. Inserting the table's frequencies into the formula for phi gives

$$\phi = \frac{ad - bc}{\sqrt{(a + c)(b + d)(a + b)(c + d)}}$$

$$= \frac{(43)(63) - (17)(77)}{\sqrt{(120)(80)(60)(140)}}$$

$$= \frac{2709 - 1309}{\sqrt{80,640,000}}$$

$$= \frac{1400}{8980}$$

$$= .156$$

This value indicates that the association between sex and the vote is weak, since it is much closer to 0 (no association) than it is to 1.0

Table 10–12

		Sex		
		Male	Female	
Vote in 1972	Nixon	43	17	60
	McGovern	77	63	140
		120	80	200

(perfect association). For phi, the sign of the coefficient, positive or negative, is, perhaps, best ignored. This is because with nominal variables the ordering of the categories has no importance and can be changed without changing the nature of the relationship between the variables. What the sign of phi indicates is whether the association tends to run along the positive diagonal of the table (from upper left to lower right) or the negative diagonal (from lower left to upper right). But changing the ordering of the categories on one of the variables can cause the sign of phi to change. And since reordering the categories is a permissible operation on nominal variables, the sign of phi only indicates how the association happens to run in a particular arrangement of a table and does not really say anything about the nature of the association between the variables. Therefore, we shall ignore the sign of any value of phi and treat it as if it varies in value from 0 to 1.0. (This point can be checked by reversing the order of the categories for one or both of the variables in Table 10–12 and recalculating the value of phi. Disregarding the sign of the coefficient, the phi value will always be the same.)

Phi is one of the measures of association for which there is no easy interpretation for numerical values between 0 and 1.0. Some insight into the meaning of this statistic, though, can be gained by examining the conditions under which it will be 1.0 or 0. The value of phi will be 1.0 when all the observations having the same score on one variable have the same score on the other variable. In terms of a 2×2 table, this means that phi will be 1.0 when all the observations are located on one or the other diagonal of the table. This kind of situation is illustrated in Table 10–13. In Table 10–13, knowing an observation's score on one variable enables us to predict, perfectly (without error), its score on the other variable. The value of phi will be 0 when the scores on one variable have the same proportional distributions in the categories of the other variable. This kind of situation is illustrated in Table 10–14. Here knowing scores on one variable does not help us in predicting scores on the other vari-

Table 10–13

		Variable X		
		X_1	X_2	
Variable Y	Y_1	50	0	50
	Y_2	0	50	50
		50	50	100

Table 10–14

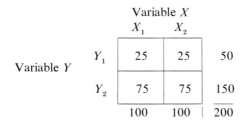

| | | Variable X | | |
		X_1	X_2	
	Y_1	25	25	50
Variable Y				
	Y_2	75	75	150
		100	100	200

able. If, for example, we were told that the X score for an observation in Table 10–14 was X_1, then the column for value X_1 on the X variable tells us that the chances are .75 ($\frac{75}{100} = .75$) that the observation's Y score is Y_1. But the marginal distribution for the Y variable (the totals at the ends of the rows in the table) tells us that the chances of any observation's Y score being Y_1 is .75 ($\frac{150}{200} = .75$), regardless of what its X score is. Thus knowing an observation's X score does not help in predicting Y scores, and exactly the same thing will be found in predicting X scores from Y scores.

Thus the meaning of phi when it is equal to 0 or 1.0 is quite clear. Intermediate values have no easily interpretable meaning. All that can be said is that the closer the value of phi is to 1.0, the nearer the association is to a perfect relationship, while the closer phi is to 0, the nearer the association is to complete independence between the variables.

It was said earlier that measures of association behave peculiarly under certain conditions. One of the most common of these peculiarities is that the maximum value of a measure of association is frequently restricted when the marginal distributions of the variables do not match. In Table 10–12, for example, the row marginals are 60 and 140, and the column marginals are 120 and 80. The marginals give the frequency distribution for each variable in the table. In Table 10–12, the row marginals give the frequency distribution for the variable "Vote in 1972." They tell us that of the 200 people in the table, 60 voted for Nixon and 140 for McGovern. We can obtain this information in percentages by dividing the marginals by the total number of observations in the table and multiplying by 100. Thus, 30% of the 200 people in the table [$\frac{60}{200}$ (100) = 30] voted for Nixon and 70% of them [$\frac{140}{200}$ (100) = 70] voted for McGovern. The distribution of the row marginals then are split 30%–70%. Again in Table 10–12, the column marginals give the frequency distribution for the variable "sex." The column marginals tell us that of the 200 people in the table, 120, or 60%, are male and that 80, or 40%,

Table 10-15

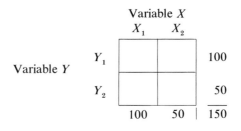

are female. Thus, the distribution of the column marginals is 60%–40%.

Because phi is 1.0 only when all the observations are located on one of the diagonals of the table, phi can only reach the value 1.0 if the row and column marginal distributions are the same. This means that the row and column marginals must split into the same proportions. It does not matter what these proportions are, only that they must match. The reason that the marginal distributions must match for the upper limit of phi to be 1.0 is that it is only under this condition that it is possible to locate all the observations on one or the other diagonal in a 2 × 2 table. Consider Table 10–15. When the distributions of the two row and column marginals match, it is possible to locate all the observations on one of the diagonals of the table and still have the cells sum up to equal the marginal values. This is illustrated in Table 10–16. Of course, it is not necessary that all the observations actually be located on one of the diagonals, only that it is possible for this to happen if phi is to have 1.0 as its upper limit. Table 10–17 illustrates a situation in which the marginal distributions are the same as those of Table 10–15 except that the observations are not all located on the diagonal.

Returning to Table 10–12, we saw there that the marginal distributions do not match. The marginals in Table 10–12 split 30%–70% and 40%–60%. In this table, or any table in which the marginal distributions

Table 10-16

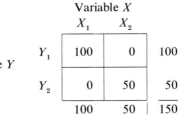

Table 10–17

	Variable X		
	X_1	X_2	
Y_1	60	40	100
Y_2	40	10	50
	100	50	150

Variable Y

Table 10–18

	Variable X		
	X_1	X_2	
Y_1			30
Y_2			70
	60	40	100

Variable Y

do not match, it is impossible to locate all the observations on one diagonal and still have the cell frequencies sum to the marginal values. Table 10–18 presents marginals in these proportions using a total of 100 observations. Examination will show that it is impossible to locate all 100 observations on either of the diagonals and still have the cells sum to the marginal values. In such situations, phi cannot even theoretically take on a value of 1.0, and consequently the upper limit of phi is restricted to some value less than 1.0.

A value of phi in a table where the upper limit of phi is restricted below 1.0 is not directly comparable to a phi value in a table in which phi can reach 1.0. In tables in which the upper limit of phi is restricted

Table 10–19

		Sex		
		Male	Female	
Vote in 1972	Nixon	0	60	60
	McGovern	120	20	140
		120	80	200

below 1.0, it is possible to calculate an adjusted phi value that is comparable to phi values where the upper limit is 1.0. This is done by first calculating the maximum value of phi possible for the set of unmatched marginal distributions in a particular table. This is done by finding the arrangement of cell frequencies for the marginals that yields the highest possible value of phi. For the marginals in Table 10–12, Table 10–19 shows the distribution of cell frequencies that yields the highest possible value of phi. This highest possible value of phi for a table with unmatched marginal distributions is called *phi maximum* (ϕ_{max}). Inserting the cell frequencies from Table 10–19 into the formula for phi gives

$$\phi_{max} = \frac{(0)(20) - (60)(120)}{\sqrt{(120)(80)(60)(140)}}$$

$$= \frac{0 - 7200}{\sqrt{80,640,000}}$$

$$= \frac{-7200}{8980}$$

$$= -.80$$

Disregarding the sign,

$$\phi_{max} = .80$$

This is the maximum possible value phi could take on in Table 10–12.

This maximum value is now used to adjust the original phi value of Table 10–12 of .156 so that it is relative to an upper limit of 1.0 instead of to .80. Rescaling the value of phi in this way will put the obtained value of phi on the standard scale of 0 to 1.0 and facilitates comparisons of phi values across tables with different marginal distributions. This rescaled value is called *phi adjusted* (ϕ_{adj}) and is found very simply as follows:

$$\phi_{adj} = \frac{\phi_{obtained}}{\phi_{max}}$$

For Table 10–12, ϕ_{adj} is

$$\phi_{adj} = \frac{\phi_{obtained}}{\phi_{max}}$$

$$= \frac{.156}{.80}$$

$$= .195$$

Phi and Chi-square

It may have been noticed that the formula for phi is very similar to the formula for chi-square for 2×2 tables. As a matter of fact, there is a close connection between the formulas. Let us take a look at this connection, for it will help to illustrate a couple of points. It will be recalled that the formula for chi-square for a 2×2 table is

$$\chi^2 = \frac{N(ad - bc)^2}{(a + c)(b + d)(a + b)(c + d)}$$

To show the relationship between chi-square and phi, we first divide both sides of the equation by N:

$$\frac{\chi^2}{N} = \frac{N(ad - bc)^2}{N(a + c)(b + d)(a + b)(c + d)}$$

Taking the square root of the expressions on both sides of the equals sign gives:

$$\sqrt{\frac{\chi^2}{N}} = \frac{ad - bc}{\sqrt{(a + c)(b + d)(a + b)(c + d)}}$$

The expression on the right side is the formula for phi. Thus,

$$\sqrt{\frac{\chi^2}{N}} = \phi \qquad \text{or} \qquad \phi = \sqrt{\frac{\chi^2}{N}}$$

It was pointed out at the beginning of the chapter that measures of association are unaffected by the size of the sample. In the case of phi, this is pointed out by the relationship between chi-square and phi. Other things being equal, the chi-square test will yield a result that will be significant at a smaller alpha level as the sample size increases. Dividing the value of chi-square by N to yield phi, though, removes the effect of different sample sizes. Thus, phi measures the association between two variables in a sample without regard to the size of the sample.

Question 10–2.

a. As shown in Chapter 9, the value of chi-square for Table 10–12 is 4.86. Use this value to verify that the formula which expresses phi as a function of chi-square yields the same value as the formula for phi that was first presented.
b. A study of local politics is undertaken to investigate why people vote in local elections. A survey of 500 people is conducted and a cross-

tabulation between whether a person owns his own home and voting produces the following results:

		Homeowner Yes	No	
Vote in Local Election	Yes	100	50	150
	No	200	150	350
		300	200	500

What is the value of phi for this table?

c. Instead of the frequencies just presented, imagine that a much larger sample was taken, which produced the following table:

		Homeowner Yes	No	
Vote in Local Election	Yes	200	100	300
	No	400	300	700
		600	400	1000

What is the value of phi for this table? Does this value equal the phi value for the table in part b? Why?

The Contingency Coefficient

The statistic phi can only be used as a measure of association on 2×2 tables. Quite often, though, nominal variables are encountered which have more than two categories. In political science and sociological research, it is not unusual for one or both variables in a cross-tabulation to consist of five or six categories. A measure of association that is frequently used in situations of this kind is the contingency coefficient (C), which is defined as

$$C = \sqrt{\frac{\chi^2}{\chi^2 + N}}$$

As can be seen, this statistic is based upon chi-square. To compute a value for the contingency coefficient, the value of chi-square for the table at hand is computed. Then the contingency coefficient is found by simply taking the square root of the value of chi-square divided by chi-square plus the sample size (N).

Table 10–20

		North	South	East	West	
			Region			
Party Identi- fication	Democrat	20	20	35	15	90
	Independent	10	5	15	20	50
	Republican	15	5	30	10	60
		45	30	80	45	200

To illustrate the computation of C, let us use the data on region and party identification that was used in Chapter 9 to illustrate the computation of chi-square for tables larger than 2×2. These data are reproduced in Table 10–20. The value of chi-square for this table was found to be 17.9. With 6 degrees of freedom, the null hypothesis of no association can be rejected at an alpha level of .01. In other words, we can be quite confident that there is an association between region and party identification. However, the value of chi-square does not serve to indicate the degree or strength of the association between region and party identification. The calculation of C will help give an indication of this. Inserting the obtained value of chi-square and the sample size into the formula for C yields

$$C = \sqrt{\frac{\chi^2}{\chi^2 + N}}$$

$$= \sqrt{\frac{17.9}{17.9 + 200}}$$

$$= \sqrt{.082}$$

$$= .286$$

Thus, the value of C for the data in Table 10–20 is .286.

It will be noted that C is always positive in value. Actually, since the last operation in computing C is the taking of a square root, C could be made either positive or negative in value by using either the positive or negative square root. But since in dealing with nominal variables it does not make sense to speak of the direction of association, the sign of C is ignored. Thus, C may vary in value from 0 to 1.0.

Unfortunately, as was the case for phi, the upper limit of the value of C can also be restricted to less than 1.0. It turns out that the upper limit of C is a function of the degrees of freedom in the table. In a square table, a table with the same number of rows and columns, the maximum value of C is given by

$$C_{max} = \sqrt{\frac{K-1}{K}}$$

where K is the number of categories for a variable. Thus, for a 3×3 table, the maximum value of C is

$$C_{max} = \sqrt{\frac{K-1}{K}}$$

$$= \sqrt{\frac{3-1}{3}}$$

$$= \sqrt{\frac{2}{3}}$$

$$= \sqrt{.67}$$

$$= .82$$

For a 5×5 table, the maximum value is

$$C_{max} = \sqrt{\frac{K-1}{K}}$$

$$= \sqrt{\frac{5-1}{5}}$$

$$= \sqrt{\frac{4}{5}}$$

$$= \sqrt{.80}$$

$$= .89$$

As can be seen, the upper limit of C gets larger as the table gets larger. The upper limit of C is 1.0 when the number of categories is infinite. The maximum upper limit for a given table can be used to adjust the value of C in the same way that we adjusted for the value of phi.

There is an additional problem here in that the exact upper limit of C is known only for square tables. Table 10–20 is, of course, not square. Based on the number of rows, the maximum value of C for Table 10–20 would be

$$C_{max} = \sqrt{\frac{r-1}{r}}$$

$$= \sqrt{\frac{3-1}{3}}$$

$$= .82$$

Based on the number of columns, the maximum value would be

$$C_{max} = \sqrt{\frac{c-1}{c}}$$

$$= \sqrt{\frac{4-1}{4}}$$

$$= .87$$

Generally, the maximum value that will provide the more conservative estimate is used. In the present case this is the maximum value based on the number of columns. The adjustment procedure now follows that which was used with phi:

$$C_{adj} = \frac{C_{obtained}}{C_{max}}$$

Inserting the appropriate values yields

$$C_{adj} = \frac{.286}{.87}$$

$$= .33$$

Thus, the adjusted value of C for Table 10–20 is .33.

In terms of meaning, this value indicates that there is a weak to moderate degree of association between region and party. Unfortunately, it is not possible to provide a more meaningful statement than this because C also lacks a straightforward, commonsense interpretation. As with phi, a value for C of 1.0 would indicate perfect association or dependence, and a value of 0 would indicate no association or independence. As for values that are between 1.0 and 0, about all that can be said is that they reflect weak, moderate, or strong association, with values below .30 usually designated as weak, .30 to .60 as moderate, and above .60 as strong.

One important difference between phi and C is that C can reach a value of 1.0 without all the observations being located along one of the diagonals of the table. Of course, a value for C of 1.0 can only be reached if the table is infinite in size. But the point is that high values of C are not dependent upon locating observations along a diagonal. What high values of C indicate is that observations which have a particular score on one variable tend to all have the same score on the other variable, regardless of where these scores are located in the table. Thus, a high value of C means that we can predict fairly well an observation's score on one variable from a knowledge of its score on the other variable. Low values of C indicate that such predictions will not be very accurate.

Question 10–3. As an exercise at the end of Chapter 9, you were asked to calculate the value of chi-square for the following table:

Religion

		Catholic	Protestant	Jewish	
	Business	5	20	5	30
Occupation	Professional	10	15	15	40
	Labor	15	15	0	30
		30	50	20	100

The value of chi-square for this table is 21.3 which, with 4 degrees of freedom, is significant at an alpha level of .001.

a. Compute the value of the contingency coefficient for this table.
b. Compute the value of C_{adj} for the table.

Lambda

For many years phi and the contingency coefficient were virtually the only measures of association available for use at the nominal level. Relatively recently, however, a statistic for nominal data called *lambda* (λ) has been developed which should replace phi and the contingency coefficient for most work in social science. Lambda has a clear, straightforward interpretation for values between 0 and 1.0 which, as has been pointed out, both phi and the contingency coefficient lack. Furthermore, lambda is appropriate for any size of table, 2 × 2 or larger. Lambda, then, possesses some desirable properties that both phi and the contingency coefficient lack. Given the superior properties of lambda, it might be wondered why so much time has been devoted to phi and the contingency coefficient. First, phi and the contingency coefficient are still used quite frequently in social science, and thus it is necessary to be familiar with these two statistics to be adequately equipped to deal with social science research materials. Second, phi and the contingency coefficient are relatively simple statistics and provide a means for coming to grips with the concept of association and with some of the difficulties of measuring association in nominal data. Finally, although lambda does have advantages over phi and the contingency coefficient, it is still not a "perfect" measure of association, and there are some situations in which it might be preferable to use phi or the contingency coefficient.

Lambda is one of a family of measures of association known as *proportional reduction in error (PRE)* statistics. Suppose that we were told that 100 people were going to walk by one at a time and that as each person passed we were to guess whether he was a Republican or a Democrat. Before the experiment starts we are told that in this group of 100, there are 70 Republicans and 30 Democrats. In other words, we are told the frequency distribution of party identification for the 100 people. Our job in guessing is to make as few errors as possible. What should our guessing strategy be? If we randomly guessed either Republican or Democrat as each person walked by, then in the long run we would be guessing Republican about 50% of the time and Democrat about 50% of the time. The best we could hope for under this strategy is that we would guess right in half the cases or correctly identify the party identification of 50 of the 100 people. (This would be true in the long run—if we repeated the experiment many times. In any one run of the experiment, of course, by chance we might correctly identify more or less than 50 out of the 100.) But another strategy can greatly improve upon this. We know from the frequency distribution of party identification that among the 100 people there are 70 Republicans. Thus, if we would guess Republican for *every* person who came by, we would be sure of correctly identifying the party identification of 70 of the people and only make 30 errors. The best guessing strategy when the frequency distribution is known, then, is to guess the *modal value* of the distribution, because this will generate the fewest number of errors.

Now suppose we are told that in the group of 100 people there are 70 men, of whom 60 are Republicans and 10 are Democrats, and 30 women, of whom 10 are Republicans and 20 are Democrats. This information can be represented in a 2 × 2 table as in Table 10–21. The experiment is now to be repeated, except that this time we will get to observe whether a person is male or is female before we guess. What should our guessing strategy be in this situation? Obviously, we would want to make use of any association between sex and party identification in order to improve our guessing. Examining Table 10–21, we see that if the person passing by is male, then 60 out of 70 times he will be a Republican. That is, in Table 10–21 the modal value of party identification for

Table 10–21

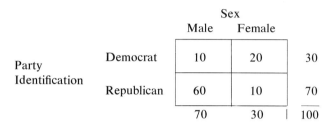

men is Republican with 60 observations. Consequently, if we always guess Republican for men, we will be correct 60 times and incorrect 10 times. For females the modal value of party identification is Democrat with 20 observations. Thus, if we always guess Democrat for women, we will be correct 20 times and incorrect 10 times. Following this strategy, we will make a total of 20 errors.

Without any information about the association between sex and party identification, our best guessing strategy resulted in 30 errors. *With* information about the association, we made 20 errors. Thus, by knowing information about the data on one variable we were able to make 10 less errors in guessing on the other variable. This information can be put in terms of a proportion by dividing the *number of errors that are saved from being made* when we have information about sex and party identification (10) by the *number of errors that are made when we do not have information about the association* between sex and party identification (30):

$$\frac{10}{30} = .33$$

Thus, the *proportional reduction in error* that is obtained from the information about the association between sex and party identification is 33%. It is exactly this proportional reduction in error that lambda measures. In general, lambda is defined as

$$\text{lambda} = \lambda = \frac{\text{number of errors saved}}{\text{number of errors without information}}$$

$$= \frac{\left(\begin{array}{c}\text{number of errors}\\\text{without information}\end{array}\right) - \left(\begin{array}{c}\text{number of errors}\\\text{with information}\end{array}\right)}{\text{number of errors without information}}$$

If two variables are completely unrelated (independent), having information about one variable will not reduce the number of errors that we would make in guessing about the other variable. We would make just as many errors in guessing as we did when we had no information about the other variable. In this situation the numerator of lambda will be 0, which will make the value of lambda 0. If two variables are perfectly related, having information on one variable will enable us to guess correctly every time on the other variable. In this situation we will be saved from making all the errors that would be made if we did not have the information and the numerator and denominator of lambda will be equal, which will make the value of lambda 1.0. The stronger the association between the two variables, the more errors that will be saved and the larger the value of lambda will be. And the nice thing about lambda is that for values between 0 and 1.0, lambda tells us the proportional reduction in errors that takes place when we have information about the other variable, as compared to the errors that are made when we do not have this information.

Since lambda deals with a comparison of the errors made in guessing about one variable when we do and do not have information about another variable, it is necessary to distinguish the variable we are guessing about and the variable from which information is obtained. By convention, this is done by designating the variable across the top of a table as A and the variable along the side of a table as B. If we are using information from variable A (sex in Table 10–21) to guess about variable B (party identification in Table 10–21), this situation is designated by labeling lambda as λ_B. If we were to reverse the process, to use information from variable B to guess about variable A, lambda is designated as λ_A. Although it may not be obvious at the moment, we will see that λ_B need not always equal λ_A in the same table.

Suppose that a survey of 200 people is conducted to see if there is an association between religion and where people live. A cross-tabulation of the data produces Table 10–22. The chi-square value for this table is 47.59, which for 6 degrees of freedom is statistically significant at an alpha level less than .001. The chi-square test tells us, then, that we can be quite confident that some sort of relationship exists between the two variables. But how strong an association exists between the variables is not specified by the significance test. If religion is designated the A variable and residential location the B variable and we are interested in predicting from religion to residential location, λ_B can be computed as a measure of the predictive strength of association between the variables. To compute λ_B two pieces of information are needed: the number of errors that would be made in predicting the residential location of people *without* having knowledge of their religion, and the number of errors that would be made in predicting the residential location of people *with* information about their religion.

Table 10-22

		Religion				
		Catholic	Protes-tant	Jew	Other	
Residential Location	Urban	30	20	10	20	80
	Suburban	15	40	15	0	70
	Rural	5	20	5	20	50
		50	80	30	40	200

The "best-guess" strategy for a nominal variable when we have no information about its relation to any other variable is to always guess the modal value of its frequency distribution. This frequency distribution is given in the marginal values for a variable in a table. In Table 10-22 the marginal distribution for residential location is given in the row marginals of the table. Examination of these marginals shows that the modal value for residential location is urban with 80 observations. Thus, we would predict urban for each person in guessing his residential location when we had no other information. Since there are 70 people who are suburban residents and 50 who are rural residents, we would make a total of 120 errors by always guessing the modal category of urban.

When information about the A variable (religion) is introduced, we look inside the table to determine the modal value on the B variable for each category of the A variable. For Catholics, the modal value on residential location is urban, with 30 observations. Since a total of 50 people are Catholics in the table (the column marginal for Catholics), we will make 20 errors (50-30) by always guessing the modal value of urban for Catholics. Moving to the second column in the table, we see that there is a total of 80 Protestants and that the modal value for Protestants is suburban, with 40 observations. By always guessing suburban for Protestants, we would make 40 errors (80-40). The marginal total for the third column indicates that 30 of the 200 people in the table are Jews. The modal value for this category is also suburban, with 15 observations. By always guessing suburban for Jews, we would make 15 errors. In the last column there are 40 people who were classified as Other for their religion. This column is bimodal: there are equal numbers of observations (20) for urban and rural. It does not matter which of these two values we would guess for those who are classified as Other. In either case we would be correct 20 times and make 20 errors. The total num-

ber of errors in predicting variable B (residential location) with information about variable A (religion) is equal to the sum of the errors that
would be made in each column, or $20 + 40 + 15 + 20 = 95$.

The two error totals are now used to calculate the value of λ_B for
Table 10–22. Using the verbal definition of lambda given above, we
have

$$\lambda_B = \frac{\left(\begin{array}{c}\text{number of errors}\\\text{without information}\end{array}\right) - \left(\begin{array}{c}\text{number of errors}\\\text{with information}\end{array}\right)}{\text{number of errors without information}}$$

$$= \frac{120 - 95}{120}$$

$$= \frac{25}{120}$$

$$= .208$$

Thus, in predicting residential location from a knowledge of religion we
would make 20.8% fewer errors than if we had to predict without knowledge about religion. Since we have reduced the errors 20.8% by knowing
religion, we can conclude that there is a mild predictive association between religion and residential location.

The procedure just used to compute lambda serves to highlight the
logic and meaning of this statistic. It is, however, too cumbersome for
efficient calculation. An easier computation formula for λ_B (when the
columns of a table are labeled A and the rows B) is

$$\lambda_B = \frac{\text{sum of the column modes} - \text{mode of the row marginals}}{N - \text{mode of the row marginals}}$$

Inserting the values from Table 10–22 into this formula gives

$$\lambda_B = \frac{(30 + 40 + 15 + 20) - 80}{200 - 80}$$

$$= \frac{105 - 80}{120}$$

$$= \frac{25}{120}$$

$$= .208$$

Thus, the computing formula gives the same answer as the definitional
procedure for lambda, with a lot less work.

Let us now reverse the direction of prediction and compute a value of lambda for predicting religion from residential location. This means that we are going to predict variable A from a knowledge of variable B, and consequently we want to compute λ_A.

As for λ_B, two pieces of information are needed to calculate λ_A. Now, however, this information will concern the number of errors made in predicting religion *without* information about residential location and the number of errors made in predicting religion *with* information about residential location. The first piece of information is obtained from the marginal values for the A variable, religion. These are the column marginals, and they show that the modal value for religion is Protestant, with 80 observations. Always guessing Protestant for religion would result in 120 errors.

Information is now introduced about the B variable, residential location. What we want now is the number of errors that would be made when we guess the modal values for each category of residential location. This information is given in the rows of the table. Looking at the first-row marginal, it can be seen that there are 80 urban residents. The mode for this category is Catholic, with 30 observations. Guessing Catholic for all urban residents, then, would result in 50 errors. The row marginal for the second row shows there are 70 suburban residents. The modal value in this row is Protestant, with 40 observations. The number of errors that would be made in always guessing Protestant for suburban residents is 30. Finally, the row marginal for the third row shows that there are 50 rural residents. This row happens to be bimodal, with 20 observations each for the values of Protestant and Other. Again, it does not matter which of these values we would guess for rural residents, as we would get the same number of errors, 30, for guessing either value. Summing the number of errors made in each row yields $50 + 30 + 30 = 110$.

Inserting the two error totals into the verbal definition of lambda yields

$$\lambda_A = \frac{\left(\begin{array}{c}\text{number of errors}\\\text{without information}\end{array}\right) - \left(\begin{array}{c}\text{number of errors}\\\text{with information}\end{array}\right)}{\text{number of errors without information}}$$

$$= \frac{120 - 110}{120}$$

$$= \frac{10}{120}$$

$$= .083$$

Thus in predicting religion from a knowledge of residential location we would make 8.3% fewer errors than if we had to predict religion without knowledge about residential location. Since the errors have been reduced only by 8.3%, we conclude that there is a weak predictive association between residential location and religion.

That λ_B and λ_A differ in value for Table 10–22 points out an important fact about what these statistics measure: They measure the extent to which knowledge of one variable improves our ability to make predictions about the other variable. These statistics, then, measure the degree of association between two variables in the sense of being able to reduce errors in prediction. The fact that λ_B and λ_A can and usually do have different values for the same table indicates that predictive association is not necessarily a symmetric property. The asymmetry of predictive association may be grasped by noting that all people who are hanged by the neck (sufficiently) will die, but not all dead people were hanged. Table 10–23 shows a situation in which it is possible to predict perfectly from variable A to variable B ($\lambda_B = 1.0$) but only imperfectly from variable B to variable A ($\lambda_A = .445$). Thus, in measuring association by lambda it is necessary to decide the direction of prediction between the variables and select λ_B or λ_A for calculation accordingly.

Just as for λ_B there is a computational formula for λ_A. It has the same form as the one for λ_B.

$$\lambda_A = \frac{\text{sum of the row modes} - \text{mode of the column marginals}}{N - \text{mode of the column marginals}}$$

Inserting the values from Table 10–23 into this formula yields

$$\lambda_A = \frac{(30 + 40 + 20) - 80}{200 - 80}$$

$$= \frac{90 - 80}{120}$$

$$= \frac{10}{120}$$

$$= .083$$

which is, of course, the same answer that was obtained using the definitional procedure for lambda.

Before moving on, it should be pointed out that a significant chi-square for a set of data does not necessarily mean that lambda will indicate a predictive association. For example, the sex and vote data that was used earlier in Table 10–12 yields a significant value for chi-square. These data are reproduced in Table 10–24. On the basis of the chi-

Table 10–23

		Variable A			
		A_1	A_2	A_3	
Variable B	B_1	40	40	0	80
	B_2	0	0	20	20
		40	40	20	100

Table 10–24

		Sex		
		Male (A_1)	Female (A_2)	
Vote in 1972	Nixon (B_1)	43	17	60
	McGovern (B_2)	77	63	140
		120	80	200

square test we can be fairly sure that some kind of relationship exists between sex and the vote. Now, let us compute $\lambda_{B'}$ and λ_A for these data. Using the frequencies in Table 10–24, we have

$$\lambda_B = \frac{(77 + 63) - 140}{200 - 140}$$

$$= \frac{140 - 140}{60}$$

$$= \frac{0}{60}$$

$$= 0$$

For λ_A,

$$\lambda_A = \frac{(43 + 77) - 120}{200 - 120}$$

$$= \frac{120 - 120}{80}$$

$$= \frac{0}{80}$$

$$= 0$$

Thus, the values of lambda would indicate that there is no predictive association between sex and vote, even though the value of chi-square for these data are statistically significant. The reason for this peculiarity is that lambda indicates association only when the modal values for columns are different (for λ_B) or when the modal values for the rows are different (for λ_A). When the modal values for columns are the same ("McGovern" in Table 10–24) and the modal values for the rows are the same ("Male" in Table 10–24), lambda will not be able to detect any association in the table, even if there is some. This is a consequence of the definition of lambda involving the reduction of errors in predicting. Thus lambda only measures a particular type of association, which allows a reduction in error in predicting. Whether the variables are associated in some other sense may be checked by calculating phi or the contingency coefficient.

Question 10–4. For the table in Question 10–3:
a. Compute λ_A.
b. Compute λ_B.

_____

ORDINAL MEASURES OF ASSOCIATION

Yule's Q

For a variable at the ordinal level of measurement, a rank order is established among observations if one observation has a higher score on an ordinal variable than another. The concept of order among observations on a variable will play an important role in our discussion of measures of association at this level of measurement.

Let us begin by considering *Yule's Q*. This statistic is designed to measure association between two ordinal variables when each variable is a dichotomy. Using the labeling indicated in Table 10–25, Yule's Q is defined as

$$Q = \frac{ad - bc}{ad + bc}$$

As can be seen from this formula, the computation of Q is quite simple. The numerator consists of the difference of the products of the cell fre-

Table 10–25

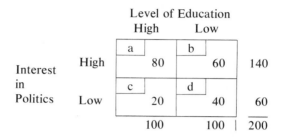

		Level of Education		
		High	Low	
Interest in Politics	High	a 80	b 60	140
	Low	c 20	d 40	60
		100	100	200

quencies on the diagonals of a 2 × 2 table. The denominator consists of the sum of these products.

Let us use the frequencies in Table 10–25 to illustrate the computation of Q:

$$Q = \frac{ad - bc}{ad + bc}$$

$$= \frac{(80)(40) - (60)(20)}{(80)(40) + (60)(20)}$$

$$= \frac{3200 - 1200}{3200 + 1200}$$

$$= \frac{2000}{4400}$$

$$= .455$$

This value indicates that for the data in Table 10–25 there is a moderate association between level of education and interest in politics.

Let us now consider interpretations for various values of Q. As for most measures of association, a Q of 0 indicates no association between the variables. Positive values of Q indicate a positive association; negative values indicate a negative association. Since order is important when dealing with ordinal variables, the sign of Q has substantive meaning. However, care must be used in interpreting the sign, since the arrangement of the variables in the table can change it. Examination of the formula for Q shows that Q will have a positive value when the quantity ad is larger than the quantity bc. This situation will occur when the observations tend to fall along the diagonal in the table that runs from the upper-left corner to the lower-right corner. Q will be negative when bc is larger than ad. This will occur when the observations tend to fall along the diagonal that runs from the lower-left corner to the upper-right corner. Whether a positive or negative association between the variables is indicated by a positive or negative value of Q will depend on how the variables were arranged in the table. Care should be taken in interpreting the sign of Q so that it is understood whether the substantive association between the variables is positive or negative in nature, not just whether the observations tend to fall along one diagonal in the table.

Unlike phi, Q can always reach a value of 1.0 or -1.0 regardless of the nature of the marginal distributions. A value of 1.0 or -1.0 for Q, though, may not always indicate a "perfect" association. If all the observations in a table fall along one of the diagonals, the two cells on the other diagonal will be empty. This means that either $a = d = 0$ or that $b = c = 0$. In either case, one of the products, ad or bc, will equal zero and the value of Q will be either 1.0 or -1.0. Since all the observations lie along one of the diagonals, an interpretation of perfect association for either value is straightforward. However, Q will also be equal to 1.0 or -1.0 when only *one* cell in the table is empty. The formula for Q contains only two quantities: ad and bc. If any one of the four cells in a table is empty, one of these quantities will be zero (any number times zero equals zero). When this happens, the remaining quantity will form both the numerator and the denominator, and since any number divided by itself equals 1, Q will equal 1.0 or -1.0. Thus, 2×2 tables that have one empty cell will always yield a Q of 1.0 or -1.0. In most situations, a table with one empty cell will not represent a case of perfect association. However, there are some situations in which a perfect relationship can be conceptualized as producing a 2×2 table with one empty cell. But regardless of what conception of a perfect relationship is used, whenever a Q of 1.0 or -1.0 is obtained, the table should be checked to see if the value was generated because of a single empty cell or because all the observations fell along one of the diagonals of the table.

An interpretation of Q has been developed which allows a very specific meaning to be attached to intermediate values of Q. This interpretation is based upon probability and involves looking at the observations in terms of pairs. Two kinds of pairs are defined: consistent pairs and inconsistent pairs. Suppose that the scores of four observations from Table 10–25 on the variables of education and political interest are as follows:

Observation Number	Education	Level of Political Interest
1	High	High
2	Low	Low
3	Low	High
4	High	Low

Observations 1 and 2 form a consistent pair. The reason for this is that the order of the two observations on each variable is consistent: On education, observation 1 ranks higher than observation 2; observation 1 also ranks higher than observation 2 on political interest. Thus, the ordering of the two observations, 1 ranking above 2, is the same on both variables. Now consider observations 3 and 4. These two observations form an inconsistent pair. On education, observation 3 ranks *below* observation 4, but on political interest observation 3 ranks *above* observation 4. Thus, the ordering of these two observations is reversed in going from one variable to the other.

Now let us locate these four observations in their appropriate cells in the table. This is done in Table 10–26. It can be seen that the consistent pair, observations 1 and 2, are located in the a and d cells of the table. All observations that fall into these two cells will form consistent pairs. Multiplying together the frequencies for cells a and d, then, will give the number of all possible consistent pairs. Thus, the product *ad* in the formula for Q is the number of consistent pairs. It measures the extent to which the variables are associated in a positive fashion so as to produce the same ordering of pairs of observations on both variables.

Again in Table 10–26, it can be seen that the observations which form the inconsistent pair, observations 3 and 4, are located in the b and c cells of the table. All observations that fall into these two cells, then, must form inconsistent pairs. Multiplying together the frequencies for cells b and c will give the number of all possible inconsistent pairs. Thus, the product *bc* in the formula for Q is the number of inconsistent pairs in the table. It measures the extent to which the variables are associated in a negative fashion so as to produce reverse orderings of pairs of observations on the variables.

The denominator of Q adds together these two products to give the total number of pairs, consistent and inconsistent, in the table. By divid-

Table 10–26

		Level of Education	
		High	Low
Interest in Politics	High	a Observation 1	b Observation 3
	Low	c Observation 4	d Observation 2

ing the products ad and bc by the denominator of Q, we can find the proportion of consistent pairs and inconsistent pairs in a table. For Table 10–25 the proportion of consistent pairs is

$$\text{proportion of consistent pairs} = \frac{ad}{ad + bc}$$

$$= \frac{(80)(40)}{(80)(40) + (60)(20)}$$

$$= \frac{3200}{4400}$$

$$= .727$$

The proportion of inconsistent pairs is

$$\text{proportion of inconsistent pairs} = \frac{bc}{ad + bc}$$

$$= \frac{(60)(20)}{(80)(40) + (60)(20)}$$

$$= \frac{1200}{4400}$$

$$= .272$$

Subtracting the proportion of inconsistent pairs from the proportion of consistent pairs yields

$$.727 - .272 = .455$$

This is the value of Q that was obtained above. What this indicates is that Q specifies the net difference in the proportions of consistent and inconsistent pairs in a table. The closer Q is to 1.0, the greater the proportion of consistent pairs to inconsistent pairs in the table, and the more positive the association between the variables. If Q is near zero, the proportions of consistent and inconsistent pairs are about equal and there is little or no association between the variables. As Q approaches -1.0, the greater the proportion of inconsistent pairs to consistent pairs and the stronger the negative association is between the variables.

Since Q gives the difference in the proportions of consistent and inconsistent pairs in a table, the value of Q indicates the *net* probability that any two observations selected at random from the table will form a consistent or inconsistent pair. This net probability is the extent to which the likelihood of selecting one type of pair exceeds the likelihood of selecting the other type of pair. Thus, a Q of .455 for Table 10–25 indicates that a pair of observations selected at random from the table has a .455 greater chance of being a consistent pair than of being an inconsistent pair. The greater the chances are of selecting a consistent pair, the stronger the positive association must be. The more equal the chances are (Q near 0) of selecting either a consistent or inconsistent pair, the weaker the association must be. Finally, the greater the chances are of selecting an inconsistent pair, the more negative association there must be. Thus, Q reflects the strength and direction of the association between the variables by indicating the net probability of selecting a consistent or inconsistent pair from the table.

Question 10–5. Suppose that in a sample of people the following frequencies for education and sense of political efficacy were found:

		Level of Education		
		High	Low	
Sense of Political Efficacy	High	84	16	100
	Low	30	20	50
		114	36	150

Compute the value of Yule's Q for these data.

Gamma

Yule's Q is designed to measure association between two ordinal variables that form a 2×2 table. For tables larger than 2×2, a different measure of association is needed. One of the most common measures that can be used on tables of any size involving two ordinal variables is *gamma* (γ). Gamma is an extension of Yule's Q and, like Q, involves the computation of consistent and inconsistent pairs. It is also related to lambda in that it is a proportional-reduction-in-error type of statistic.

Consider the data shown in Table 10–27. Here we wish to know the extent to which social status is associated with political participation. As can be seen, a person can be "high," "medium," or "low" on social status and also be "high," "medium," or "low" on political participation. As in the case of Yule's Q, the definition of a consistent or an inconsistent pair depends upon the ordering of observations on the two variables. An observation that scores "high" on one of the variables ranks above an observation that scores *either* "medium" or "low" on that variable. And an observation that scores "medium" ranks above an observation that scores "low." What we are interested in when we consider a pair of observations is whether the ordering of the two observations on

Table 10–27

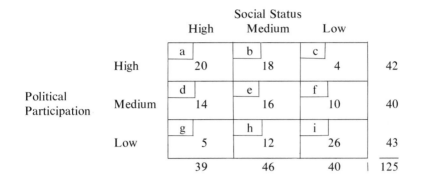

		Social Status			
		High	Medium	Low	
Political Participation	High	a 20	b 18	c 4	42
	Medium	d 14	e 16	f 10	40
	Low	g 5	h 12	i 26	43
		39	46	40	125

both variables is or is not the same. If one observation ranks above another observation on both variables, regardless of what the particular scores are, the observations form a consistent pair. Consider the following observations and their scores:

Observation Number	Social Status	Political Participation
1	High	High
2	Low	Low
3	High	Medium
4	Low	High

For observations 1 and 2, observation 1 (high) ranks above observation 2 (low) on social status and observation 1 (high) ranks above observation 2 (low) on political participation. Thus, the ordering of the two observations is the same on both variables and they form a consistent pair. Observations 2 and 3 also form a consistent pair. Observation 3 (high) ranks above observation 2 (low) on social status and observation 3 (medium) also ranks above observation 2 (low) on political participation. Thus, even though the values of the scores are different on the two variables, the *ordering* of observations 2 and 3 on the variables is the same and therefore they form a consistent pair. An inconsistent pair is formed only when the ordering of two observations reverses itself on the variables. Observations 3 and 4 form an inconsistent pair.

Examination of the four observations will show that it is not possible to determine an ordering of certain pairs on one of the variables.

Observations 1 and 3 have the same score (high) on social status. Observations 2 and 4 are both "low" on social status and observations 1 and 4 are both "high" on political participation. When two observations have the same score on one or both of the variables, they are said to be *tied*. When ties occur, it is not possible to determine an ordering for the observations involved and they cannot be classified as either consistent or inconsistent pairs. Gamma avoids this problem by ignoring all tied pairs, dealing only with untied pairs.

The formula for gamma is

$$\text{gamma} = \gamma = \frac{P_c - P_i}{P_c + P_i}$$

where P_c is the number of consistent pairs in the table and P_i is the number of inconsistent pairs in the table. The computation of gamma, then, requires that we first determine P_c and P_i.

Let us illustrate the process of computing gamma by using the data in Table 10–27. To find the number of consistent pairs, we must determine what pairs of observations will have consistent orderings on both variables. Consider the observations in cell a of Table 10–27. The observations in this cell have a score of "high" on both variables. The observations in cells b and c have a score of "high" on political participation and the observations in cells d and g have a score of "high" on social status. Thus, the observations in cells b, c, d, and g form tied pairs with the observations in cell a. Since ties cannot be classified as forming consistent or inconsistent pairs, the observations in these cells cannot be classified as forming consistent pairs with the observations in cell a. The observations in cells e, f, h, and i all have scores on both variables which rank below the scores of the observations in cell a. Thus, the observations in these cells form consistent pairs with the observations in cell a.

To find the number of consistent pairs that can be formed with the observations in cell a, the number of observations in cells e, f, h, and i are summed and multiplied by the number of observations in cell a. Thus, the number of consistent pairs in the table involving the observations in cell a is

$$a\,(e + f + h + i) = (20)(16 + 10 + 12 + 26)$$

$$= (20)(64)$$

$$= 1280$$

Now let us see if we can derive a general rule for computing consistent pairs. If any cell in Table 10–27 is considered, the observations in the cells which lie in the same row or column as that cell must form tied pairs with the observations in that cell. For cell a, these were cells b, c, d, and g. For cell e, they would be cells d, f, b, and h. Therefore, such cells must be eliminated from the computation of consistent pairs. Observations that lie in cells *below and to the right* of a particular cell will always form consistent pairs with the observations in that cell because their scores on both variables must always be lower than the scores of the observations in the selected cell. Therefore, the number of consistent pairs in the table can be found in the following way. Beginning with the upper-left cell, multiply the number of observations in the cell by the sum of all the observations that lie in the cells below and to the right. We then proceed through the table in a similar fashion, taking each cell and multiplying the number of observations in it by the sum of the observations that lie in the cells below and to the right of it. There will be some cells in the table for which there are no cells which lie below and to the right, and they will not enter into the calculation of consistent pairs. For cell b, cells f and i lie below and to the right, so the number of consistent pairs formed with the observations in cell b is: $b(f + i) = (18)(10 + 26) = (18)(36) = 648$. For cell d, cells h and i lie below and to the right, so the number of consistent pairs formed with the observations in cell d is: $d(h + i) = (14)(12 + 26) = (14)(38) = 532$. Finally, for cell e, cell i is the only one that lies below and to the right, so: $e(i) = (16)(26) = 416$. The total number of consistent pairs in the table is found by summing the number of consistent pairs formed with each cell. The whole process then is

Consistent Pairs

$$a(e + f + h + i) = (20)(16 + 10 + 12 + 26) = (20)(64) = 1280$$

$$b(f + i) \qquad = (18)(10 + 26) \qquad = (18)(36) = \quad 648$$

$$d(h + i) \qquad = (14)(12 + 26) \qquad = (14)(38) = \quad 532$$

$$e(i) \qquad = (16)(26) \qquad = (16)(26) = \quad \underline{416}$$

$$2876$$

A similar process is followed to find the number of inconsistent pairs in the table. If cell c in Table 10–27 is considered, it is seen that the observations in this cell score "low" on social status and "high" on political participation. Examination of the table will show that only observations that lie in cells *below and to the left* of cell c form inconsistent pairs with the observations in cell c. The observations in cell e, for example, score "medium" on social status (which ranks them *above* the observations in cell c on social status) and "medium" on political participation (which ranks them *below* the observations in cell c on political participation). Thus, the ordering of observations in cells c and e reverses itself from one variable to the other, and consequently they form inconsistent pairs. This will be true of observations in all cells below and to the left of cell c. Thus, to find the number of inconsistent pairs in the table, we begin with the upper-right cell and multiply the number of observations in that cell by the sum of all the observations which lie in cells below and to the left. We then proceed through the table, taking each cell and multiplying the number of observations in it by the sum of the observations in the cells lying below and to the left. The total number of inconsistent pairs in Table 10–27 is found, then, as follows:

Inconsistent Pairs

$$c(e + d + h + g) = (4)(16 + 14 + 12 + 5) = (4)(47) \quad = 188$$
$$b(d + g) \qquad\quad = (18)(14 + 5) \qquad\quad = (18)(19) = 342$$
$$f(h + g) \qquad\quad = (10)(12 + 5) \qquad\quad = (10)(17) = 170$$
$$e(g) \qquad\qquad\quad = (16)(5) \qquad\qquad = (16)(5) \quad = \underline{\quad 80}$$
$$780$$

Once the number of consistent and inconsistent pairs is found, we insert these values into the formula for gamma. For Table 10–27,

$$\gamma = \frac{P_c - P_i}{P_c + P_i}$$

$$= \frac{2876 - 780}{2876 + 780}$$

$$= \frac{2096}{3656}$$

$$= .57$$

As a proportional-reduction-in-error statistic, gamma can be interpreted along similar lines as lambda. But the interpretation of gamma does not involve single observations, but *pairs* of observations, and their order relative to each other on each of the two variables. Let us call the variable for which we are to predict the order of observations the dependent variable and the variable information about which we use in making the predictions, the independent variable. What gamma tells us is the proportional reduction in error obtained in guessing the ordering of pairs of observations on the dependent variable *with* information about their ordering on the independent variable compared to the errors made in randomly guessing the ordering of pairs on the dependent variable when we do not have information about their order on the independent variable. The sign of gamma tells us how to guess. If the sign of gamma is positive, we should guess the *same* order for the pair on the dependent variable as they have on the independent variable. By doing so, we will reduce our errors by a percentage given by the value of gamma. If the sign of gamma is negative, we should guess the *reverse* order for the pair on the dependent variable as they have on the independent variable. In doing so we will reduce our errors by a percentage given by the numerical value of gamma.

In Table 10–27, if we were told to guess the order of a selected pair of observations on political participation and given no other information we would have to guess randomly as to which observation in the pair scores higher and which lower. If we were now asked to guess their order on political participation but were first told that observation 1 scored higher on social status than observation 2, we should guess that observation 1 ranks higher on political participation than observation 2 because the gamma for this table is positive in value, indicating that we should guess the same order for a pair on the dependent variable as the pair had on the independent variable. The value of gamma of .57 tells us that we would reduce our errors by 57% when we followed this guessing strategy over the number of errors we would make when we had no information about social status and had to guess randomly. Since we could reduce our errors by more than half, this value of gamma indicates a moderate positive association between social status and political participation.

Values of gamma range from −1.0 to 0 to 1.0. A value of 0 means that consistent pairs and inconsistent pairs are equally likely to be found in the table and that there is no advantage to be gained in guessing the order of pairs one way or the other over random guessing. Consequently, there would be no association in the table. A value of 1.0 for gamma means that there is a perfect relationship between the variables *in the sense* that for untied pairs, if we know that one observation ranks higher on the independent variable than another observation and we guess the *same* ordering of the observations on the dependent variable, we will never make a mistake in our predictions. This means that all untied pairs in the table form only consistent pairs. A value of −1.0 for gamma means that for untied pairs, if we guess the *reverse* order on the dependent variable from the order of the observations on the independent variable, we will never make a mistake in our predictions. This means that all untied pairs in the table form only inconsistent pairs. The stronger the association between two variables, the more one type of pair, consistent or inconsistent, will dominate over the other type, and the more accurately we could predict the ordering of pairs on the dependent variable from a knowledge of their order on the independent variable. Gamma indicates the increasing strength of association by moving closer to 1.0 or −1.0. And the numerical value of gamma indicates the proportional reduction in error in predicting the order of untied pairs on the dependent variable obtained from a knowledge of their order on the independent variable over random guessing.

Gamma is not restricted in value by unmatched marginal distributions and thus can always reach a value of 1.0 or −1.0. But gamma may reach a value of 1.0 or −1.0 in situations that would probably not be considered perfect association. Such situations occur when certain arrangements of empty cells appear in a table. A related problem is that different arrangements of data in a table may yield the same value for gamma. This happens because gamma only considers consistent and inconsistent pairs as has been defined and not the precise differences in rankings. Gamma also only detects an association between variables that runs consistently in the same direction throughout the table. If the association changes direction in the middle of the table, gamma could yield a value of 0 even though there is a very strong U-shaped association between the variables. What these problems indicate is that, in determining the nature and strength of association between two variables, the arrangement of the data in the table must be considered as well as the numerical value of gamma.

A final problem of gamma is that its value is computed only on the untied pairs in a table. If a table contains many ties, gamma ignores them and its value may be misleading. The best solution to this problem seems to be to retain as many categories for each variable as possible so as to reduce the number of ties in the table. If this is not possible and the proportion of ties in a table is great, it is probably best not to use gamma to measure the association.

Question 10−6. Compute the value of gamma for the following tables:

a.

	H	M	L	
H	25	0	25	50
L	0	50	0	50
	25	50	25	100

b.

	H	L	
H	30	10	40
M	20	20	40
L	10	30	40
	60	60	120

EXERCISES FOR CHAPTER 10

Each of the following problems presents a contingency table between two variables. For each table determine which measure of association discussed in this chapter is appropriate and specify the reasons for your selection. Compute the value of the selected measure of association for the table and interpret what this value indicates about the association between the variables. If more than one measure of association is applicable, calculate the values for all measures that are applicable.

1.

		Region of Country				
		North	South	East	West	
Favor Amnesty for Vietnam War Draft Evaders	Yes	30	15	60	40	145
	No	50	35	40	30	155
		80	50	100	70	300

2.

		Sex		
		Male	Female	
Attention to Political News	High	150	100	250
	Low	110	140	250
		260	240	500

3.

		Age			
		Young (18–35)	Medium (36–55)	Old (56 and over)	
Are You in Favor of Capital Punishment?	Yes	40	40	30	110
	No	30	40	20	90
		70	80	50	200

4.

		Year in College				
		Fresh-man	Sopho-more	Junior	Senior	
Scale of Conserva-tiveness of Politi-cal Attitude	Liberal	40	50	50	60	200
	Middle of Road	30	40	40	20	130
	Conserva-tive	80	40	30	20	170
		150	130	120	100	500

5.

		Occupation				
		Profes-sional	White Collar	Farmer	Manual Worker	
Ever Write Letter to Congressman?	Yes	50	60	40	40	190
	No	30	40	20	120	210
		80	100	60	160	400

6.

		Sex		
		Male	Female	
Vote in Local Election?	Yes	50	30	80
	No	40	80	120
		90	110	200

7.

		Level of Education		
		Low	Medium	High	
Size of Political Contribution	None	50	70	10	130
	$1 – $25	40	50	20	110
	Over $25	10	30	20	60
		100	150	50	300

8.

		Age		
		Young	Old	
Level of Political Information	High	50	100	150
	Low	150	100	250
		200	200	400

9.

		Union Membership		
		Yes	No	
Do You Feel Your Interests Are Being Represented in Washington?	Yes	40	50	90
	No	20	40	60
		60	90	150

ANSWERS TO QUESTIONS IN CHAPTER 10

10–1:

		Age		
		Young	Medium	Old
Party Identification	Democrat	33.3%	40.0%	46.7%
	Independent	46.7%	35.0%	13.3%
	Republican	20.0%	25.0%	40.0%
		100%	100%	100%

The pattern of association is that among these respondents, as age increases, the proportion of people who have *either* a Democratic or Republican party identification increases. Thus, the proportion of respondents who give Independent as their party identification decreases as age increases, whereas the proportions of people saying they are Democrats or Republicans both increase as age increases.

10–2: a. $\phi = \sqrt{\dfrac{\chi^2}{N}} = \sqrt{\dfrac{4.86}{200}} = \sqrt{.0244} = .156.$

  b. $\phi = \dfrac{ad - bc}{\sqrt{(a+c)(b+d)(a+b)(c+d)}} = \dfrac{(100)(150) - (50)(200)}{\sqrt{(300)(200)(150)(350)}}$

  $= \dfrac{15,000 - 10,000}{\sqrt{3,150,000,000}} = \dfrac{5000}{56,124.86} = .09.$

c. $\phi = \dfrac{(200)(300) - (100)(400)}{\sqrt{(600)(400)(300)(700)}} = \dfrac{60{,}000 - 40{,}000}{\sqrt{50{,}400{,}000{,}000}}$

$= \dfrac{20{,}000}{224{,}499.44} = .09.$

10–3: a. $C = \sqrt{\dfrac{\chi^2}{\chi^2 + N}} = \sqrt{\dfrac{21.3}{21.3 + 100}} = \sqrt{.175} = .42.$

b. $\sqrt{\dfrac{K-1}{K}} = \sqrt{\dfrac{3-1}{3}} = \sqrt{.67} = .82,$

$C_{adj} = \dfrac{.42}{.82} = .51.$

10–4: a. $\lambda_A = \dfrac{(20 + 15 + 15) - 50}{100 - 50}$

$= \dfrac{50 - 50}{50}$

$= 0.$

b. $\lambda_B = \dfrac{(15 + 20 + 15) - 40}{100 - 40}$

$= \dfrac{50 - 40}{60}$

$= \dfrac{10}{60}$

$= .17.$

10–5: $Q = \dfrac{(84)(20) - (30)(16)}{(84)(20) + (30)(16)}$

$= \dfrac{1680 - 480}{1680 + 480}$

$= \dfrac{1200}{2160}$

$= +.56.$

10–6: a.

Consistent	Inconsistent
$(25)(50 + 0) = 1250$	$(25)(50) = 1250$
$(0)(0) = 0$	$(0)(0) = 0$

$\gamma = \dfrac{1250 - 1250}{1250 + 1250}$

$$= \frac{0}{2500}$$

$$= 0.$$

b.   Consistent   Inconsistent

$$(30)(20 + 30) = 1500 \qquad (10)(30) = 300$$

$$(20)(30) = \frac{600}{2100} \qquad (20)(10) = \frac{200}{500}$$

$$\gamma = \frac{2100 - 500}{2100 + 500}$$

$$= \frac{1600}{2600}$$

$$= +.615.$$

11
Correlation and Regression

In this chapter we consider the Pearson product-moment correlation coefficient. This coefficient is designed to measure association among variables at the interval or ratio level of measurement. It is named after an English statistician, Karl Pearson, and is one of the most widely used measures of association in social science. In fact, when social scientists use the term "correlation" in an unqualified way, they are usually referring to the Pearson product-moment correlation, even though the term "correlation" refers in a general way to association or dependence among variables.

In this chapter we shall also consider a technique known as *regression analysis*. It will be recalled from the discussion of association in Chapter 9 that when two variables are related, there are two things we want to know about the relationship between them. One is the strength or degree of association between the variables. For variables at the interval or ratio level, the Pearson product-moment correlation coefficient provides a means to obtain this information. The second thing we wish to know about the relationship between two variables is the nature of the association. For interval or ratio level variables, regression analysis provides a way to specify a mathematical formula which describes the nature of the association between two variables.

Suppose that at college X the numbers 4, 3, 2, 1, and 0 are used to indicate the grades A, B, C, D, and F. At college Y, however, these grades are indicated by the numbers 10, 8, 6, 4, and 2. Suppose now that we wish to convert the grade-point averages of students at college X into the grade-point averages used at college Y. This can be done by using the formula

$$Y = 2 + 2X$$

where X is the grade-point average at college X and Y will be the equivalent grade-point average in the system used by college Y. If a student had a B average at college X (a grade-point average at college X of 3.0), his corresponding grade-point average in the system used at college Y could be found as follows:

$$Y = 2 + 2X$$
$$= 2 + (2)(3.0)$$
$$= 2 + 6$$
$$= 8$$

From this formula, the corresponding grade-point average at college Y could be found for the grade-point average of any student at college X. The formula precisely specifies the relationship between the two grade-point systems. Besides describing the relationship between the two systems, the formula also gives a rule for determining scores on one variable *(Y)* from a knowlege of scores on the other variable *(X)*. Since the X variable is usually called the independent variable and the Y variable the dependent variable, the formula gives a way to find or predict scores on the dependent variable from a knowledge of scores on the independent variable.

In mathematics such rules are called *functions*. When the terms involving the variable X in a function are raised only to the first power (X^1), it is called a *linear function*. That is, a linear function has no terms that involve X^2 or X^3 or higher powers of X. When only two variables are involved in a linear function *(X and Y)*, the function can be represented graphically by a straight line.

The formula for converting grade-point averages is a linear function. In that situation it was fairly easy to determine the (linear) function which converts grade-point averages from one system to the other. But in dealing with more realistic problems of association, the determination of a function that describes the nature of the association is not so obvious. The technique of regression provides a means for doing this.

A major premise in regression of the type which will concern us here is that the association between two variables can be properly described by a straight line. This is why the technique is called *linear regression*. Of course, there is no reason why an association has to be linear in nature. It is possible to have nonlinear or curvilinear associations. The association between age and strength mentioned in Chapter 10, where strength increases up to the late thirties or early forties and then declines, is an example of a nonlinear relationship between two variables. If the technique of linear regression were applied to such situations, it would produce misleading results because the technique describes the extent of linear association and ignores nonlinear association. Despite this possibility, there are many variables which are linearly related (or which closely approximate a linear relationship), and thus the technique of linear regression is very useful in social science.

Regression analysis has a special importance in social science because, by specifying the nature of the relationship between variables, it strongly contributes to the development of theory. Commonsensically, we know that there is a relationship between the distance a falling object covers and the amount of time it has been falling. But a much more powerful and useful form of this knowledge is obtained when the knowledge is specified by the function $s = \frac{1}{2}gt^2$, which tells us how far in feet (s) an object will fall as a function of time (t) and the gravitational constant (g). Of course, it must be admitted that at present there are precious few situations in which social science knowledge can be specified by so precise a function. Regression analysis, however, will play an important part in the discovery and formulation of such functions.

As will be seen later, there is a close relationship between linear regression and the Pearson product-moment correlation coefficient. In fact, an adequate understanding of the correlation coefficient is dependent upon an understanding of linear regression. Consequently, we shall begin by discussing regression and then return later to the topic of correlation.

REGRESSION

As was just seen, *regression* involves the idea of finding or predicting scores on one variable from a knowledge of scores on another variable. Consequently, it is necessary to distinguish between the variable about which predictions are to be made and the variable from which we make the predictions. This means that the investigator must designate one variable as the independent or "causal" variable and the other as the dependent or "effect" variable. Ideally this is done on the basis of information outside the data that are of immediate consideration, such as previous work, or on the basis of a theory that is being developed or tested. For example, a researcher might be working with a theory of economic determinism which assumes that economic structures in a society determine the political structures. A hypothesis derivable from this theory would be that economic status yields political power. In using regression analysis to investigate this proposition, economic status would be the independent variable and political power the dependent variable. Researchers often find themselves in situations wherein they want to use regression analysis but do not have a theory that can serve as a basis for determining which variable is independent and which is dependent. The only general rule that is available for guidance in such situations is that the dependent variable cannot temporally occur prior to the occurrence of the independent variable. Other than this, the investigator must rely on his common sense and on a substantive knowledge of the area in which he is working to properly specify which variable is to be treated as independent and which as dependent. In the discussion that follows, we use the convention of representing the indepedent variable by X and the dependent variable by Y.

As we said earlier, the type of regression that will be considered here assumes that the association between two variables can properly be represented by a straight line. From Chapter 3 it will be recalled that the general formula for a straight line is $Y = a + bX$. The purpose of the technique of linear regression is to extract information from a set of data to determine the values of a and b so as to specify that particular straight line which best describes the linear association in the data.

To illustrate the meaning of these terms in the equation for a straight line and to show how such an equation is used to describe the association between two variables, let us consider the following problem. Suppose that we are interested in formulating a general rule by which to predict the gas mileage of a particular automobile at various speeds. Data on the problem are gathered by determining the gas mileage (miles per gallon) of the car at four different speeds (miles per hour), as follows:

Observation Number	Speed (mph)	Gas Mileage (mpg)
1	30	13.5
2	50	10.5
3	60	9.0
4	80	6.0

These data can be graphed to obtain a visual presentation such as that in Figure 11–1. In graphing such information, it is customary to use the horizontal axis to represent the independent variable and the vertical axis to represent the dependent variable. In this problem the horizontal axis represents speed and the vertical axis gas mileage. Once the axes are marked off in units of their respective variables, the observations are plotted on the graph. To plot the first observation, for example, the point 30 units is found on the horizontal axis and the point 13.5 units is found on the vertical axis. The intersection of lines drawn through these points perpendicular to each axis marks the location of this observation on the graph. Repeating the process for all the observations results in the graph

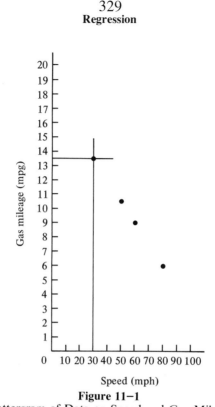

Figure 11–1
Scattergram of Data on Speed and Gas Mileage

presented in Figure 11–1. Such a graph is called a *scatter diagram* or *scattergram*. The construction of a scattergram is often the first step in regression analysis because it allows a visual inspection of the data to be made. From it a researcher can see whether or not the data form a configuration which at least roughly corresponds to a straight line. If the configuration is U- or S-shaped or some other form which is obviously not linear, then linear regression is not applicable and the researcher should not proceed with the technique.

Having plotted the points and seen that they form at least a rough linear configuration, we can turn to the problem of obtaining a formula for a straight line that best describes the association between the variables. The "best" line is that line which will yield *predicted values* for the Y variable which are as accurate as possible. In other words, the best line is that line which will give predicted values of Y which are as close as possible to the actual Y values of the observations. Finding such a line for the data in Figure 11–1 is quite easy because all the points fit exactly onto a single straight line, as shown in Figure 11–2. This line is the best line that describes the association between speed and gas mile-

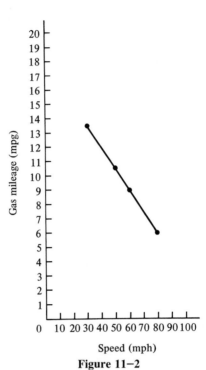

Figure 11–2
Straight Line Connecting Data Points

age for our car because it will yield predicted Y values that are perfect-ly accurate; the predicted Y values obtained from the line will exactly match the actual Y values of the observation. This can be seen through the following process. Locate any X value (speed) on the horizontal axis, say, 50 mph. Through this point draw a line perpendicular to the X axis upward until it meets the line drawn through the data points. From this point draw a line perpendicular to the Y axis. The point at which the last line intercepts the Y axis is the predicted value of Y [represented as Y' (called "Y prime")] that the line gives for an X value of 50. The entire process is illustrated in Figure 11–3, where a value of 10.5 is obtained for Y'. It can be seen from observation 2 in the data given above that when the car was driven at 50 mph, the actual, observed value for gas mileage was 10.5 (Y). Thus, the actual or obtained value (Y) exactly matches the predicted value given by the line (Y'). This is because the data point sits precisely on the line. Had not the line exactly passed through the point for observation 2, there would have been some dis-crepancy between what the line predicts as the Y value when the X val-ue is 50 and the actual Y value, the gas mileage the car was observed to get at 50 mph. Since all the data points sit precisely on the line, all pre-

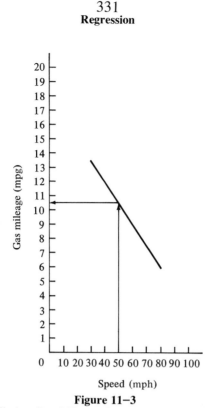

Figure 11–3
Predicting Gas Mileage for a Speed of 50 mph.

dicted values for Y obtained from the line will precisely match the actual Y values.

If on the basis of our data (four observations are not a lot of data), we were to accept the line as a good description of the nature of the association between speed and gas mileage, we can use the line to predict the gas mileage of the car at any given speed. Suppose that we want to determine the gas mileage the car would get at 40 mph. Following the same process of drawing lines on the graph as before, we obtain a pre-dicted value for gas mileage of 12 mpg. This predictive process is il-lustrated in Figure 11–4.

Question 11–1. Using the prediction line in Figure 11–2, determine the predicted gas mileage for the car at the following speeds:

a. 45 mph.　　b. 75 mph.
c. 68 mph.　　d. 57 mph.　　　　——

Let us now determine a formula for the prediction line in Figure 11–2. The general formula for a straight line is $Y = a + bX$. The formula

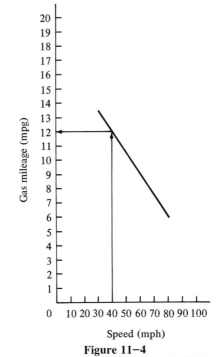

Figure 11−4
Predicting Gas Mileage for a Speed of 40 mph.

for a particular straight line, such as that in Figure 11–2, is determined by specifying the values of the constants a and b. This can be done directly from the prediction line as illustrated in Figure 11–5. The quantity a in the formula for a line is the Y intercept. It is the point at which the line intercepts the Y axis. By extending the prediction line it can be seen in Figure 11–5 that the line intercepts the Y axis at point 18. Thus, the value of a in the formula for the prediction line will be 18.

In Chapter 3 it was shown that b in the equation for a straight line is its slope. The slope specifies the angle of the line on the graph. From Figure 11–5, the slope of the line can be found as follows:

$$b = \frac{\Delta y}{\Delta x} = \frac{9 - 12}{60 - 40} = \frac{-3}{20} = -.15$$

The equation for the prediction line, then, is

$$Y = 18 - .15X$$

The values of a and b in the formula for the prediction line usually have important substantive meaning. The value of a gives the predicted Y value when X equals zero. But because a is the value of the Y inter-

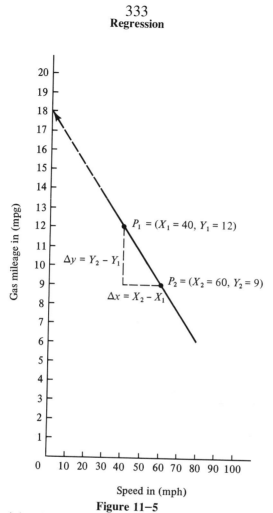

Figure 11–5
Graphically Obtaining the Slope and the Y Intercept for the Line Passing through the Data Points

cept for the prediction line, in some situations it is determined by extending the prediction line beyond values which the observations can realistically assume. Thus, in the present problem, the a value says that if the car were not moving it would get 18 mpg! Such a statement, of course, is nonsensical. The reason is that it is impossible to observe gas mileage in terms of miles per gallon unless the car moves. We might, perhaps, interpret the a value as follows: that if the car could move when the engine is at idle speed, the mileage would be 18 mpg. In most problems a will be substantively meaningful, although it is not uncommon to run into situations that present difficulties similar to the present one in the interpretation of a. The important point is that a gives the predicted Y value when the X variable equals zero.

The value of b or the slope tells us how much change in the Y variable will be caused by a 1-unit increase in value of the X variable. The b value of $-.15$ for the present problem, then, tells us that for each increase in speed of 1 mph, the gas mileage of the car will decrease by .15 mpg. The value of the slope is very important because it precisely specifies how a change in the X variable will affect the Y variable.

Having determined the values of a and b for the prediction line, we can now use the formula for the line to generate predicted values of Y. Since predicted Y values are represented as Y', the formula used to generate predicted Y values is written

$$Y' = a + bX$$

For a speed of 40 mph, the predicted gas mileage will be

$$Y' = 18 + (-.15)(40)$$
$$= 18 - 6.0$$
$$= 12$$

Thus, the predicted value of Y at 40 mph is 12 mpg, which is the same value derived from the graphical method of prediction used earlier.

Question 11-2. Using the prediction equation $Y' = a + bX$, where $a = 18$ and $b = -.15$, calculate the predicted Y values for the following values of X. Do the predicted Y values match those obtained in Question 11-1?

a. 45 mph.   b. 75 mph.
c. 68 mph.   d. 57 mph.   _____

LEAST SQUARES

In the gas-mileage example, when the observations were plotted on a graph, all the points sat exactly on one straight line. This made it very easy to determine the best prediction line, for all that we had to do was draw a straight line connecting the points. When dealing with real problems, however, it will rarely, if ever, happen that the data points will all fit precisely on a single straight line. Rather, there is likely to be some scattering among the points, owing to the effect of measurement error and/or other variables. Thus, it will not usually be possible to draw a single straight line that will connect all the points. Table 11-1, for example, lists scores for education and income for a sample of people. When these observations are plotted, the scattergram of Figure 11-6 is obtained. When such scattergrams are encountered, it is not as obvious that the variables are linearly related, as was the case in Figure 11-1. In

Table 11−1
Scores on Income and Education
for 15 People

Observation Number	Years of Education (X)	Income in Thousands of dollars (Y)
1	7	9
2	12	16
3	8	14
4	12	12
5	14	11
6	9	16
7	18	19
8	14	13
9	8	13
10	12	14
11	17	14
12	10	16
13	16	15
14	10	10
15	13	18

such cases we look to see if there is a *linear trend* in the data as opposed to nonlinear trends. Since in Figure 11–6, the trend of Y values is steadily upward for increases in the X values, we can conclude that there is a linear trend among the points. Figure 11–7 presents some scattergrams where there are nonlinear trends among the points.

Returning to Figure 11–6, the problem is to find a straight line that best describes the linear trend among the points. One way to do this might be to take a ruler and draw a line through the points which best seems to match the trend. But with many data points this method is subject to error. What is needed is a mathematical method that will tell us exactly where to locate the line. Earlier it was said that the best prediction line would be the one that yielded predicted Y values that were as close as possible to the actual Y values. This criterion can be developed to give a mathematical method for determining the best prediction line. The actual Y values of the points appear in Table 11–6. When a line is selected to fit the points, a predicted Y value can be calculated from the line for the X value of each point. These X values are also given in Table 11–1. Thus for each observation we would have its actual Y value (Y) and its predicted Y value (Y'). The best line will be the one that makes

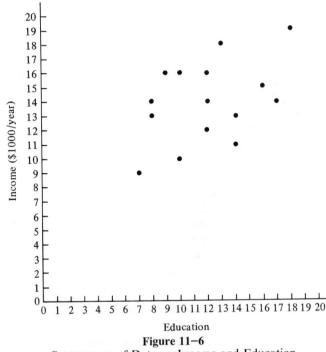

Education

Figure 11–6

Scattergram of Data on Income and Education

the differences between Y and Y' as small as possible. Thus, it would seem that the criterion for the line of best fit is the one that makes the sum of all the differences between Y and Y' [$\Sigma (Y - Y')$] as small as possible. Unfortunately, we cannot work with this criterion as it stands. This is because for some points Y will be larger than Y', making the difference positive; for other points, Y' will be greater than Y, making the difference negative. Consequently, just as for the sum of the distances of scores about their mean, the quantity $\Sigma (Y - Y')$ would equal zero for the line of best fit. To avoid this problem, the differences between Y and Y' are squared to get rid of the minus signs. Thus, the criterion for the line of best fit is that the sum of the squared differences between Y and Y' for each observation be as small as possible, or:

$$\Sigma (Y - Y')^2 = \text{a minimum}$$

This criterion, that the sum of the squared differences be as small as possible, gives the name *least-squares regression* to the technique of finding the best prediction line.

In this technique the prediction line that best fits the data according

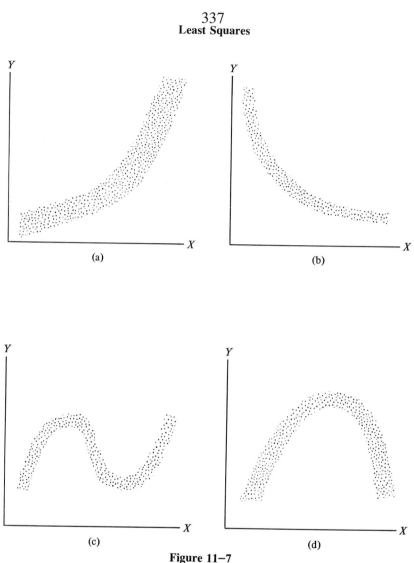

Figure 11–7
Examples of Nonlinear Trends

to the least-squares criterion is called the *regression line*. The least-squares procedure yields a unique regression line for a particular set of points. Thus, the method provides a means to determine a line of best fit that is not open to errors of judgment in finding its location.

We now want to use the principle of least squares to develop a way to compute the values of a and b so that we can find the equation for the straight line which best fits the data points (the regression line). A complete derivation of the formulas which are used to calculate the values of

a and b requires the use of calculus, which is beyond the scope of this text. Consequently, the derivations are not done here and the final formulas are just presented without proof.

The least-squares principle yields the following formula for the slope of the regression line:

$$b_{yx} = \frac{\Sigma(X - \overline{X})(Y - \overline{Y})}{\Sigma(X - \overline{X})^2}$$

The symbol $\overline{Y}$ indicates the mean of the Y scores, just as $\overline{X}$ indicates the mean of the X scores. And since x is the deviation of an X score from the X mean $(x = X - \overline{X})$ and y is the deviation of a Y score from the Y mean $(y = Y - \overline{Y})$, the formula for the slope can also be written

$$b_{yx} = \frac{\Sigma xy}{\Sigma x^2}$$

The symbol b_{yx} is called the *regression coefficient*. The placement of the subscripts in the regression coefficient serves to indicate which variable is being treated as the dependent and which the independent, with the symbol for the dependent variable being listed first. Since y is in the first position here, this means that this formula is for the slope of the line for the regression of the Y variable (the dependent one) on the X variable (the independent one). There is a special verbal convention for reading these subscripts which goes like this: b_{yx} refers to the regression of Y *on* X.

The least-squares principle yields the following formula for a:

$$a_{yx} = \overline{Y} - b_{yx}\overline{X}$$

This value is the Y intercept and tells us where the regression line of Y on X crosses the Y axis. The subscripts here again indicate that Y is being treated as the dependent variable and X as the independent variable.

Let us now take a closer look at the formulas for b_{yx} and a_{yx}. We will start with the formula for the regression coefficient. The numerator of b_{yx} consists of the sum of the products of the X and Y deviations for each observation. If for a particular observation its X score is far from the X mean and its Y score is far from the Y mean, the product of the deviations will be large. If, however, for a particular observation the X score is far from the X mean but the Y score is *close* to the Y mean, the product of the deviations will be small. Thus if in a particular set of data the observations which have an X score that is far from the X mean also have a Y score that is far from the Y mean, the products of the deviations for these observations will be large and will also make the sum of all the products large in value. If, however, observations that have an X

score far from the X mean have a Y score that is *close* to the Y mean (or vice versa), the products of the deviations will be relatively small, and as a result the sum of all the products will be small in value. The sum of the products of the deviations, then, is a measure of the extent to which observations tend to have X and Y scores that are similar in their distances from their means or dissimilar in their distances from their means. Another way of saying this is that the sum of the products measures the extent to which observations that have extreme X scores also have extreme Y scores, where extreme means far from the mean. The numerator of b_{yx}, then, indicates the extent to which X and Y scores *vary together.* Hence the quantity in the numerator of b_{yx} is called the *covariance.*

When it happens that the X scores which are above $\overline{X}$ are usually accompanied by Y scores which are above $\overline{Y}$ and that the X scores which are below $\overline{X}$ are usually accompanied by Y scores which are below $\overline{Y}$, the covariance will be positive in value because positive X deviations are multiplied by positive Y deviations and negative X deviations are multiplied by negative Y deviations. In this type of situation, the variables are related in such a way that increases in the value of X are accompanied by increases in the value of Y. This type of situation is illustrated in Figure 11–8a. In Figure 11–8b the opposite type of situation is presented. Here, X scores below $\overline{X}$ are usually accompanied by Y scores *above* $\overline{Y}$ and X scores above $\overline{X}$ are usually accompanied by Y scores *below* $\overline{Y}$. This means that the products of the deviations will be predominantly negative, making the value of the covariance negative. This, in turn, will make the sign of the regression coefficient negative. In this type of situation, increases in the value of X are accompanied by decreases in the value of Y. The covariance, then, measures the extent to which X and Y vary together and the sign of the covariance indicates whether X and Y vary together in a positive or negative way.

Finally, let us look at two other types of situations. In Figure 11–8c the X scores above $\overline{X}$ are accompanied by Y scores that are about equally likely to be above or below $\overline{Y}$. The same is also true for X scores below $\overline{X}$. Thus, when the deviations of the X scores and the Y scores are multiplied and added, the result will be about zero, because the products of deviations will be about equally positive and negative and thus cancel each other out. In this type of situation, the regression coefficient will be zero, indicating that changes in the independent variable are not accompanied by any consistent type of change in the dependent variable.

In Figure 11–8d it can be seen that each Y score is equal in value to $\overline{Y}$. Consequently, there is no dispersion or variation present in the Y scores. Under this condition the covariance between X and Y will also be zero because the Y scores all equal $\overline{Y}$, and therefore the deviation of

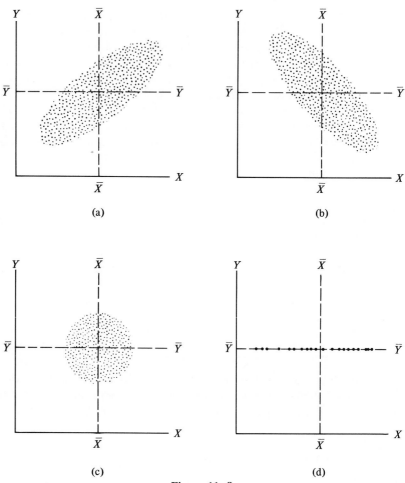

Figure 11–8

Examples of (a) Positive Covariation; (b) Negative Covariation; (c) Random Covariation; (d) No Covariation among Data Points

each Y score from $\overline{Y}$ will be zero. Here, then, the regression coefficient is zero because changes in the independent variable are accompanied by *no* changes at all in the dependent variable.

Let us now take a closer look at the formula for a_{yx}; it is: $a_{yx} = \overline{Y} - b_{yx}\overline{X}$. This formula may be manipulated to obtain: $\overline{Y} = a_{yx} + b_{yx}\overline{X}$. This form says that when X is equal to $\overline{X}$, Y will be equal to $\overline{Y}$. What this means is that the statistic a_{yx} sets the height of the regression line on the

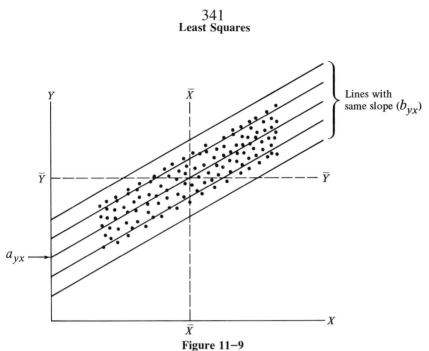

Figure 11-9
Determination of the Value of the Y Intercept so that the Regression Line Passes
through the "Middle" of the Data Points

graph so that it passes through the point on the graph whose coordinates
are ($X = \bar{X}$, $Y = \bar{Y}$). This is illustrated in Figure 11-9. Since the middle
of a scatter of points may be defined as the point at which lines drawn
through the values of $\bar{X}$ and $\bar{Y}$ intersect, it would seem reasonable to expect that the regression line should pass through this point. The value of
a_{yx}, then, selects to be the regression line that line which runs precisely
through the middle of the scatter of points out of all lines that have a
slope equal to b_{yx}.

Having presented and interpreted the formulas for a_{yx} and b_{yx}, the
computation of these statistics will be illustrated on the data for education and income in Table 11-1. As for many other statistics, a computational formula has been developed for b_{yx} which is more convenient to
use than the definitional formula presented above. This formula is algebraically equivalent to the definitional formula. It is

$$b_{yx} = \frac{N \Sigma XY - (\Sigma X)(\Sigma Y)}{N \Sigma X^2 - (\Sigma X)^2}$$

Once b_{yx} is computed, a_{yx} can easily be found from the formula given
earlier.

Table 11–2

Obser-vation Number	Educa-tion (X)	X^2	In-come (Y)	Y^2	XY
1	7	49	9	81	63
2	12	144	16	256	192
3	8	64	14	196	112
4	12	144	12	144	144
5	14	196	11	121	154
6	9	81	16	256	144
7	18	324	19	361	342
8	14	196	13	169	182
9	8	64	13	169	104
10	12	144	14	196	168
11	17	289	14	196	238
12	10	100	16	256	160
13	16	256	15	225	240
14	10	100	10	100	100
15	13	169	18	324	234
	$\Sigma X = 180$		$\Sigma Y = 210$		$\Sigma XY = 2577$

$$\Sigma X^2 = 2320 \qquad \Sigma Y^2 = 3050$$

$$\bar{X} = \frac{180}{15} = 12 \qquad \bar{Y} = \frac{210}{15} = 14$$

$$s_X = \sqrt{\frac{2320 - [(180)^2/15]}{15}} \qquad s_Y = \sqrt{\frac{3050 - [(210)^2/15]}{15}}$$

$$= 3.27 \qquad\qquad\qquad = 2.71$$

From the computational formula for b_{yx}, it can be seen that it is necessary to first calculate the quantities ΣX, ΣY, ΣX^2, and ΣXY. The calculation of these quantities is set out in Table 11–2. (The value of ΣY^2 will be of use later so it is also calculated in the table.)

Inserting the appropriate values from Table 11–2 into the computational formula for b_{yx} yields

$$b_{yx} = \frac{N\Sigma XY - (\Sigma X)(\Sigma Y)}{N\Sigma X^2 - (\Sigma X)^2}$$

$$= \frac{(15)(2577) - (180)(210)}{(15)(2320) - (180)^2}$$

$$= \frac{38{,}655 - 37{,}800}{34{,}800 - 32{,}400}$$

$$= \frac{855}{2400}$$

$$= .35625 \quad \text{or about} \quad .356$$

The data for income are given in terms of thousands of dollars. Therefore, the value for the regression coefficient of .356 means that an increase of 1 year of education produces (or is associated with) an increase in income of .356 of $1000, or $356. Thus, for these data, an increase of 1 year in education increases income by $356. The value of a_{yx} is now easily calculated as follows:

$$a_{yx} = \overline{Y} - b_{yx}\overline{X}$$
$$= 14 - (.356)(12)$$
$$= 14 - 4.275$$
$$= 9.725$$

Thus, the point at which the regression line would intercept the Y axis is 9.725 thousands of dollars.

The term a_{yx} gives the value of the Y variable (income) when the X variable (education) equals zero. In the present problem, it says that a person with no education earns $9725 per year. But it would probably be misleading to accept such a statement. The data contain no observations which have a level of education that is even close to zero (the lowest score on education is 7). Thus, the data contain no information about income levels for people with very low levels of education. The value of a_{yx} is determined by extending the regression line for data about moderate and highly educated people until it hits the Y axis. And the value of a_{yx} is chosen so that the regression line passes through the middle of these data. When the data do not contain observations with values near zero on the X variable, a_{yx} must be interpreted not as a realistic statement about what the Y score is of an observation with a zero X score, but as the value of the Y intercept necessary to make the regression line pass through the middle of the observed data points.

Having determined the values of b_{yx} and a_{yx}, the formula for the line which best describes the linear trend in the data for education and income (the regression line) can be specified as follows:

$$Y = 9.725 + .356X$$

This line is plotted on the data points in Figure 11-10.

The equation for the regression line can now be used to derive pre-

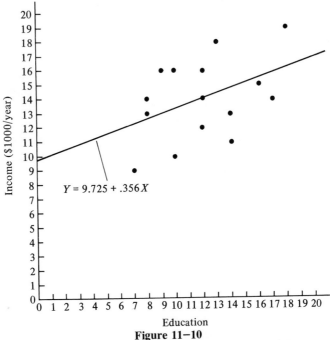

Figure 11–10
Regression Line for Data Points on Education and Income

dicted Y scores (Y') for various values of X. In calculating predicted Y scores, the equation is written as

$$Y' = 9.725 + .356X$$

Suppose, for example, that we wish to determine what the regression line predicts as the income for a person with 16 years of education. The calculation is

$$Y' = 9.725 + .356X$$

$$= 9.725 + (.356)(16)$$

$$= 9.725 + 5.700$$

$$= 15.425$$

Since this is in terms of thousands of dollars, the regression line predicts that a person with 16 years of education will have an income of $15,425.

Question 11–3. Using the equation for the regression line found above for the data on education and income, find predicted values of Y for the following values of X:

a. 8 years of education. b. 10 years of education.
c. 14 years of education. d. 18 years of education.

THE ACCURACY OF PREDICTING
Y VALUES FROM THE REGRESSION LINE

The equation for the regression line describes the linear trend of the as-sociation between X and Y. But how strong the linear trend is in the data is *not* indicated by the regression equation. Figure 11–11a and b present scattergrams of two sets of data with the same linear trend. Consequent-ly, the two sets of data have the same regression line. But it can be seen that in Figure 11–11b, the data points are tightly clustered about the regression line, while in Figure 11–11a the points are widely dispersed about the regression line. Obviously there is an important difference in these two situations, a difference we would like to measure.

The more tightly the data points cluster about the regression line, the better the regression line is as a description of the configuration among the data points. One way to measure how well the regression line functions as a description of the configuration among the data points is to ask how accurately the regression line (or equation) can predict the Y scores of observations from their X scores. The height of the regression line above a particular X score is the value that the regression line pre-dicts for the corresponding Y score. This same value, of course, can be calculated from the regression equation. The distance of a point from the regression line is the amount of discrepancy or error between what the

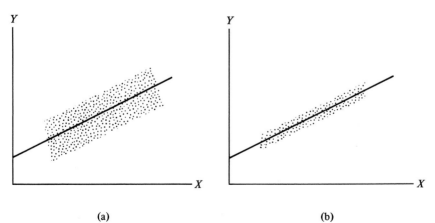

(a)   (b)

Figure 11–11
Sets of Data with Different Dispersions about the Same Regression Line

regression line predicts as the Y score of an observation and its actual Y value. The farther a point lies from the regression line, the less accurate the predicted Y score for the observation will be. Thus, we can use the discrepancies between what the regression line (or equation) predicts as the Y scores for observations (Y') and their actual Y scores (Y) to measure how closely the data points cluster about the regression line. This then will provide us with a measure of how well the regression line functions as a description of the configuration among the data points.

Let us consider observation 1 in the data for education and income. Its X score (education) is 7. What the regression line predicts as its Y score can be calculated from the regression equation:

$$Y' = 9.725 + .356X$$
$$= 9.725 + (.356)(7)$$
$$= 9.725 + 2.494$$
$$= 12.219$$

But from Table 11–1, it can be seen that the actual Y score of this observation is 9. Thus, the predicted Y score (Y') deviates from the actual Y score (Y) by 3.19 units $(Y - Y')$. Repeating this procedure for each observation we can see how well the regression equation predicts the Y

Table 11–3

Obser- vation Number	Educa- tion (X)	In- come (Y)	Predicted Y Value (Y')	Y − Y'	(Y − Y')²
1	7	9	12.219	−3.219	10.362
2	12	16	14.000	2.000	4.000
3	8	14	12.575	1.425	2.031
4	12	12	14.000	−2.000	4.000
5	14	11	14:713	−3.713	13.786
6	9	16	12.931	3.069	9.412
7	18	19	16.137	2.863	8.197
8	14	13	14.713	−1.713	2.934
9	8	13	12.575	.425	.181
10	12	14	14.000	.000	.000
11	17	14	15.781	−1.781	3.172
12	10	16	13.288	2.712	7.355
13	16	15	15.425	−.425	.181
14	10	10	13.288	−3.288	10.811
15	13	18	14.356	3.644	13.279
					$\Sigma(Y - Y')^2 = 89.701$

score for each observation. And by subtracting Y' from Y we can obtain the amount of discrepancy between the predicted Y score and the actual Y score. This is done in Table 11–3.

At this point we have measured the discrepancy between the predicted Y score and the actual Y score for each observation. What we now need is some sort of measure that will summarize the amount of discrepancy between predicted Y scores and actual Y scores among all the observations. The first thing that might come to mind is to add up the discrepancies for each observation and divide the sum by N (the number of observations) to obtain the average amount of discrepancy between the predicted Y scores and the actual Y scores. In Table 11–3 the column labeled "$Y - Y'$" contains the discrepancies between the actual Y score and the predicted Y score for each observation. When the values in this column are summed, though, the result is zero. That is, $\Sigma (Y - Y') = 0$. This will always be true, because the distance of points from the regression line will be equally distributed above and below the regression line. Thus, the regression line will underpredict Y scores for points that lie above it, overpredict for points that lie below it, and correctly predict the Y scores only for points that lie precisely on the line. And the total amounts of underprediction and overprediction will be equal, so when the discrepancies are summed, the result is zero.

This problem is avoided in the usual way by squaring the value of $Y - Y'$ for each observation before summing. This is done in the last column of Table 11–3. In the present problem the sum of the squared differences between the actual Y score and the predicted Y score for each observation $[\Sigma (Y - Y')^2]$ is 89.701. Now if this quantity is divided by N and the square root taken, a measure is formed called the *standard error of estimate*. It is symbolized as $s_{y \cdot x}$. Thus,

$$s_{y \cdot x} = \sqrt{\frac{\Sigma (Y - Y')^2}{N}}$$

This formula is similar to the formula for the ordinary standard deviation and is interpreted along similar lines. The symbol $s_{y \cdot x}$ is to be read as the standard error in predicting Y scores from the X scores. The standard error of estimate is a measure of how tightly points cluster about the regression line. Consequently, it is a measure of the amount of error that would be made in predicting the Y scores of the observations from the regression equation. If the distribution of Y scores is normal in shape, the standard error of estimate tells us that 68.26% of the points lie within a distance equal to $1s_{y \cdot x}$ of the regression line and, consequently, that 31.74% of the points lie farther than this distance from the regression line. For the present problem the value of the standard error of estimate is

$$s_{y \cdot x} = \sqrt{\frac{(Y - Y')^2}{N}}$$

$$= \sqrt{\frac{89.701}{15}}$$

$$= \sqrt{5.980}$$

$$= 2.445$$

What this value tells us is that about 68.26% of the points lie within two parallel lines drawn at a distance of 2.445 Y units above and below the regression line. This is illustrated in Figure 11–12. In data where the points lie close to the regression line, the predicted Y scores will closely approximate the actual Y scores, and the value of $s_{y \cdot x}$ will be small. In data where the points tend to lie far from the regression line, the predicted Y scores will deviate a good deal from the actual Y scores and the value of $s_{y \cdot x}$ will be large.

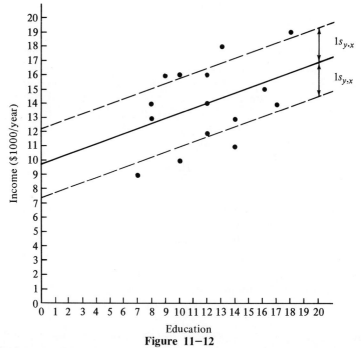

Figure 11–12
Standard Error of Estimate as a Measure of the Dispersion of Points about the Regression Line.

What we have seen so far, then, is that the regression equation specifies the nature of the linear trend between two variables in the sense of specifying that line which is best able to predict values on the dependent variable from a knowledge of scores on the independent variable. The standard error of estimate measures how close the points lie to this line, and consequently measures how well the regression equation can predict the scores on the dependent variable. Thus, whereas the regression equation indicates the nature of the linear trend, the standard error of estimate indicates how well the regression equation functions as a description of the association between the variables.

Before concluding this section it should be pointed out that the regression line for X on Y is usually not the same as the regression line for Y on X. For the problem on education and income, the equation for predicting income from education $(Y$ on $X)$ was

$$Y = 9.725 + .356X$$

If we were to reverse the roles of the variables and wanted to predict education (X) from income (Y), we would want the regression of X on Y. This can be done simply by reworking the problem, treating income as the independent variable and education as the dependent variable. We would find that the equation for the regression line would differ from the one given above. Most importantly, we would find that b_{yx} does not equal b_{xy}. The slope b_{yx} gives the change in income for an increase of 1 unit (1 year) of education and was .356 thousand dollars. But b_{xy} will be the change in education for a 1-unit increase ($1000) in income. This will obviously be a value different than .356 and, consequently, result in a different regression line. Thus, the slope is an asymmetric measure. Its value depends upon the direction of prediction: that is, whether we are predicting from X to Y or from Y to X. The computation of b_{xy} in the present problem is left as an exercise at the end of the chapter.

Question 11–4. Suppose that you are studying rates of political participation during a campaign. You measure political participation by counting how many of the following six types of political activity a person engaged in during the campaign: talking to friends about the candidates, attending campaign speeches, wearing a campaign button, giving money to a candidate or party, writing a letter about the campaign to a newspaper, and voting. Your independent variable is social status, which is measured on a scale from 1 (low) to 6 (high). You are interested in determining if there is a linear association between social status and the number of different types of political activity people engage in. In a sample of 10 people, the following scores are obtained:

Observation Number	Social Status (X)	Political Participation (Y)
1	2	3
2	4	3
3	2	1
4	3	2
5	5	4
6	3	4
7	2	2
8	5	5
9	4	5
10	3	4

a. Plot a scattergram of these data. Is there a linear relationship between the two variables?
b. Calculate the value of b_{yx}.
c. Calculate the value of a_{yx}.
d. Write the equation for the regression line.
e. From the regression equation, calculate what the regression line predicts as the Y score for each variable.
f. Calculate the value of $s_{y \cdot x}$. _____

CORRELATION

Regression coefficients are among the most important of all statistical measures. By themselves, however, they are not adequate to fully describe the relationship between two variables because they do not take into account the degree or strength of association. The standard error of estimate gets at this problem, but the information is presented in an awkward form. What is needed is a measure of association along the lines of the coefficients considered in Chapter 10, which follows the conventional forms for measuring association. That is, a coefficient is needed that varies from 0 for no association to 1.0 for a perfect positive association or to −1.0 for a perfect negative association. In social science the statistic most often used for this purpose is the *Pearson product-moment correlation coefficient*. This coefficient, hereafter referred to just as the *correlation coefficient,* yields a value of 0, indicating no linear relationship between the variables and a value of 1.0 or −1.0 when all the points lie precisely on the regression line.

The formula for the correlation coefficient (symbolized as *r*) is

$$r = \frac{\Sigma(X - \bar{X})(Y - \bar{Y})}{\sqrt{[\Sigma(X - \bar{X})^2][\Sigma(Y - \bar{Y})^2]}}$$

Since $x = (X - \bar{X})$ and $y = (Y - \bar{Y})$, this formula can also be written

$$r = \frac{\Sigma xy}{\sqrt{(\Sigma x^2)(\Sigma y^2)}}$$

As can be seen, the numerator for the correlation coefficient is the co-variance, just as was the numerator for the regression coefficient. The denominator, however, is different from the denominator of the regression coefficient and will give it a quite different meaning. To explore the meaning of r, it will be useful to manipulate the formula for r and obtain it in a different form.

To begin this manipulation, the numerator and denominator are divided by N, the sample size:

$$r = \frac{\dfrac{\Sigma xy}{N}}{\sqrt{\dfrac{(\Sigma x^2)(\Sigma y^2)}{N}}}$$

$$r = \frac{\dfrac{1}{N}\Sigma xy}{\sqrt{\left(\dfrac{\Sigma x^2}{N}\right)\left(\dfrac{\Sigma y^2}{N}\right)}}$$

It will be remembered that the formula for the variance is

$$s^2 = \frac{\Sigma x^2}{N}$$

If the variance of the X scores is represented as s_X^2 and the variance of the y scores as s_Y^2 and their standard deviations as s_X and s_Y, then

$$= r = \frac{\dfrac{1}{N}\Sigma xy}{\sqrt{s_X^2 s_Y^2}}$$

$$= \frac{\dfrac{1}{N}\Sigma xy}{s_X s_Y}$$

The denominator now consists of the product of the standard deviation of X and the standard deviation of Y. Since these values are constants within a particular set of data, the last formula can be rewritten as

$$r = \frac{1}{N}\Sigma\left(\frac{x}{s_X}\right)\left(\frac{y}{s_Y}\right)$$

Finally, it will be recalled that the raw scores on a variable can be transformed to "standardized" scores without losing any information present in the data. When scores are standardized, they have a mean of zero and a standard deviation of 1.0. (A set of standardized scores does not necessarily have a normally shaped distribution. Standardized scores will be normally distributed only if the raw scores from which the standardized scores are derived are normally distributed.) A standardized score is represented by the symbol z and we shall attach the subscripts X or Y to denote a standardized score obtained from the X variable or the Y variable. The standardized scores for variables X and Y are calculated from the formulas

$$z_X = \frac{X - \overline{X}}{s_X} = \frac{x}{s_X}$$

$$z_Y = \frac{Y - \overline{Y}}{s_Y} = \frac{y}{s_Y}$$

Substituting into the last equation for the correlation coefficient yields

$$r = \frac{1}{N}\Sigma z_X z_Y$$

$$= \frac{\Sigma z_X z_Y}{N}$$

This formula tells us that if the X scores and Y scores for two variables are put in terms of standardized scores, ones with a mean of zero and a standard deviation of 1.0, then the correlation coefficient is equal to the covariance of these standardized scores divided by N. The correlation coefficient, then, is the average amount of covariance in the standardized scores of the variables.

Earlier it was said that the correlation coefficient may vary in value from -1.0 to $+1.0$. From the last formula it might not be evident how the result of the covariance of the standardized scores divided by N is constrained within these limits. It is the standardizing process which insures that these bounds cannot be exceeded. If there is a perfect positive linear association between two variables, then all the data points must lie precisely on the regression line. Because the data points all lie on the same straight line, the X and Y scores of all the points must be related to each other in the same way. This relation is as follows: The X score of a point lies the same distance in units of standard deviation of the X variable (s_X) from $\overline{X}$ as the Y score of the point lies from $\overline{Y}$ in units of standard deviation of the Y variable (s_Y). If, for example, the X score for a point is a distance of $2s_X$ from $\overline{X}$, then the Y score for that point will be a distance of $2s_Y$ from $\overline{Y}$. It is only when this relation holds that all the points will lie on the same straight line.

When the X and Y variables of a set of points which lie on a straight line are standardized, then the z_X score of each point must be equal to the z_Y score of the point. This follows from the fact that a standardized score gives the distance of a score from the mean in units of standard deviation. And therefore a point whose X score lies the same distance in units of standard deviation of the X variable from $\overline{X}$ as its Y score lies from $\overline{Y}$ in units of standard deviation of Y will have the same value for its z_X and z_Y scores.

Now, by definition a standardized variable is one with a mean of 0 and a variance (and standard deviation) of 1.0. The formula for calculating the variance of a standardized variable, z, would be

$$s^2 = \frac{\Sigma(z - \bar{z})^2}{N}$$

where $\bar{z}$ stands for the mean of the standardized scores. But since the mean of a set of standardized scores is 0, the formula becomes

$$s^2 = \frac{\Sigma(z - 0)^2}{N}$$

$$= \frac{\Sigma z^2}{N}$$

Looking back at the formula for the correlation coefficient, it can be seen that to multiply a point's z_X and z_Y scores together when all the data points lie on the same straight line is going to mean multiplying the same value by itself for each data point. This is equivalent, of course, to squaring the value. Thus when a perfect positive linear relationship exists among a set of points the numerator of the formula for r above is equivalent to the numerator of the formula for the variance of a set of standardized scores. And since a set of standardized scores must always have a variance of 1.0, the value for r when all the data points lie on a straight line must also be 1.0. Of course, a similar argument also applies when a perfect negative linear association is present: If the z_X and z_Y values match but are opposite in sign, then the limit of r will be -1.0.

The values of 1.0 and -1.0 form the limits of the correlation coefficient because it is only when z_X and z_Y values for points are matched in size that the maximum value of the covariance $(\Sigma z_X z_Y)$ is reached. This may be seen intuitively from the following points. If the data points all lie on a single straight line, then the largest values of z_X will be multiplied by the largest values of z_Y (since z_X values and z_Y values for points in this situation must be equal), which results in their products being large in value and which in turn makes the value of the covariance large. But if the data points do not all lie precisely on the same straight line, then the z_X and z_Y values for many of the points will not be matched in size. When this happens some of the points will have large values of z_X but relatively small values of z_Y, and other of the points will have large values of z_Y but relatively small values of z_X. Consequently the products of the z_X and z_Y scores for these points will be relatively small and will make the sum of the products, the covariance, smaller in value than if the z_X and z_Y scores of points were matched in size. In such situations the value of r will be smaller in magnitude than 1.0 or -1.0.

Let us standardize the data on education and income presented in Table 11-1. In Table 11-2 it was found that $\overline{X} = 12$, $s_X = 3.27$, $\overline{Y} = 14$, and $s_Y = 2.71$. For the first observation, then.

$$z_{X_1} = \frac{7 - 12}{3.27} \qquad z_{Y_1} = \frac{9 - 14}{2.71}$$

$$= -1.53 \qquad = -1.84$$

Table 11-4

Obser-vation Number	Standardized Score for Education (z_X)	Standardized Score for Income (z_Y)	$z_X z_Y$
1	−1.53	−1.84	2.81
2	.00	.74	.00
3	−1.22	.00	.00
4	.00	−.74	.00
5	.61	−1.11	−.68
6	−.92	.74	−.68
7	1.83	1.84	3.37
8	.61	−.37	−.23
9	−1.22	−.37	.45
10	.00	.00	.00
11	1.53	.00	.00
12	−.61	.74	−.45
13	1.22	.37	.45
14	−.61	−1.48	.90
15	.31	1.48	.46
			$\Sigma z_X z_Y = 6.40$

When this procedure is repeated for each of the 15 observations, the standardized scores for education and income are obtained as in Table 11–4. Now, if the standardized scores for education and income (z_X and z_Y) are plotted, the scattergram in Figure 11–13 is obtained. Notice that the configuration of points in Figure 11–12 is exactly the same as the original configuration of points presented in Figure 11–6. Thus, translating the raw scores of the variables into standardized scores does not affect the relationship between the variables. Since the mean of the standardized variable is 0 and its standard deviation 1.0, what has been done is to transfer the location of points so that the middle of the points sits over the origin of the graph and to measure the dispersion in the points in units of standard deviation.

The sign of r indicates whether there is a positive or negative association between the variables given the way the scores of the variables are set out for calculation. How r measures the direction of the association is easily seen from Figure 11–13. Points located in quadrants I and III of the graph have the same sign for their z_X and z_Y values. The product of numbers with the same sign (both positive or both negative) will be positive. Thus, if most of the points appear in quadrants I and III, r will be positive in sign, indicating a positive relationship between the variables — that increasing values on one variable go with increasing values on the other. Points located in quadrants II and IV have different signs for their z_X and z_Y values, and thus the product of the standardized scores

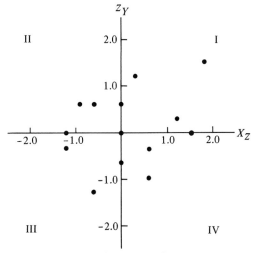

Figure 11–13
Standardized Score Scattergram for Data on Education and Income

for points in these quadrants will be negative. If most of the points appear in quadrants II and IV, the sign of *r* will be negative, indicating a negative relationship between the variables—that increasing values on one variable go with decreasing values on the other. In Figure 11–13 most of the points appear in quadrants I and III, and therefore we should expect *r* to be positive for these data.

It must be emphasized that the sign of *r* arises by definition from the way the scores are set out for computation. This occurs because the correlation coefficient only deals with the numerical values of scores, not with their substantive meaning. In the present example, the nature of the variables of education and income is such that a positive sign of the correlation coefficient will correspond to a substantive positive association between the variables. But suppose that we had correlated parents' income with class standing among a sample of high school students, where a 1 means the top student in the class, a 2 means the second top student, and so on. When these variables are correlated, a minus sign for the correlation coefficient would be obtained. What this sign means is that as the numbers on one variable increase (parents' income), the numbers on the other variable (class standing) decrease. But, substantively, what the sign indicates is that as parents' income increases, students do better in school. Care, then, needs to be taken in translating the sign of the correlation coefficient into a description of the substantive association between the variables.

Let us now calculate the correlation between education and income. The last column in Table 11–4 contains the product of z_X and z_Y for each observation. Summing these products yields 6.40. Therefore,

$$r = \frac{z_X z_Y}{N}$$

$$= \frac{6.40}{15}$$

$$= .43$$

While the standardized score formula for the correlation is useful for explaining the correlation coefficient, it is not very convenient for computational purposes. The following computational formula for the correlation coefficient may seem formidable, but is much easier to use, especially when a large number of observations are involved:

$$r = \frac{N\Sigma XY - (\Sigma X)(\Sigma Y)}{\sqrt{[(N)(\Sigma X^2) - (\Sigma X)^2][(N)(\Sigma Y^2) - (\Sigma Y)^2]}}$$

In Table 11–2, all the quantities in this formula have been calculated for education and income. Inserting them into the formula yields

$$r = \frac{(15)(2577) - (180)(210)}{\sqrt{[(15)(2320) - (180)^2][(15)(3050) - (210)^2]}}$$

$$= \frac{38,655 - 37,800}{\sqrt{[34,800 - 32,400][45,750 - 44,100]}}$$

$$= \frac{855}{\sqrt{(2400)(1650)}}$$

$$= \frac{855}{\sqrt{3,960,000}}$$

$$= \frac{855}{1989.98}$$

$$= .43$$

This is the same value that was obtained through the standardized score formula for r.

Question 11–5. Use the data in Question 11–4 to:

a. Calculate ΣX and ΣY.
b. Calculate ΣX^2 and ΣY^2.
c. Calculate ΣXY.
d. Use the quantities just calculated to compute the value of the correlation coefficient between social status and political participation.

INTERPRETATIONS OF
THE CORRELATION COEFFICIENT

Figure 11–13 presented the scattergram for the data on education and income in terms of standardized scores. An interesting question is, Where would the regression line fall on this graph? It has been seen that the regression line must pass through the point $(X = \bar{X}, Y = \bar{Y})$. When the raw scores are translated into standardized scores, this point coincides with the origin of the graph. Therefore, the regression line must pass through the origin in Figure 11–13. This will always be true for any set of data since the mean of a standardized variable is zero.

The formula for the regression line when the variables are in their raw score form was

$$Y = 9.725 + .356X$$

Previously, we used this regression equation to calculate the predicted Y value for a person with 16 years of education and found it to be 15.425.

Since this predicted Y value is equal to the height of the regression line above the X score of 16, the regression line must pass through the point $(X = 16, Y = 15.425)$. Now let us find the coordinates of this point in terms of standardized scores:

$$z_{(X=16)} = \frac{16 - 12}{3.27} \qquad z_{(Y=15.425)} = \frac{15.425 - 14}{2.71}$$

$$= \frac{4}{3.27} \qquad\qquad = \frac{1.425}{2.71}$$

$$= 1.22 \qquad\qquad\quad = .53$$

Since the regression line passed through the point $(X = 16, Y = 15.425)$ on the raw score scattergram of the data, the regression line must pass through the point $(z_x = 1.22, z_y = .53)$ on the standardized score scattergram of the data. And we have already seen that the regression line must pass through the origin of the standardized score scattergram since it corresponds to the point $(X = \overline{X}, Y = \overline{Y})$. These two points then must lie on the regression line and we can use them to draw the regression

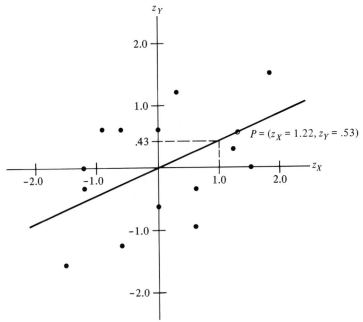

Figure 11–14
Regression Line for Education and Income Drawn onto Standardized Score Scattergram

line onto the standardized score scattergram of the data points as in Figure 11–14. As can be seen, this line again is the line of best fit to the data, and thus nothing has been lost in translating the data and the regression line into the standardized score framework.

Now we may ask what is the slope of this regression line on the standardized score scattergram? Examining Figure 11–14 it can be seen that if we travel on the regression line from the origin up to the right a distance of 1 unit on the z_X axis, we will in the process have covered a vertical distance of .43 unit on the z_Y axis. The slope of this regression line on the standardized score scattergram then is

$$\text{slope} = \frac{\Delta z_Y}{\Delta z_X} = \frac{.43}{1.00} = .43$$

Interestingly enough, this is the value of the correlation coefficient for the data. What this means is that if the variables are put in terms of standardized scores, the correlation coefficient is the value of the slope of the regression line. Since the variables have been standardized, r is *the standardized regression coefficient.*

When the variables are standardized, the equation for the regression line will have a different form from the equation for the raw-score regression line. Since the regression line must pass through the origin of the graph when the variables are standardized, the z_Y intercept (a in the equation) will always be zero. The slope of the regression line (b) will always equal r. And since the values of the variables are given in terms of z scores, the equation for the *standardized* regression line will be

$$z_Y = 0 + r z_X \qquad \text{or} \qquad z_Y = r z_X$$

The correlation coefficient may now be interpreted as the slope of a regression line when the variables have been translated into standardized scores. The r of .43 for the data on education and income, then, tells us that when education increases by 1 standard deviation (1 unit of z_X), income increases by .43 standard deviation (.43 unit of z_Y). If two variables were perfectly related, a 1-standard-deviation change in the independent variable would be associated with a 1-standard-deviation change in the dependent variable and r would equal 1.0. If two variables would be perfectly unrelated, a 1-unit change in the independent variable would be associated with no consistent change in the dependent variable and r would be 0. The sign of r, just as for any regression coefficient, indicates the direction of the association. Thus, r can be interpreted as indicating the degree and direction of change in the dependent variable that is associated with a change in the independent variable when the variables are put in standardized form.

The correlation coefficient as a standardized regression coefficient, though, says something more than just describing the nature of the linear trend in the data. Figure 11–15a presents a random scatter of points. Here there is no linear trend in the data, and thus no regression line can be drawn. Notice that if a particular X score is selected, there is no single particular value for the Y scores of the points that lie above that X score. That is, for any X score there is a scatter among the Y scores of the points which have that particular X score, and the vertical range and

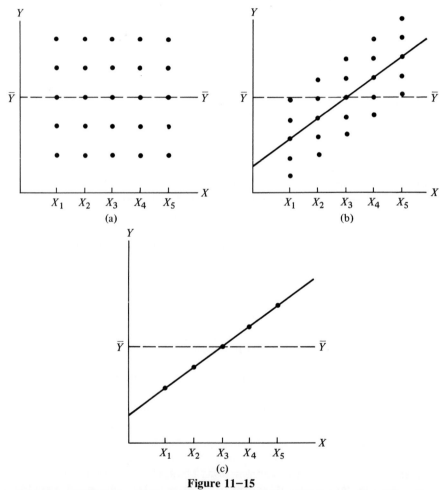

Figure 11–15
Accuracy of Predicting Y Scores from a Knowledge of X Scores when There Is
(a) A Random Scatter Among the Points; (b) A Linear Trend among the Points;
(c) A Perfect Linear Relationship among the Points

location of this scatter is pretty much the same for any X score that is selected. Thus, if we were asked to predict the Y score of a point, knowing its X score would be of no help. The best prediction under these conditions would be to predict the mean value of the Y scores $(\overline{Y})$ for the Y score of any point. And this is the best prediction we could make for any point, regardless of its X score.

If the data in Figure 11–15a were translated into z scores, the X variable would become z_X, the Y variable z_Y, and the mean of Y $(\overline{Y})$ zero. If the best prediction for the Y score of each point is $\overline{Y}$, then the best prediction for the z_Y score of *each* point is zero. Because the standardized regression equation is $z_Y = rz_X$, the value of r for these data must be zero for this prediction to be made.

In Figure 11–15b, there is a linear trend among the points and a regression line can be drawn. If a particular X score is selected on this graph, there is still a vertical scatter among the Y values of the points which have the selected X score. But now the location of the range of the scatter of Y values for each X score is different. Consequently, unlike Figure 11–15a, the Y values tend to be different for different X scores. The regression line follows the changes in location of the vertical scatters of the points by passing through the middle of the ranges of the Y scores for the different X scores. Thus, instead of always guessing $\overline{Y}$ as the Y score for all the points, we can increase our accuracy of prediction by using the regression equation to make different predictions according to the value of the X score for a point. Although there is still a vertical scatter among the Y scores of the points which lie above any one X score, and thus some error in prediction, the total amount of error will be smaller than if we had guessed $\overline{Y}$ for each point.

If the data in Figure 11–15b are translated into standardized scores, the value of r in the standardized regression equation must be some nonzero value in order to be able to arrive at different z_Y scores according to the z_X scores. Thus the value of r for these data must be different from zero.

In Figure 11–15c, all the points lie precisely on the regression line. Points that have the same X score must all have the same Y score, whose value is the height of the regression line above that X score. What the regression equation predicts as a point's Y score, then, will coincide perfectly with the point's actual Y score. By knowing the X score of a point the regression equation can be used to predict its Y score without error. As was said earlier, when all the data points lie precisely on the regression line, the z_X score for any point must be equal in size to the z_Y score for the point. Since the standardized regression equation is $z_Y = rz_X$, it is obvious that in this situation r must equal 1.0 (or -1.0) if the regression equation is to predict the z_Y scores perfectly.

We have seen, then, that when there is a random scatter of points, a knowledge of the X scores is of no help in predicting the Y scores and r is equal to zero. When a linear *trend* is present in the data, a knowledge of the X scores will improve our ability to predict Y scores (through the regression equation) and r will differ from zero (but not reach 1.0 or -1.0). When a perfect linear relationship is present, a knowledge of the X scores will enable us to predict perfectly the Y scores (again through the regression equation) and r will equal 1.0 or -1.0. In other words, *the correlation coefficient is an indicator of the extent to which Y scores can be predicted from X scores.*

The correlation coefficient, then, can be interpreted as an index of the extent to which the Y scores of observations in a set of data can be predicted through the regression equation from their X scores. The more accurate the predictions, the closer r will be to 1.0 or -1.0.

This interpretation of r is quite similar to the meaning of the standard error of estimate, $s_{y \cdot x}$, discussed earlier. The standard error of estimate indicates how much error can be expected in predicting Y scores from the regression equation. And it presents this information in the actual units of measurement of the Y variable. The correlation coefficient, though, ignores the units of measurement and presents this information in terms of a standardized scale from 0 to 1.0 (or -1.0). Thus, r facilitates comparisons of the predictive power of regression equations in different sets of data. The standard error of estimate indicates in absolute terms how much error there will be in using the regression equation to predict the Y scores within one particular set of data.

Although it is useful to have a measure of how accurately the regression equation can predict Y scores, it is still not an easily interpretable meaning for the correlation coefficient as a measure of association. Suppose that we had a set of data and the scores of the observations on two variables. One variable we denote as X and the other as Y. Assume now that we wish to predict the Y scores of the observations without using any information about their X scores. As we have seen, the mean of a variable is the value "closest" to all the scores. (This follows from the fact that the sum of the squared deviations from the mean is smaller than the sum of squared deviations about any other possible value.) Thus if we guess the mean $(\overline{Y})$ in predicting Y scores when we have no information about the X scores, the total amount of error we would make would be smaller than if we would guess in any other way. Figure 11 – 16a shows the scattergram for the data on education and income in Table 11 – 1. The line drawn through the points in Figure 11 – 16a is the value of the Y mean. The vertical distance from any point to this line is the amount of error that we would make in guessing $\overline{Y}$ as the value of the Y score for a point. This distance is equal to $Y - \overline{Y}$ or y. Since to

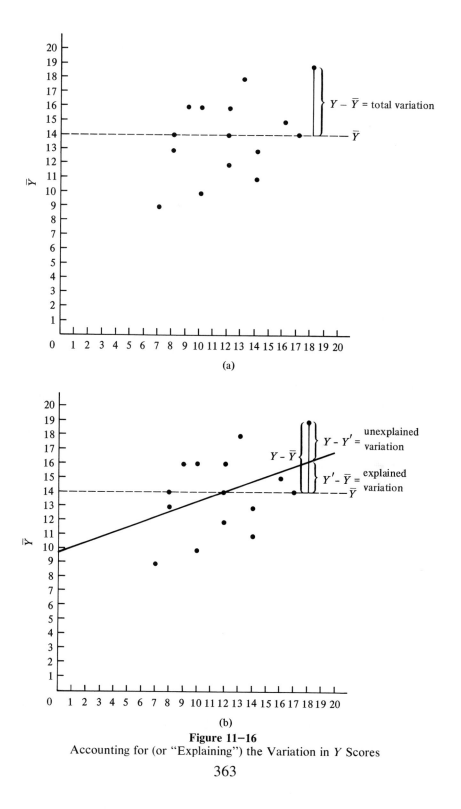

Figure 11-16
Accounting for (or "Explaining") the Variation in Y Scores

363

sum the deviations of scores about their mean will always result in a value of zero, we must square each deviation to obtain a measure of the total amount of error that is made in predicting $\overline{Y}$ as the Y score for each point. Thus

$$\text{total error} = \Sigma(Y - \overline{Y})^2 = \Sigma y^2$$

In Chapter 5 it was said that this quantity was called the sum of squares. For convenience we will refer to it as the *variation,* since it is also a measure of the total amount of variation among the Y scores. (Notice that this is not the same thing as the variance.) Therefore,

$$\text{total variation} = \Sigma(Y - \overline{Y})^2$$

Now let us introduce information about the X scores. If we know the X score for an observation, our best prediction for its Y score will be obtained from the regression line. In Figure 11 – 16b, the regression line for these data, $Y = 9.725 + .356X$, is drawn through the points. The predicted Y score given by the regression line for a given X score is indicated by Y'. The value of Y' is given on the graph by the height of the regression line above the X score.

The distance from $\overline{Y}$ to Y' is the increase in accuracy obtained in predicting Y scores using the regression line over the prediction using only the Y mean. The distance from Y' to Y is the error that still remains in using the regression line to predict Y scores. Thus, the distance from a point to $\overline{Y}$ can be split into two parts: the improvement in prediction over $\overline{Y}$ made when using the regression line to predict $(Y' - \overline{Y})$, and the error that still remains in predicting Y scores when using the regression line $(Y - Y')$. Therefore,

$$Y - \overline{Y} = (Y' - \overline{Y}) + (Y - Y')$$

These distances are shown for one point in Figure 11 – 16b.

In order to show the relation to the total variation, this equation must be squared:

$$(Y - \overline{Y})^2 = [(Y' - \overline{Y}) + (Y - Y')]^2$$

This yields

$$(Y - \overline{Y})^2 = (Y' - \overline{Y})^2 + 2(Y' - \overline{Y})(Y - Y') + (Y - Y')^2$$

This equation shows the relation of one squared deviation to its two component parts. To show the relation for the *sum* of all the squared deviations of the points, each term must be summed over all the observations:

$$\Sigma(Y - \overline{Y})^2 = \Sigma(Y' - \overline{Y})^2 + 2\Sigma(Y' - \overline{Y})(Y - Y') + \Sigma(Y - Y')^2$$

Let us look at the middle term on the right side of the equation: 2Σ $(Y' - \bar{Y})(Y - Y')$. It turns out that this term is always equal to zero. Why this is so can be seen through the following illustration. Figure 11–17 presents a set of data which for purposes of illustration is subdivided into four groups according to whether observations score X_1, X_2, X_3, or X_4 on the X variable. Thus, all points belonging to the same group have the same X score. The regression line for *all* the data points passes through the average Y score for each group. Thus, the regression line predicts an observation's Y score as the mean Y score for the group to which it belongs. For any observation scoring X_1 on the X variable, for example, the regression equation predicts its Y score (Y') as the mean Y score of all points that have a score of X_1 on the X variable. This value is indicated in Figure 11–17 as Y'_{X_1}. Similar statements hold for observations scoring X_2, X_3, or X_4 on the X variable.

Let us now focus our attention upon (any) one group of points. Since the regression equation predicts the same Y' score for all points belonging to the same group, within that group of scores the difference between Y' and $\bar{Y}$ will be a constant. Thus, for points that belong to the same group, the quantity $2\Sigma(Y' - \bar{Y})(Y - Y')$ can be written as $2(Y' - \bar{Y})\Sigma(Y - Y')$. The term $(Y' - \bar{Y})$ can be pulled outside the summation sign because its value is constant for all points belonging to the same group.

Keeping our attention upon a particular group of scores, let us now look at the term $\Sigma(Y - Y')$. Y' is what the regression equation yields as the predicted Y score for an observation. For a group of points that have

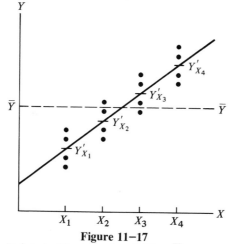

Figure 11–17
Data Points to Illustrate that $\Sigma(Y' - \bar{Y})(Y - Y') = 0$

the same X score, Y' is the mean of the Y scores for the group. Thus, the term $\Sigma(Y - Y')$ is the sum of the deviations of the Y scores of a group of observations about the mean Y score for that group. And we know that the sum of the deviations of scores about their mean always equals zero. Thus, for each group of scores $\Sigma(Y - Y') = 0$ and makes the quantity $2(Y' - \overline{Y})\Sigma(Y - Y') = 0$ for a group of scores, since zero times anything is zero. This means that the quantity $2\Sigma(Y' - \overline{Y})(Y - Y') = 0$ for each group taken separately. And when the results for each group are summed to give the result for all points in the set of data, the quantity $2\Sigma(Y' - \overline{Y})(Y - Y')$ will equal zero, since it is the result of adding up the zeros for each group. Of course in real problems it is unlikely that scores would lump themselves neatly into discrete groups. But the differences in predicted Y values must be balanced about the regression line and always results in $2\Sigma(Y' - \overline{Y})(Y - Y')$ equaling zero.

Dropping out the term $2\Sigma(Y' - \overline{Y})(Y - Y')$ from the last equation above leaves

$$\Sigma(Y - \overline{Y})^2 = \Sigma(Y' - \overline{Y})^2 + \Sigma(Y - Y')^2$$

The term on the left side of the equation is the total variation. The first term on the right side is that part of the total variation which is accounted for or "explained" by the regression equation. That is, it is the distance from the Y mean to the regression line for a point, squared and summed over all the points. It is called the *explained variation*. (This use of "explained" is in a statistical and not causal sense; it is best interpreted as "associated with.") The second term on the right side of the equation is that part of the total variation which is unaccounted for or "unexplained" by the regression line. It is the distance from the regression line to the point, squared and summed over all the points. This term is called the *unexplained variation*. In verbal terms, then,

$$\frac{\text{total}}{\text{variation}} = \left(\frac{\text{explained}}{\text{variation}}\right) + \left(\frac{\text{unexplained}}{\text{variation}}\right)$$

In Table 11–3, we calculated $(Y - Y')^2$ for each observation in the data for education and income. In Table 11–5, this information is repeated and, in addition, the values for $(Y' - \overline{Y})^2$ and $(Y - \overline{Y})^2$ are presented. For the first observation in Table 11–5, for example, Y is 9. The mean of the Y scores is 14, so $Y - \overline{Y} = 9 - 14 = -5$ and $(Y - \overline{Y})^2 = 25$. For this observation, Y' is equal to 12.219. Therefore, $Y' - \overline{Y} = 12.219 - 14 = -1.781$ and $(Y' - \overline{Y})^2 = 3.172$. When the quantities $(Y' - \overline{Y})^2$ and $(Y - Y')^2$ are summed over all 15 observations, we obtain

$$\left(\frac{\text{explained}}{\text{variation}}\right) + \left(\frac{\text{unexplained}}{\text{variation}}\right) = 20.304 + 89.701 = 110.005$$

Table 11–5

Observation Number	Education (X)	Income (Y)	Predicted Y value (Y')	(Y − Y')	(Y − Y')²	(Y' − Ȳ)	(Y' − Ȳ)²	(Y − Ȳ)	(Y − Ȳ)²
1	7	9	12.219	−3.219	10.362	−1.781	3.172	−5	25
2	12	16	14.000	2.000	4.000	.000	.000	2	4
3	8	14	12.575	1.425	2.031	−1.425	2.031	0	0
4	12	12	14.000	−2.000	4.000	.000	.000	−2	4
5	14	11	14.713	−3.713	13.786	.713	.508	−3	9
6	9	16	12.931	3.069	9.412	−1.069	1.143	2	4
7	18	19	16.137	2.863	8.197	2.137	4.567	5	25
8	14	13	14.713	−1.713	2.934	.713	.508	−1	1
9	8	13	12.575	.425	.181	−1.425	2.031	−1	1
10	12	14	14.000	.000	.000	.000	.000	0	0
11	17	14	15.781	−1.781	3.172	1.781	3.172	0	0
12	10	16	13.288	2.712	7.355	−.712	.507	2	4
13	16	15	15.425	−.425	.181	1.425	2.031	1	1
14	10	10	13.288	−3.288	10.811	−.712	.507	−4	16
15	13	18	14.356	3.644	13.279	.356	.127	4	16
		$\Sigma Y = 210$ $\bar{Y} = 14$			$\Sigma(Y - Y')^2 = 89.701$		$\Sigma(Y' - \bar{Y})^2 = 20.304$		$\Sigma(Y - \bar{Y})^2 = 110$

This, within rounding error, equals the total variation of 110, which is also calculated in Table 11 – 5.

The breakdown in the total variation can be used to provide a meaningful interpretation for the correlation coefficient as a measure of association. It turns out that the explained variation is equal to r^2 times the total variation:

$$\text{explained variation} = (r^2)(\text{total variation})$$

or

$$\Sigma(Y' - Y)^2 = (r^2)[\Sigma(Y - \overline{Y})^2]$$

Manipulating this equation, r^2 is found to be equal to the ratio of the explained variation to the total variation:

$$r^2 = \frac{\Sigma(Y' - \overline{Y})^2}{\Sigma(Y - \overline{Y})^2} = \frac{\text{explained variation}}{\text{total variation}}$$

Thus, r^2 is the proportion of the total variation in the Y scores that can be accounted for by the regression line. In the data for education and income, the ratio of the explained to total variation is

$$\frac{\text{explained variation}}{\text{total variation}} = \frac{20.304}{110}$$

$$= .185$$

The correlation for these data found earlier is .43. Squaring this value we obtain .185, which matches the value of the ratio of the explained to total variation. What this value indicates is that 18.5% of the variation in income can be accounted for by education in this set of data.

By squaring the value of r, a very meaningful interpretation of the correlation coefficient is obtained. This method of interpreting the correlation coefficient is so important that r^2 is given a special name: *coefficient of determination.* It varies in value from 0 to 1.0. If r^2 equals zero, the independent variable accounts for none of the variation in the dependent variable. When r^2 equals 1.0, the independent variable accounts for 100% of the variation in the dependent variable. For intermediate values, r^2 specifies the percentage of the variation in the dependent variable that can be accounted for by the independent variable. It must be emphasized that since correlation is based upon linear regression, the coefficient of determination specifies the strength of linear association in the data. Any sort of nonlinear association is not detected by either regression or the coefficient of determination. Thus, r^2 tells us how strongly two variables are linearly associated by indicating the percentage of variation in one variable that can be accounted for by the other. While regression describes the nature of the linear trend in the data and hence how much change in the dependent variable will be caused by a change in the independent variable, the coefficient of determination tells us how important one variable is in accounting for scores on the other variable. Obviously, the greater the percentage of variation in the scores of one variable that is accounted for by another variable, the more important this last variable must be in accounting for scores on the first variable.

Unlike the slope of the regression line, the coefficient of determination is a symmetric measure. Its interpretation, then, does not depend upon which variable is independent and which dependent. It is often more appropriate, therefore, to say that r^2 specifies the extent to which the variation in one variable is "associated with" the variation in another. The coefficient of determination, then, provides an easily interpretable measure of the strength of linear association between two interval or ratio level variables.

We have seen that while r can vary from -1.0 to 0 to 1.0, r^2 varies only from 0 to 1.0. This is because squaring the value of r must always result in a positive value. This makes sense in terms of what r^2 means. The correlation coefficient indicates the direction of association, besides indicating how well the regression line can predict Y scores. The coefficient of determination, though, specifies the proportion of the total variation in one variable that is "explained" or associated with the other. Obviously, to make sense this proportion must either be zero or positive in value.

As with some of the nominal and ordinal measures of association discussed in Chapter 10, the value of the correlation coefficient (and hence the coefficient of determination) can be affected if certain peculiarities are present in the data. As was said earlier, the correlation coefficient as a measure of association measures the strength of linear association in a set of data. If nonlinear trends are present, the correlation coefficient would present a misleadingly low value for the degree of association between the variables. To emphasize this point, Figure 11–18 presents scattergrams of some nonlinear trends. In Figure 11–18a and b, the value of r would be about zero. It would obviously be incorrect to conclude from this value of r that there is no association between the variables in these situations. In Figure 11–18c, the value for r would be moderately high. Yet the actual degree of association is considerably greater than what r would indicate, because the relationship is again curvilinear in form.

Another point is that the value of r can be misleading if the data present values for X or Y only over a small range of the values that X and Y really take on. In Figure 11–19, there is obviously a fair-sized association between the variables overall. But if data were selected so

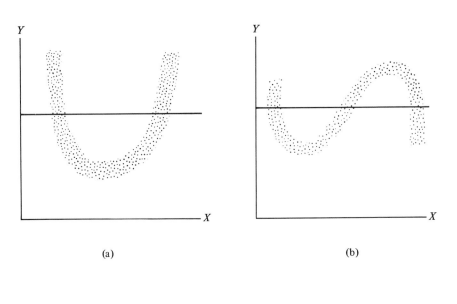

(a)   (b)

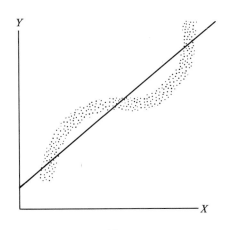

(c)
Figure 11–18
Examples of Nonlinear Trends where the Correlation Coefficient Would Be Misleading as a Measure of Association

that only points within a restricted range were used to compute r, such as those in the box in Figure 11 – 19, the value of r would be small.

The correlation coefficient is also very sensitive to extreme values. In Figure 11 – 20a, the effect of the three extreme cases is to produce a moderate correlation between the variables, when among the majority of

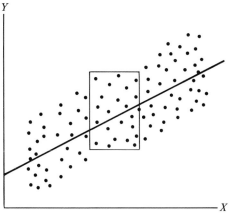

Figure 11–19
Illustration of Points Showing No Relationship within a Limited Range of Values
but a Positive Relationship Overall

points there is no relationship at all. In Figure 11–20b, there is a strong
linear relationship among most of the points but the value of *r* will be
depressed because of the three extreme cases that are out of line. In
such situations it might be appropriate to exclude the few extreme cases
in calculating the value of *r*.

Finally, the value of *r* is affected by mismatches in the marginal distributions of the variables. Ideally, the distribution of *X* scores and the

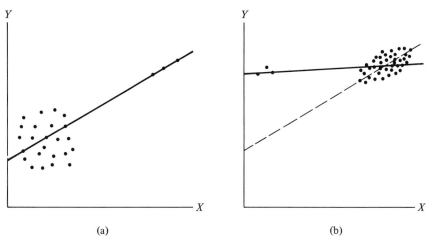

(a)   (b)

Figure 11–20
Examples of the Effect of Extreme Values on the Correlation Coefficient

distribution of Y scores should both be normal. But for describing the amount of linear association in a set of data, no problem arises if the distributions of the variables are not normal as long as both distributions have the same form (match). But if the marginal distributions of scores on the X and Y variables do not match, the upper limit of r is depressed below 1.0. There are some methods available for coping with such situations which are presented in more advanced books on correlation and regression.

Perhaps the best way to check to see if any of the problems mentioned above are present in a set of data is to construct a scattergram of the points and frequency distributions of the X and Y values. It is then possible to visually inspect the data for curvilinear trends, extreme values, and skewed or nonnormal distributions.

Before closing this chapter, a few words should be said about statistical inference. In this chapter we have focused on the problem of describing the nature and strength of linear association between two variables as it exists in a set of data. If the data consist of an entire population, no inference problem exists, as whatever linear association found by correlation and regression is a fact for the data. If, however, the data represent a random sample drawn from a population, the question arises as to whether the linear association described by correlation and regression is also present in the population or whether the linear association in the sample is a result of chance because of the random-selection process used in drawing the sample from the population. There are methods available for constructing confidence intervals about the statistics a_{yx}, b_{yx}, $s_{y \cdot x}$, and r. By constructing such confidence intervals, a researcher could find the limits within which the values for these statistics fall in the population at a certain level of confidence (that is, the 95% level). In this way the researcher could infer the characteristics of the linear association between two variables in a population from a randomly selected sample.

These inference methods are not considered here. The main reason for this is because in most social science research at this time, the question is not what the precise limits of correlation and regression are in the population but whether the amount of association between two variables is large enough to be of substantive interest. For a random sample of 100 observations, for example, a value for r of .195 is statistically significant (not equal to zero) at an alpha level of .05. But a correlation of this size says that only 3.8% of the variation in one variable can be accounted for by another variable. Such a small amount of association, despite being statistically significant, is usually not large enough to be of substantive interest. And in larger samples, which are quite frequent in social science research, even smaller values for the correlation coefficient will be statistically significant.

Thus, the question is whether the amount of correlation is great enough to say that there is an important substantive relationship between the variables and not just whether the relationship is statistically significant. Exactly what size the correlation must be before it is of substantive interest must be determined by the nature of the problem under consideration. In most cases it will need to be considerably larger than the minimum size needed for statistical significance.

Similar considerations surround the values of the slope and Y intercept in the regression equation. Again because most samples in social science are fairly large, confidence intervals for these statistics will usually be fairly narrow. Under such conditions it is usually substantively more important to obtain a rough idea as to how changes in the independent variable affect the dependent variable than it is to know the precise values of the confidence intervals for these statistics in the population. Of course, if the sample size is small, the inferential problems should be considered and a more advanced text on correlation and regression should be consulted.

Another way of saying all this is that many areas of social science are at the state of development where the major questions involve discovering what variables are associated with each other rather than specification of the precise degree to which they are associated. Correlation and regression, then, are most often used as tools for detecting and describing substantively important associations. When sample sizes are fairly large, this criterion is usually more important than questions of statistical significance.

EXERCISES FOR CHAPTER 11

1. A researcher constructs an index of social status (based upon education, income, and occupation) and an index of liberal–conservative political attitude. The index of social status runs from 1 (low) to 5 (high) and the index of liberal–conservative attitude from 1 (strong conservative) to 7 (strong liberal). The scores for 12 people on these indices are:

Index of Social Status	Index of Liberal–Conservative Political Attitude
4	3
2	5
3	4
3	3
1	6
2	4
5	3
4	2
3	5
4	4
2	3
5	2

Construct a scattergram of these data. Is there a linear trend among the data points? Find the equation for the regression line of political attitude on social status and draw it on the scattergram. What is the direction of the relationship between the variables? How would you describe this relationship? Find the predicted Y score for each observation. What is the value of the standard error of estimate? How would you interpret the value for this statistic? How much of the variation in political attitude can be accounted for by social status? (That is, compute the value of the correlation coefficient.)

2. In Table 11–1, the scores for 15 observations on education and income are given. Below are the scores of these same observations on age and income:

Age	Income (thousands of dollars)
20	9
30	16
50	14
25	12
35	11
40	16
55	19
20	13
30	13
50	14
30	14
45	16
60	15
25	10
50	18

Construct a scattergram for the data. Find the equation for the regression line for income on age. Compute the value of the correlation coefficient for the data. How much of the variation in income is accounted for by age? Which variable do you think is more important in determining income: education or age? What problems do you find in trying to answer this question with the information you presently have?

3. For the data on education and income given in Table 11–1, treat income as the X variable and education as the Y variable. Find the regression line for education on income. Construct a scattergram of the data and plot this regression line on it. Is it the same as the regression line for income on education?

4. Use the information on education in Table 11–1 and on age in Exercise 2 to compute the correlation between age and education. What does the value of the correlation coefficient tell you about the relationship between the two variables?

ANSWERS TO
QUESTIONS IN CHAPTER 11

11–1: a. About 11.2 mpg. b. About 6.8 mpg.
      c. About 7.6 mpg.   d. About 9.4 mpg.

11–2: a. 11.25 mpg.   b. 6.75 mpg.
      c. 7.8 mpg.   d. 9.45 mpg.

11–3: a. $Y' = 12.573$.   **b.** $Y' = 13.285$.
    c. $Y' = 14.709$.   **d.** $Y' = 16.133$.

11–4: a. Yes.   **b.** $b_{yx} = .835$.
    c. $a_{yx} = .545$.   **d.** $Y = .545 + .835X$.
    e. $Y'_1 = 2.215$, $Y'_6 = 3.05$,   **f.** $s_{y \cdot x} = .876$.
       $Y'_2 = 3.885$, $Y'_7 = 2.215$,
       $Y'_3 = 2.215$, $Y'_8 = 4.72$,
       $Y'_4 = 3.05$, $Y'_9 = 3.885$,
       $Y'_5 = 4.72$, $Y'_{10} = 3.05$.

11–5: a. $\Sigma X = 33$, $\Sigma Y = 33$.   **b.** $\Sigma X^2 = 121$, $\Sigma Y^2 = 125$.
    c. $\Sigma XY = 119$.   **d.** $r = .724$.

12
Analysis of Variance

Chapter 8 discussed how the difference between the means of two groups may be tested to determine if the difference is statistically significant. The problems in these situations involved a dichotomous nominal level variable such as sex or urban–rural residence and a ratio level variable such as income or age. In these problems, two independent random samples are drawn, one for each category of the nominal variable (such as males and females) and the means of the two random samples on the ratio level variable compared. But there is no reason why a nominal variable has to be dichotomous (only two values). Religion, for example, is often used as a nominal variable and may be categorized to have five values: (1) Protestant, (2) Catholic, (3) Jewish, (4) other, and (5) none. Or employment status might be classified as: (1) employed, (2) unemployed, (3) retired, and (4) housewife. Given nominal variables of this sort, problems might arise wherein a number of random samples, one for each category of the nominal variable, is selected and it is desired to see if there is a statistically significant difference among all the sample means on a ratio level variable.

It might be thought that it would be possible to determine if three or more sample means differ by conducting a *t* test on each possible pair of sample means. But there are a number of problems with such a strategy. First, the number of tests increases rapidly as the number of samples increases. For three samples there are three different pairs of means to test. With five samples, there are 10 tests and with 10 samples there would be 45 significance tests to be made. Aside from the work involved, this would be dangerous because whenever a large number of significance tests are made, there is a high probability that one or more of the tests would turn out to be statistically significant by chance alone (a Type I error). Second, in a series of such tests, the repeated use of the same means results in the tests not being independent of each other and thus violates a basic assumption of the testing procedure. Finally, because each test involves only two samples, it is not possible to take into account the information present in all the samples in any single test. The statistical technique of analysis of variance provides a way to overcome these difficulties.

Analysis of variance (often abbreviated as *Anova*) is a technique that has developed into a highly sophisticated and complicated subject. Its greatest application in the social sciences is in the area of psychology. This is primarily because it is most useful in experimental situations where the experimenter can exercise great control over the subjects in the experiment and over the "treatments" that are administered to them. In political science and sociology, which rely more on survey rather than experimental data, the more sophisticated forms of analysis of variance are less likely to be used. We will, therefore, focus on the basic application of analysis of variance to the problem of determining whether there is a statistically significant difference among the means of three or more independent random samples.

THE LOGIC OF
ANALYSIS OF VARIANCE

Suppose that a researcher is interested in seeing if there is a difference in U.S. voter turnout rates in local elections by geographical region: North, South, East, and West. He draws four separate (independent) random samples of counties in each geographical region and obtains the percentage of voter turnout in the last local election for each county. He then calculates the mean turnout rate in the counties for each of the four samples $(\overline{X}_N, \overline{X}_S, \overline{X}_E, \overline{X}_W)$. In this situation, geographical region is a nominal variable and the percentage of voter turnout in the counties is a ratio level variable.

The null hypothesis for this situation is that there is no difference in the mean population voter turnout rates of the four regions. Symbolically, this is stated as

$$H_o: \mu_N = \mu_S = \mu_E = \mu_W$$

The alternative hypothesis is that some sort of difference does exist in the population means. The nature of the difference is not specified but only that the means differ in some way. Figure 12–1 graphically presents the distributions of county turnout rates under the null hypothesis and under *one* possible alternative hypothesis.

Each sample of counties is conceptualized as being randomly selected from a population. Analysis of variance assumes that these populations are normally shaped, at least if the sample sizes are small, and that the variances of the populations (σ^2) are all equal. The property of equal variances is called *homoscedasticity*. Homoscedasticity is an essential

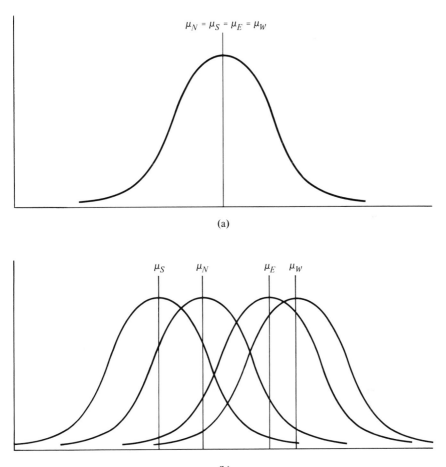

$$\mu_N = \mu_S = \mu_E = \mu_W$$

(a)

(b)

Figure 12-1
Population Distributions: (a) Distribution of Voter-Turnout Rate of Counties under the Null Hypothesis; (b) One Possible Set of Distributions of Voter-Turnout Rates of Counties under the Alternative Hypothesis

assumption to analysis of variance. If the researcher is uncertain if it holds in his data, it is possible to test the data for it and a more advanced text on analysis of variance should be consulted. It will be assumed here that the data meet this requirement. If all the populations, then, are normally shaped and have equal variances, they can differ only in the value of their means. And if the null hypothesis is true (the population means are equal), the populations are identical or all the groups really belong to the same population and not different ones. For the problem of geographical region and turnout rates, this would mean that all the counties belong to the same population as far as turnout rates are concerned. This is the situation depicted in Figure 12–1a.

If the null hypothesis is false, however, there is not a single population for all turnout rates but a number of different ones according to region. Figure 12–1b presents one way in which the null hypothesis could be false. But there are a variety of other ways that the population means could differ besides that shown in Figure 12–1b. It is possible that μ_S is greater than μ_N, or μ_E could be less than μ_S, and so on. The problem, then, is to find a way to test for a significant difference among the sample means which allows for all possible ways in which the population means might differ.

The sampling distribution of the mean was discussed in Chapter 8. It was seen that this sampling distribution is obtained by selecting a random sample of observations from an original population, calculating the mean of the sample and plotting the value of the sample mean $(\overline{X})$. When this process is repeated over and over, a distribution of sample means results, which is called the *sampling distribution of the mean*. The standard deviation of this sampling distribution $(\sigma_{\overline{x}})$ is a measure of the dispersion among all possible sample means. That is, $\sigma_{\overline{x}}$ is a measure of how much variation there is among sample means which are all obtained from random samples from the same population.

Now, if the null hypothesis is true $(\mu_N = \mu_S = \mu_E = \mu_W)$, the samples for the four regions are really four independent random samples from the same population. If this is the case, the means of the four samples $(\overline{X}_N, \overline{X}_S, \overline{X}_E,$ and $\overline{X}_W)$ can be conceived of as four randomly selected sample means taken from the distribution of all possible sample means, which is the sampling distribution of the mean. And since $\sigma_{\overline{x}}$ is the standard deviation of this distribution, then, if *the standard deviation of the four sample means* is calculated, its value should differ from $\sigma_{\overline{x}}$ only by an amount that can be expected to occur by chance through sampling error. If the null hypothesis is false, however, the samples will *not* be four independent samples from the same population. In this case, the standard deviation of the sample means will be *larger* than $\sigma_{\overline{x}}$ by an amount that is unlikely to occur by chance if the null hypothesis were true. This is because if the means of the populations differ, the sample means will reflect this and will differ among themselves *more* than they would if the samples had all come from the same population. Thus, if the value of $\sigma_{\overline{x}}$ were known or could be estimated, the value of the standard deviation of the sample means could be compared to it. And if the standard deviation of the sample means deviated from $\sigma_{\overline{x}}$ by more than what could be expected to occur by chance under the null hypothesis, the null hypothesis could be rejected and we could say that there is a statistically significant difference of some sort among the sample means.

In this account, the standard deviation has been used as the measure of dispersion among the sample means. Analysis of variance, though, uses the variance as the measure of dispersion. But since the variance is merely the square of the standard deviation, the reasoning is exactly the same. In analysis of variance, then, we want to compare the variance of the sample means to $\sigma_{\bar{X}}^2$ or an estimate of it.

Since only sample information is available in real problems, $\sigma_{\bar{X}}^2$ remains unknown to us. It will be recalled, however, that through the central limit theorem, there is a relationship between the standard deviation of the sampling distribution of the mean ($\sigma_{\bar{x}}$) and the standard deviation of the original population (σ_o):

$$\sigma_{\bar{x}} = \frac{\sigma_o}{\sqrt{N}}$$

If σ_o were known, the value of $\sigma_{\bar{x}}$ and $\sigma_{\bar{X}}^2$ could be calculated. But again because only sample information is available in an actual problem, the value of σ_o is also unknown to us. It was said before, though, that if the sample size is moderately large, then $\hat{s}$ provided a good estimate of σ_o and $\sigma_{\bar{x}}$ could, in turn, be estimated by $\hat{s}_{\bar{x}}$, as follows:

$$\hat{s}_{\bar{x}} = \frac{\hat{s}}{\sqrt{N}}$$

Squaring both sides of this equation yields

$$\hat{s}_{\bar{x}}^2 = \frac{\hat{s}^2}{N}$$

From this equation an estimate of $\sigma_{\bar{X}}^2$, $\hat{s}_{\bar{X}}^2$, is obtained against which we can compare the variance of the sample means. To use the equation, $\hat{s}^2$ must be calculated. Since the assumption of homoscedasticity (equal variances) has been made, the variance in *each* sample is an estimate of the same population variance. Therefore, the variances from each of the samples may be combined into one common estimate of the population variance. This is done through the pooling procedure for the variance that was discussed for the case of two samples in Chapter 8 extended now to the case of more than two samples. From this pooled estimate ($\hat{s}^2$) of the original population variance (σ_o^2), an estimate of the standard error can be obtained through the last formula given above against which the variance of the sample means can be compared. If the variance of the sample means deviates more than can be expected by chance from this estimate of the variance of the sampling distribution of the mean, the null hypothesis can be rejected.

One last complication in the logic of analysis of variance needs to be discussed. Since

$$\hat{s}_{\bar{X}}^2 = \frac{\hat{s}^2}{N}$$

then multiplying both sides of the equation by N yields

$$N\hat{s}_{\bar{X}}^2 = \hat{s}^2$$

If the null hypothesis is true, then if the unbiased variance for the sample means is calculated, it will be an estimate of $\sigma_{\bar{X}}^2$. This last equation says that if the variance of the sample means is multiplied by N, the size of the samples, it will provide (under the null hypothesis) an estimate of the original population variance (σ_o^2). The quantity $N\hat{s}_{\bar{X}}^2$, then, is an estimate of the original population variance *obtained from the variance of the sample means.* In analysis of variance this estimate of the original population variance is compared to the estimate of the original population variance *obtained from the variances within each sample.* Thus, rather than comparing two estimates of the variances of the sampling distribution of the mean, analysis of variance compares two estimates of the original population variance. One estimate is obtained from the variances of the *observations* within each of the samples, and the other estimate is obtained from the variance among the sample *means.* The basic point, though, still holds: If the population means really do differ in some way, the estimate of the variance obtained from the sample means will deviate from the estimate of the variance obtained from the observations within the samples more than could be expected by chance if the null hypothesis were true.

Question 12–1. A researcher is interested in the effects of different types of news media on the formation of people's attitudes toward political candidates. He randomly assigns 15 college sophomores to three groups. Each group then receives the same campaign speech through a different type of media. One group reads the campaign speech in a newspaper. Another hears it over the radio, and the third group watches it on television. Immediately after the speech, each subject in the experiment rates how much he likes or dislikes the candidate on a scale from 1 (strong dislike) to 10 (strong like). The ratings were:

Newspaper Group (1)	Radio Group (2)	Television Group (3)
3	9	7
2	6	6
1	10	6
5	7	9
4	8	7

a. Calculate the mean $(\overline{X})$ and the unbiased variance $(\hat{s}^2)$ for each group.
b. Calculate the unbiased variance $(\hat{s}_{\overline{X}}^2)$ among the *sample means* found in part a. Multiply $\hat{s}_{\overline{X}}^2$ by the size of the samples (N_s).
c. Calculate a pooled estimate of the variance from the variances within each sample according to the formula

$$\hat{s}^2 = \frac{(N_1 - 1)\hat{s}_1^2 + (N_2 - 1)\hat{s}_2^2 + (N_3 - 1)\hat{s}_3^2}{N_1 + N_2 + N_3 - 3}$$

PARTITIONING
THE SUM OF SQUARES

The purpose of the discussion so far has been to indicate how the analysis of variance provides a test of whether the differences among three or more sample means is statistically significant. The technique of analysis of variance, though, does more than provide a significance test. It is also able to analyze how much of the variation on the ratio level variable in the observations of all the samples can be accounted for by the nominal level variable. It is this ability to analyze the variation in the observations from which the technique gets its name.

Suppose that the researcher interested in voter-turnout rates selects a random sample of 10 counties from each of the four regions of the country. He obtains the voter turnout for each county in the last local election and lists the information as in Table 12–1. In the table the mean voter-turnout rate for each sample is calculated. It is also possible to calculate the mean rate for all the counties (disregarding their regions). This mean is found by simply summing the raw scores for all the observations (counties) and dividing by the total number of observations. This mean is called the *grand mean* and is labeled $\overline{X}_G$. Analysis of variance analyzes the deviation of the observations from the grand mean.

In order to conduct the analysis of these deviations, some new terminology is needed. We are accustomed to representing the score of an observation as X_i, where the subscript i indicates the number of the observation in the sample. But analysis of variance deals with more than one sample. Consequently, it is necessary to indicate not only what number an observation is in a sample, but also to what sample an observation belongs. This is done by using the letter j as a second subscript to refer to the number of the sample. The score of a particular observation is now represented as X_{ij}, indicating the ith observation in the jth sample. If the samples in Table 12–1 are numbered as North = 1, South = 2, East = 3, and West = 4, the score of the fifth observation in the North sample is represented as X_{51}. And the score of the fifth observation in

Table 12–1

Voter-Turnout Rates for Randomly Selected Counties in the Four Regions of the United States

North	South	East	West	Total
38	24	31	44	
40	36	40	27	
28	35	42	34	
39	23	33	40	
25	27	45	32	
29	37	32	31	
28	30	30	36	
34	29	39	41	
33	26	35	37	
36	33	43	38	
$\sum_i X_{i1} = 330$	$\sum_i X_{i2} = 300$	$\sum_i X_{i3} = 370$	$\sum_i X_{i4} = 360$	$\sum_j \sum_i X_{ij} = 1360$
$\overline{X}_1 = 33$	$\overline{X}_2 = 30$	$\overline{X}_3 = 37$	$\overline{X}_4 = 36$	$\overline{X}_G = 34$
$N_1 = 10$	$N_2 = 10$	$N_3 = 10$	$N_4 = 10$	$N = 40$

the West sample is represented as X_{54}. In Table 12–1, $X_{51} = 25$ and $X_{54} = 32$.

Since j refers to the number of a sample, the mean of any sample is represented as $\overline{X}_j$. The sample means for the North and West samples, then, are represented as $\overline{X}_1$ and $\overline{X}_4$. In Table 12–1, $\overline{X}_1 = 33$ and $\overline{X}_4 = 36$.

In order to compute the mean for a particular sample, it is necessary to obtain the sum of all the scores in that sample. Since the scores all belong to the same sample, the j subscript will be the same for all these scores. Thus, the sum of all the scores in the North sample is represented as ΣX_{i1} and the sum of the scores in the West sample is represented as ΣX_{i4}. In Table 12–1, $\Sigma X_{i1} = 330$ and $\Sigma X_{i4} = 360$. The formulas for the sample means of the North and West samples will now be

$$\overline{X}_1 = \frac{\Sigma X_{i1}}{N_1} = \frac{330}{10} = 33$$

$$\overline{X}_4 = \frac{\Sigma X_{i4}}{N_4} = \frac{360}{10} = 36$$

where N_1 is the number of observations in the first (North) sample and N_4 is the number of observations in the fourth (West) sample.

To obtain the value of the grand mean, it is first necessary to obtain the sum of the scores of all observations in all samples. This requires two summations: (1) summing the raw scores in each sample, and (2) summing the results of step (1) across the samples. This process, therefore, is indicated by a double summation:

$$\sum_j \sum_i X_{ij}$$

In Table 12–1, $\sum_j \sum_i X_{ij} = 1360$. It can also be seen from Table 12–1 that

$$\sum_j \sum_i X_{ij} = \Sigma X_{i1} + \Sigma X_{i2} + \Sigma X_{i3} + \Sigma X_{i4}$$

$$= 330 + 300 + 370 + 360$$

$$= 1360$$

The grand mean $(\overline{X}_G)$, then, is equal to this sum divided by the total number of observations in all samples, N (with no subscript):

$$\overline{X}_G = \frac{\sum_j \sum_i X_{ij}}{N}$$

$$= \frac{1360}{40}$$

$$= 34$$

It was said that analysis of variance analyzes the deviations of observations from the grand mean. The deviation of any observation from the grand mean may be divided into two parts: (1) the deviation of the observation from the mean of the particular sample to which it belongs, plus (2) the deviation of that sample mean from the grand mean. In symbols, this statement is

$$X_{ij} - \overline{X}_G = (X_{ij} - \overline{X}_j) + (\overline{X}_j - \overline{X}_G)$$

or

$$\left(\begin{array}{c}\text{raw} \\ \text{score}\end{array} - \begin{array}{c}\text{grand} \\ \text{mean}\end{array}\right) = \left(\begin{array}{c}\text{raw} \\ \text{score}\end{array} - \begin{array}{c}\text{sample} \\ \text{mean}\end{array}\right) + \left(\begin{array}{c}\text{sample} \\ \text{mean}\end{array} - \begin{array}{c}\text{grand} \\ \text{mean}\end{array}\right)$$

If we look at the first observation in the second (South) sample in Table 12–1, $X_{12} = 24$ and $\overline{X}_2 = 30$. Then

$$X_{ij} - \overline{X}_G = (X_{ij} - \overline{X}_j) + (\overline{X}_j - \overline{X}_G)$$

$$24 - 34 = (24 - 30) + (30 - 34)$$

$$-10 = (-6) + (-4)$$

When analyzing deviations from the grand mean, the problem arises that the sum of deviations about a mean is always equal to zero. To avoid this problem it is necessary to work with squared deviations. When the squared deviations of observations from a mean are summed, the result is called the *sum of squares*. This quantity has been encountered before in the formula for the variance:

$$\hat{s}^2 = \frac{\Sigma x^2}{N-1} = \frac{\Sigma (X - \bar{X})^2}{N-1}$$

The numerator of the formula for the variance is the sum of the deviations of the raw scores from the mean in a distribution or the sum of squares. In analysis of variance, the sum of squares of all the observations about the grand mean is called the *total sum of squares* (abbreviated *total SS*) and is written

$$\text{total SS} = \sum_j \sum_i (X_{ij} - \bar{X}_G)^2$$

Analysis of variance analyzes the variation among the scores in the samples by determining how much of the total SS can be "accounted for" by the nominal variable. And at the same time, the analysis of the sum of squares provides the two estimates of the variance necessary in determining if there is a statistically significant difference among the means of the samples. We shall now consider how the total SS is broken down for analysis.

The components of a squared deviation of an observation from the grand mean may be found by squaring both sides of the equation, which divided the deviation from the grand mean of an observation into two parts. That equation was

$$X_{ij} - \bar{X}_G = (X_{ij} - \bar{X}_j) + (\bar{X}_j - \bar{X}_G)$$

Squaring both sides of this equation yields

$$(X_{ij} - \bar{X}_G)^2 = [(X_{ij} - \bar{X}_j) + (\bar{X}_j - \bar{X}_G)]^2$$

When the right side of the equation is multiplied out, the formula becomes

$$(X_{ij} - \bar{X}_G)^2 = (X_{ij} - \bar{X}_j)^2 + 2(X_{ij} - \bar{X}_j)(\bar{X}_j - \bar{X}_G) + (\bar{X}_j - \bar{X}_G)^2$$

This formula breaks down the squared deviation for one observation from the grand mean. The breakdown for the sum of the squared deviations of all observations in all samples can be found by summing the terms in the equation across all the observations. This requires two summations, first within the samples and then across the samples:

$$\sum_j \sum_i (X_{ij} - \bar{X}_G) = \sum_j \sum_i (X_{ij} - \bar{X}_j)^2 + 2\sum_j \sum_i (X_{ij} - \bar{X}_j)(\bar{X}_j - \bar{X}_G)$$
$$+ \sum_j \sum_i (\bar{X}_j - \bar{X}_G)^2$$

This equation is simplified by the fact that the middle term on the right-hand side of the equation must always be equal to zero. This can be seen as follows. When the observations *within* any sample are summed, the quantity $(\overline{X}_j - \overline{X}_G)$ of the middle term will be a constant. This quantity, then, can be moved outside of the $\sum\limits_{i}$ summation as follows:

$$2\sum_{j}\sum_{i}(X_{ij} - \overline{X}_j)(\overline{X}_j - \overline{X}_G) = 2\sum_{j}(\overline{X}_j - \overline{X}_G)\sum_{i}(X_{ij} - \overline{X}_j)$$

The term $\sum\limits_{i}(X_{ij} - \overline{X}_j)$ is the sum of the (unsquared) deviations of observations from their sample mean. But the sum of the deviations of the scores in a distribution from the mean of that distribution must always equal zero. Therefore, the whole term will always equal zero, since any number times zero equals zero. The formula then reduces to

$$\sum_{j}\sum_{i}(X_{ij} - \overline{X}_G)^2 = \sum_{j}\sum_{i}(X_{ij} - \overline{X}_j)^2 + \sum_{j}\sum_{i}(\overline{X}_j - \overline{X}_G)^2$$

The left-hand side of this equation is the total SS as defined previously. The formula says that the total SS, the sum of the squared deviations of all observations about the grand mean, may be broken into two parts. The first part $\left[\sum\limits_{j}\sum\limits_{i}(X_{ij} - \overline{X}_j)^2\right]$ is the sum of all the squared deviations of observations *from their particular sample mean*. This is a measure of the variation of observations within the samples and is called the *within sum of squares* (within SS). The second part $\left[\sum\limits_{j}\sum\limits_{i}(\overline{X}_j - \overline{X}_G)^2\right]$ is the sum of the squared deviations of *sample means from the grand mean* weighted by the size of the samples. It is a measure of the variation between the sample means and is called the *between sum of squares* (between SS). Consequently,

total SS = within SS + between SS

When independent random samples are drawn for each category of a nominal variable, we want to know if the different categories affect the scores of the observations. The total SS measures the variation present in all the observations. The within SS tells us how much of the total variation occurs within the categories. Since within a category, the value of the nominal variable is the same for all observations, the variation measured by the within SS cannot be accounted for by differences in the nominal variable. It is therefore called in statistics the *unexplained* or *error variation*. The between SS, though, tells us how much of the total variation is due to differences between the categories of the nominal variable. Therefore, the nominal variable accounts for or "explains" this variation. Statistics, then, calls the between SS the *explained variation*. This use of "explained" and "unexplained" is by convention. It does not mean that the nominal variable is explaining something in a causal sense.

The use of "explained" in analysis of variance is best interpreted as "associated with." Whether the nominal variable explains something in a causal sense depends upon the substantive, not statistical, interpretation of the researcher.

How the nominal variable "explains" variation may be illustrated by two extreme situations. Suppose that the means of all the samples had exactly the same value. In such a case the variance among the sample means would be zero, and there would be no between or explained variation. This means that the categories of the nominal variable make no difference in the scores of the observations on the ratio level variable. Knowing from which category of the nominal variable an observation comes does not help in predicting its score on the ratio level variable. Thus, the nominal variable is not able to "explain" anything about the variation that occurs among the observations.

Now suppose that the sample means have different values but that all the observations in a sample have the same score (the mean of that sample). In this situation there is no variation among observations belonging to the same sample and there would be no within or unexplained variation. There is variation among the sample means, however. And by knowing from which category of the nominal variable an observation comes we could perfectly predict its score on the ratio level variable (the sample mean). Thus, all the variation among the observations is "explained" or accounted for by the differences in the categories of the nominal variable.

Question 12–2. For the problem in Question 12–1:

a. Find $\sum_j \sum_i X_{ij}$.

b. Find $\bar{X}_G$.

c. Find the total SS according to the formula
$$\text{total SS} = \sum_j \sum_i (X_{ij} - \bar{X}_G)^2$$

d. Find the within SS according to the formula
$$\text{within SS} = \sum_j \sum_i (X_{ij} - \bar{X}_j)^2$$

e. Let J represent the number of independent samples in the problem. Divide the within SS by $N - J$. Does this value equal the pooled estimate of the variance found in Question 12–1c?

f. Find the between SS according to the formula
$$\text{between SS} = \sum_j \sum_i (\bar{X}_j - \bar{X}_G)^2$$

g. Divide the between SS by $J - 1$. Does this value equal the value of $N_s \hat{s}_{\bar{X}}^2$ found in Question 12–1b?

OBTAINING ESTIMATES OF
THE ORIGINAL POPULATION VARIANCE

The between and within sum of squares are also used to form the two estimates of the variance which are compared to see if there is a statistically significant difference among the sample means. One of these estimates, it was said, is obtained by pooling the variances of the observations within each of the samples. In Chapter 8, a pooled estimate of the variance was obtained from two samples through the formula

$$\hat{s}_p^2 = \frac{(N_1 - 1)\hat{s}_1^2 + (N_2 - 1)\hat{s}_2^2}{N_1 + N_2 - 2}$$

Since

$$\hat{s}_1^2 = \frac{\Sigma x_1^2}{N_1 - 1} \quad \text{and} \quad \hat{s}_2^2 = \frac{\Sigma x_2^2}{N_2 - 1}$$

then

$$\hat{s}_p^2 = \frac{(N_1 - 1)\dfrac{\Sigma x_1^2}{N_1 - 1} + (N_2 - 1)\dfrac{\Sigma x_2^2}{N_2 - 1}}{N_1 + N_2 - 2}$$

$$= \frac{\Sigma x_1^2 + \Sigma x_2^2}{N_1 + N_2 - 2}$$

In other words, in the case of two samples, the numerator of a pooled estimate of the variance is obtained by adding together the sum of squares of the two samples.

It will be recalled that the number of degrees of freedom in a problem is equal to the number of total observations in the samples minus the number of means that are calculated. Since there will be one mean for each sample, the degrees of freedom in a problem will be equal to the total number of observations in all the samples minus the number of samples. In the formulas for the pooled variance for two samples above, it can be seen that the denominator of the formulas is the number of degrees of freedom. For pooled estimates of the variance from more than two samples, the denominator will simply be the number of degrees of freedom for the problem.

A pooled estimate of the variance may be formed from any number of samples, then, by dividing the sum of the sum of squares within each sample by the number of degrees of freedom. In analysis of variance, the within SS is the sum of the sum of squares within each sample. If the number of independent samples in a problem is J, there will be J means and the number of degrees of freedom will be equal to N (the total number of observations) minus J. The estimate of the variance obtained from pooling the variances within each sample then will be

$$\hat{s}^2_{within} = \frac{within\ SS}{N - J}$$

The second estimate of the variance is obtained from the variance of the sample means about the grand mean. If the number of sample means is J, the variance of the sample means ($\hat{s}^2_{\bar{X}}$) will be equal to the sum of the squared deviations of the sample means from the grand mean divided by $J - 1$.

$$\hat{s}^2_{\bar{X}} = \frac{\sum_j (\bar{X}_j - \bar{X}_G)^2}{J - 1}$$

The formula relating the variance of the sample means to the original population variance is

$$\hat{s}^2 = N\hat{s}^2_{\bar{X}}$$

Thus, if the number of observations in each sample is the same (represented by N_s), the second estimate of the original population variance is obtained by multiplying the variance of the sample means by N_s:

$$\hat{s}^2_{between} = N_s\hat{s}^2_{\bar{X}} = \frac{N_s \sum_j (\bar{X}_j - \bar{X}_G)^2}{J - 1}$$

But to multiply $\sum_j (\bar{X}_j - \bar{X}_G)^2$ by N_s is equivalent to adding $\sum_j (\bar{X}_j - \bar{X}_G)^2$ to itself as many times as there are observations in the sample. (To multiply a number by 10 is the same as adding the number to itself 10 times.) Thus,

$$\hat{s}^2_{between} = N\hat{s}^2_{\bar{X}} = \frac{N_s \sum_j (\bar{X}_j - \bar{X}_G)^2}{J - 1} = \frac{\sum_i \sum_j (\bar{X}_j - \bar{X}_G)^2}{J - 1}$$

In the last term of this equation it does not matter whether the i or j summation is performed first, so it can be written

$$\hat{s}^2_{between} = \frac{\sum_j \sum_i (\bar{X}_j - \bar{X}_G)^2}{J - 1}$$

The numerator of this term is the between SS. Therefore, the second estimate of the original population variance is simply the between SS divided by $J - 1$:

$$\hat{s}^2_{between} = \frac{between\ SS}{J - 1}$$

In finding this second estimate of the original population variance, the sample means are treated as scores about the grand mean. Since one mean is calculated for these scores, the number of degrees of freedom in this calculation is the number of sample means (J) minus 1. Consequently, *the two estimates of the original population variance are obtained simply by dividing the within and between sum of squares by their appropriate degrees of freedom.*

For the purposes of this discussion, it has been convenient to have the number of observations in the samples equal. This is not a necessary requirement, however. When different sample sizes are involved, the between SS will automatically take this into account, so the second estimate of the variance will still be equal to the between SS divided by its degrees of freedom.

The comparison of the two variances is made by finding the ratio of the estimate of the variance based on the between SS to the estimate of the variance based upon the within SS. This ratio is called the F ratio:

$$F = \frac{\hat{s}^2_{\text{between}}}{\hat{s}^2_{\text{within}}}$$

The F ratio is a test statistic and statisticians have calculated distributions for it. If there really is a difference among the sample means, the estimate of the variance based upon the between SS will be larger than the estimate based upon the within SS. When this happens, the value of F will be substantially larger than 1.0. If there is no significant difference among the sample means, the two estimates will be approximately equal and the value of F will be close to 1.0.

Distributions of F, like the chi-square and t distributions, vary according to the parameter of degrees of freedom. In the case of the F distribution, though, there are 2 separate degrees of freedom involved and the distribution of F changes as *either* of the 2 degrees of freedom change. The degrees of freedom involved are those used in finding the two estimates of the variance. They are the degrees of freedom associated with the numerator of the F ratio, $J - 1$, and the degrees of freedom associated with the denominator of the F ratio, $N - J$.

Because the F distribution changes according to a change in either of the degrees of freedom associated with the F ratio, it would be impractical to construct a table that would give the probability of all possible F values. In Table A–6 of Appendix A, the critical values of F are given for alpha levels of .05 and .01. The critical value of F in a particular problem is found by locating the degrees of freedom associated with the numerator of the F ratio across the top of the table and the degrees of freedom associated with the denominator of the F ratio down the side of the table. Once this cell in the table is found, the critical value of F for an alpha of .05 is given in light type and the critical value for an alpha level of .01 is given in dark type. If the obtained F ratio from the problem exceeds the critical value, the null hypothesis can be rejected and it can be concluded that there is a statistically significant difference among the sample means.

CALCULATIONS IN
THE ANALYSIS OF VARIANCE

Although the calculations in analysis of variance involve only arithmetic, a large amount of computation needs to be done and it is easy to get lost in the detail. The goal of the computations is to obtain the values of the within and between sum of squares. Computational formulas have been worked out which are of aid in calculating these values. These computational formulas are for the total SS and the between SS. Once these are found, the within SS is obtained by subtracting the between SS from the total SS.

The computational formula for the total SS is

$$\text{total SS} = \sum_j \sum_i X_{ij}^2 - \frac{\left(\sum_j \sum_i X_{ij}\right)^2}{N}$$

Except for the double summation signs, this is the same formula that is used in the numerator of the computational formula for the ordinary variance.

The computational formula for the between SS looks more formidable:

$$\text{between SS} = \sum_j \frac{\left(\sum_i X_{ij}\right)^2}{N_j} - \frac{\left(\sum_j \sum_i X_{ij}\right)^2}{N}$$

The second term of this formula is the same as the second term in the computational formula for the total SS. The first term is simpler than it looks. What it indicates is to sum the scores in a sample, square the result, and then divide by the number of observations in the sample. This operation is repeated for each of the samples, and then the results from all the samples are summed.

Table 12-2 presents the computations for an analysis of variance for the data on voter turnout in the sample of four regions of the United

Table 12-2
Calculations for Analysis of
Variance for Data in Table 12-1

	North		South		East		West		Total
	X_{i1}	X_{i1}^2	X_{i2}	X_{i2}^2	X_{i3}	X_{i3}^2	X_{i4}	X_{i4}^2	
	38	1444	24	576	31	961	44	1936	
	40	1600	36	1296	40	1600	27	729	
	28	784	35	1225	42	1764	34	1156	
	39	1521	23	529	33	1089	40	1600	
	25	625	27	729	45	2025	32	1024	
	29	841	37	1369	32	1024	31	961	
	28	784	30	900	30	900	36	1296	
	34	1156	29	841	39	1521	41	1681	
	33	1089	26	676	35	1225	37	1369	
	36	1296	33	1089	43	1849	38	1444	

$$\sum_i X_{i1} = 300 \qquad \sum_i X_{i2} = 300 \qquad \sum_i X_{i3} = 370 \qquad \sum_i X_{i4} = 360 \qquad \sum_j \sum_i X_{ij} = 1360$$

$$\sum_i X_{i1}^2 = 11,140 \qquad \sum_i X_{i2}^2 = 9230 \qquad \sum_i X_{i3}^2 = 13,958 \qquad \sum_i X_{i4}^2 = 13,196 \qquad \sum_j \sum_i X_{ij}^2 = 47,624$$

$$\left(\sum_i X_{i1}\right)^2 = (330)^2 \quad \left(\sum_i X_{i2}\right)^2 = (300)^2 \quad \left(\sum_i X_{i3}\right)^2 = (370)^2 \quad \left(\sum_i X_{i4}\right)^2 = (360)^2 \quad \left(\sum_j \sum_i X_{ij}\right)^2 = (1360)^2$$

$$= 108,900 \qquad\qquad = 90,000 \qquad\qquad = 136,900 \qquad\qquad = 129,600 \qquad\qquad = 1,849,600$$

States. Substituting the values in Table 12–2 into the formula for the total SS, we have

$$\text{total SS} = \sum_j \sum_i X_{ij}^2 - \frac{\left(\sum_j \sum_i X_{ij}\right)^2}{N}$$

$$= 47{,}524 - \frac{(1360)^2}{40}$$

$$= 47{,}524 - \frac{1{,}849{,}600}{40}$$

$$= 47{,}524 - 46{,}240$$

$$= 1284$$

For the between SS, we have

$$\text{between SS} = \sum_j \frac{\left(\sum_i X_{ij}\right)^2}{N_j} - \frac{\left(\sum_j \sum_i X_{ij}\right)^2}{N}$$

$$= \frac{\left(\sum_i X_{i1}\right)^2}{N_1} + \frac{\left(\sum_i X_{i2}\right)^2}{N_2} + \frac{\left(\sum_i X_{i3}\right)^2}{N_3}$$

$$+ \frac{\left(\sum_i X_{i4}\right)^2}{N_4} - \frac{\left(\sum_j \sum_i X_{ij}\right)^2}{N}$$

$$= \frac{108{,}900}{10} + \frac{90{,}000}{10} + \frac{136{,}900}{10}$$

$$+ \frac{129{,}600}{10} - \frac{(1360)^2}{40}$$

$$= 10{,}890 + 9000 + 13{,}690 + 12{,}960 - 46{,}240$$

$$= 46{,}540 - 46{,}240$$

$$= 300$$

The within SS is

$$\text{within SS} = \text{total SS} - \text{between SS}$$

$$= 1284 - 300$$

$$= 984$$

Table 12–3

Summary of Analysis of Variance
for the Problem of Voter Turnout
in Counties by Region

Source	Sums of Squares	Degrees of Freedom	Estimate of Variance	F
Total	1284	$N - 1 = 39$		
Between	300	$J - 1 = 3$	100	3.66
Within	984	$N - J = 36$	27.33	

The degrees of freedom associated with the between SS is $J - 1$, and the degrees of freedom associated with the within SS is $N - J$. Therefore, the estimated variances from the between and within sum of squares are

$$\hat{s}^2_{between} = \frac{between\ SS}{J - 1} = \frac{300}{3} = 100$$

$$\hat{s}^2_{within} = \frac{within\ SS}{N - J} = \frac{984}{36} = 27.33$$

The F ratio will be

$$F = \frac{\hat{s}^2_{between}}{\hat{s}^2_{within}} = \frac{100}{27.33} = 3.66$$

It is customary to present the results of an analysis of variance in the form of a summary table as is done in Table 12–3.

We are now ready to see if the obtained value of F for this problem is statistically significant. Suppose that the researcher chose an alpha of .05. We therefore want to see if the obtained value of F exceeds the critical value of F for an alpha of .05. Since there are 3 degrees of freedom associated with the numerator and 36 degrees of freedom associated with the denominator of the obtained F ratio, it will be found from Table A–6 that the critical value of F is 2.86. The obtained value exceeds this critical value, so the null hypothesis may be rejected and it can be concluded that there is a difference among the means. Interpreting this in terms of the substantive problem, the researcher concludes that there is a difference by region in the United States in the percentage of voter turnout in counties in the last local election.

Question 12–3. Here again are the data for the experiment on the news media:

	Newspaper Group (1)		Radio Group (2)		Television Group (3)	
	X_{i1}	X_{i1}^2	X_{i2}	X_{i2}^2	X_{i3}	X_{i3}^2
	3		9		7	
	2		6		6	
	1		10		6	
	5		7		9	
	4		8		7	

a. Find $\sum_i X_{i1}$, $\sum_i X_{i2}$, $\sum_i X_{i3}$, $\sum_j \sum_i X_{ij}$, and $\left(\sum_j \sum_i X_{ij}\right)^2$.

b. Find $\sum_i X_{i1}^2$, $\sum_i X_{i2}^2$, $\sum_i X_{i3}^2$, and $\sum_j \sum_i X_{ij}^2$.

c. Find $(\sum X_{i1})^2/N_1$, $(\sum X_{i2})^2/N_2$, $(\sum X_{i3})^2/N_3$, $\sum_j \dfrac{(\sum X_{ij})^2}{N_j}$

d. Find the total SS according to the formula

$$\text{total SS} = \sum_j \sum_i X_{ij}^2 - \frac{\left(\sum_j \sum_i X_{ij}\right)^2}{N}$$

Does this value equal the value of the total SS found in Question 12-2c?

e. Find the between SS according to the formula

$$\text{between SS} = \sum_j \frac{\left(\sum_i X_{ij}\right)^2}{N_j} - \frac{\left(\sum_j \sum_i X_{ij}\right)^2}{N}$$

Does this value equal the value of the between SS found in Question 12-2f?

f. Find the within SS by subtracting the between SS from the total SS found in parts d and e. Does this value equal the value of the within SS found in Question 12-2d?_____

ADVANCED CAPABILITIES
OF ANALYSIS OF VARIANCE

The discussion of analysis of variance in this chapter presents only the simplest type of situation to which this technique may be applied. It is possible to use analysis of variance to analyze the effect of two or more nominal variables upon a single ratio level variable simultaneously. In such situations the total sum of squares is partitioned into the within or error SS and a series of between SS associated with each of the nominal variables. It is then possible to tell if any one or more of the nominal variables is significantly associated with the ratio level variable. When two or more nominal variables are thought to affect a given ratio variable, it is possible that certain combinations of the nominal variables have an effect upon the ratio variable that is unexpectedly large given the separate effects of the variables. Such combination effects are called *interactions*. Suppose, for example, it is found that education and ambition are modestly related to income. If it were then found that a combination of high education and high ambition led to a higher level of income than should have been expected just from adding together the separate effects of education and ambition, there would be present an interaction effect between education and ambition. Analysis of variance allows for the testing of such interactions.

As was mentioned earlier, analysis of variance is most frequently used in experimental situations. Analysis of variance is so powerful in these situations that experiments are designed around its capabilities. Quite complex experimental designs have been developed to take advantage of the capabilities of the technique. In fact, experiments and analysis of variance have become so intertwined that an experiment would probably be inadequate or wasteful unless the plan of the experiment fully anticipated the type of analysis of variance that is to be used to analyze the results.

The purpose of this section has been to indicate that analysis of variance has some powerful and useful capabilities that are beyond the scope of an introductory book such as this. A lack of knowledge of these capabilities could lead to wasteful or inefficient use of time and resources. If it is thought that something more than the simple analysis of variance presented here might be appropriate to a particular situation, it is probably worthwhile to consult a more advanced text on analysis of variance or an expert in the area.

Question 12–4. From Questions 12–3e and f:

a. Compute the two estimates of the variance according to the formulas

$$\hat{s}^2_{within} = \frac{within\ SS}{N - J}$$

$$\hat{s}^2_{between} = \frac{between\ SS}{J - 1}$$

b. Compute the F ratio according to the formula

$$F = \frac{\hat{s}^2_{between}}{\hat{s}^2_{within}}$$

c. What are the degrees of freedom associated with the numerator and denominator of the F ratio?
d. Is the obtained value of the F ratio for the problem statistically significant at an alpha of .05?
e. Interpret the results of the statistical test in terms of the original experiment.

_____

EXERCISES FOR CHAPTER 12

1. A researcher wishes to see if type of high school — urban, suburban, small city, or rural — affects students' success in college. He selects random samples of students at a college for each of the four types of high schools and obtains their cumulative grade-point averages at the college:

Cumulative College Grade-Point Averages by Type of High School Attended

Urban	Suburban	Small City	Rural
3.0	2.4	2.7	2.2
2.2	2.8	2.0	2.4
2.0	2.6	2.3	3.0
1.8	3.4	1.7	2.8
2.6	3.3	2.8	3.2
2.4	2.3	2.1	2.0
2.8	2.8	1.8	2.6

Is there a statistically significant difference among the grade-point averages of the four samples at an alpha level of .01? Use the computational formulas for the sum of squares in your calculations.

2. Repeat Exercise 1, but this time use the "definitional" formulas for the between and within SS.

3. A researcher wishes to see if the political party composition of counties affects the amount of money that a county spends on education. Three independent samples of three types of counties are drawn and the expenditure per pupil for education in each county is obtained as follows:

Educational Expenditure Per Pupil

Dominant Democratic Counties	Dominant Republican Counties	Two-Party Counties
200	300	225
250	250	300
220	280	200
300	325	250
280	225	275
250	270	310

Use analysis of variance to determine if there is a statistically significant difference at an alpha level of .01 among the three samples.

4. A company wishes to know if different kinds of organization in its work groups affects the absenteeism rate among workers. It randomly assigns workers to one of three groups: (1) an "equalitarian" group, where the workers decide among themselves how work shall be done; (2) a "hierarchical" group, where each worker is assigned a specific task; and (3) a "foreman" group, where one worker is given the authority to tell the others what to do. The number of days each worker is absent over a period of 6 months is recorded:

Number of Days Absent

Equalitarian Group	Hierarchical Group	Foreman Group
2	2	3
3	3	2
0	4	4
4	5	2
1	4	5
3	6	2
1	4	3

If an alpha level of .05 is selected, is there a statistically significant difference in the average number of days workers are absent in the groups? What conclusions do you draw from the results?

ANSWERS TO
QUESTIONS IN CHAPTER 12

12–1: a. $\overline{X}_1 = 3$, $\hat{s}_1^2 = 2.5$; $\overline{X}_2 = 8$, $\hat{s}_2^2 = 2.5$; $\overline{X}_3 = 7$, $\hat{s}_3^2 = 1.5$.
   b. $\hat{s}_{\overline{X}}^2 = 7.0$; $N_s \hat{s}_{\overline{X}}^2 = (5)(7) = 35$.
   c. 2.17.

12–2: a. 90.   b. 6.0.
   c. 96.   d. 26.
   e. $\frac{26}{12} = 2.17$; yes.   f. 70.
   g. $\frac{70}{2} = 35$; yes.

12–3: a. 15, 40, 35, 90, 8100.   b. 55, 330, 251, 636.
   c. 45, 320, 245, 610.   d. $636 - \frac{8100}{15} = 96$; yes.
   e. $610 - \frac{8100}{15} = 70$; yes.   f. $96 - 70 = 26$; yes.

403
Answers to Questions in Chapter 12

12–4: a. $\hat{s}^2_{\text{within}} = \frac{26}{12} = 2.17$; $\hat{s}^2_{\text{between}} = \frac{70}{2} = 35.$ b. $F = \frac{35}{2.17} = 16.13.$
c. 2, 12.   d. Yes.
e. The use of the different media to transmit the campaign speech makes a difference in the way subjects rate the candidate on the scale of like and dislike.

13
Introduction to Multivariate Analysis

In Chapter 1 it was said that an important goal of social science is the development of true generalizations. Generalizations are statements which assert that one or more variables are related to another one or more variables. From true generalizations, explanations and predictions can be derived that contribute to our knowledge and can be used to make recommendations about practical social problems.

So far we have been concerned with univariate and bivariate (one or two variable) techniques. Such techniques are used to describe the characteristics of one variable or the relationship between two variables in a sample or a population and to make inferences from a sample to a population about the characteristics of one variable or the relationship between two variables. Univariate and bivariate statistics are fundamental tools in the search for true generalizations. In the social world, however, they are not usually adequate to the task. This inadequacy does not stem from any flaw in these statistical techniques but rather is a consequence of the complexity of social and political life.

When any social problem is discussed, it is often common to hear statements which assert that a single variable is "the cause" of the problem. Continued discussion of the problem, however, will usually lead to consideration of other variables, until eventually it is said that the original variable offered as "the cause" is just one of "many factors" that are involved.

Suppose that in a discussion of the high rates of juvenile delinquency in ghetto areas, someone says that the cause is a lack of strong parental discipline. Someone else might say that, yes, that is true, but the reason for poor parental discipline is the lack of decent job opportunities and thus adequate income and self-respect for the family. Someone else might say that these things might be involved, but the cause is really poor schools and a lack of recreational facilities in the ghetto. And finally someone else might offer that it is due to racism and the resulting discrimination that young people experience as they are growing up. Or consider voting behavior. Do people cast their ballot for the candidate who they perceive best represents their own positions on the issues? Do they vote for a candidate because of loyalty to the candidate's party? Do people feel loyalty to a party because they like the party's position on issues, or do people adjust their position on issues to correspond to those of the party to which they feel loyal? What effect does the reporting of public opinion polls have on voting? What effect does party organization through canvassing have on the vote? Does the incumbent have an advantage or do people vote against the "ins"? Or is none of this important and people vote for the candidate who has been able to present the best image in the media?

In both illustrations more variables could be named that are involved in the problem. But from these illustrations we can note the following points. First, no single variable can be said to be "the cause." Except for very isolated problems, there are likely to be a number of variables that affect any social situation. What this means is that the social universe is *multivariate,* that many variables are involved in a particular situation. Sometimes the question of "the cause" is restated so as to ask what is the most important variable. This requires assessing the degree to which independent variables are related to dependent variables (voting, juvenile delinquency) and determining which independent variable is most strongly related to the dependent variable. This leads to two additional points. First, even a strong association does not necessarily mean causation. Causation is not established just by statistical association. Geographic location in a city is probably strongly related to delinquency rates. But it is not geographic location which causes delinquency but other social conditions that happen to be concentrated within certain geographic areas. Second, just looking for associations between two variables may result in concealing the actual sequence of events involved in a problem. Thus, having a good job may affect juvenile delinquency *through* the mediating variable of parental discipline: A good job may lead to more self-respect, which, in turn, may lead to more discipline, which then might affect delinquency. Simply looking at two variables at a time might miss such a chain or web of relations. The assessment of which variable is most important in a problem requires the untangling of webs of interrelationships among variables.

In the social sciences it is very difficult to isolate one variable from all the others that are involved in a problem and manipulate it systematically to see how it affects a dependent variable. Psychologists are able to approach this ideal in their experiments. But with more complex problems such as juvenile delinquency and voting, experiments are usually too far removed from reality to be of much use. Consequently, it is necessary to have techniques that can examine the interrelationships among more than two variables simultaneously. Thus, because of the multivariate nature of the social and political world, univariate and bivariate statistics are limited in their ability to uncover generalizations in these realms. In this chapter we shall consider some basic multivariate techniques that are very useful in the search for generalizations in the social sciences.

CONTROLLING
FOR THIRD VARIABLES

The advantage of multivariate techniques over univariate and bivariate statistics is in their capacity to untangle the web of relationships that may exist among variables. The most basic multivariate technique is that of controlling for a third variable. Suppose that a survey is conducted to examine the relationship between people's level of political information and whether they have isolationist or interventionist attitudes on foreign policy.

Table 13 – 1 presents the data from the survey. As can be seen, 60% of the respondents who are high on political information have an interventionist foreign policy attitude while only 40% of those who are low on political information are interventionist. Thus, there is an association between level of political information and foreign policy attitude: People with a high level of political information are more likely to have an interventionist foreign policy attitude than people who are low on political information. The chi-square value for this table is 8.00, which is statistically significant at an alpha level of .01. (This is indicated in the table by the statement "$p < .01$," which says that the probability of a chi-square value of this size or larger occurring by chance is less than .01. Results

Table 13–1

Level of
Political Information

		High	Low	
	Interventionist	60 60%	40 40%	100
Foreign Policy Attitude	Isolationist	40 40%	60 60%	100
		100	100	200
		$x^2 = 8.0$	$p < .01$	

are often reported in this way rather than saying that the result was statistically significant at a particular alpha level because it allows the reader to select his own alpha level in interpreting the results.)

Because of the large association, as indicated by the difference in
the percentages (20 points across the columns) and the fact that the table
is statistically significant, we might be tempted to form a generalization
that higher levels of political information lead people to hold more interventionist attitudes on foreign policy. Before we accept this conclusion,
however, suppose it is suggested that women tend to have lower levels
of political information than men and perhaps that women have more
isolationist foreign policy attitudes than men. And, consequently, the
association in Table 13–1 is in some way the result of a third variable,
sex. The problem now is to find a way to enter this third variable into
the analysis to check out this suggestion.

The effect of the third variable, sex, on the association between level of political information and foreign policy attitude may be examined
through a "controlling" process. This is done by dividing the respondents into two groups, males and females, and examining the association
between level of political information and foreign policy attitude within
each group. This is done in Tables 13–2 and 13–3. Each of these tables
examines the relationship between level of political information and foreign policy attitude. Table 13–2 contains only males, while Table 13–3
contains only females. In each table, then, the variable sex cannot be a
factor, since everyone in a table has precisely the same value on this
variable. Stated another way, in each table the value of the third variable, sex, is held constant and any effects it might be causing have been
removed or "controlled."

The idea of controlling, then, is to reexamine the original associa-

Table 13–2
Males
Level of
Political Information

		High	Low	
	Interventionist	57 85.1%	35 85.4%	92
Foreign Policy Attitude				
	Isolationist	10 14.9%	6 14.6%	16
		67	41	108

Table 13–3
Females
Level of
Political Information

		High	Low	
		High	Low	
Foreign Policy Attitude	Interventionist	3 9.1%	5 8.5%	8
	Isolationist	30 90.9%	54 91.5%	84
		33	59	92

tion between two variables while holding the value of a third variable constant. In the present example the third variable has only two values, male and female. If religion were being used, it might have four values: Catholic, Protestant, Jew, and other. To hold religion constant, four separate tables would be necessary, one for each "value" of religion. This would require dividing the original sample into four separate groups according to the value people have on the variable of religion. It is perfectly possible to control for a third variable which has more than two values, but for convenience and simplicity we will use dichotomous variables in illustrations.

Sometimes it is desired to control for a third variable which is continuous, such as age or years of education. If we desired to precisely control for a continuous third variable, say age, we would be required to construct a different table for each value of age. If the possible range of values of age were from 18 to 75, there could be 57 different tables! This would require splitting the sample into 57 different parts, one for each year of age. The likely result is that there would be so few people in each table that no conclusions could be drawn from any one of them. This problem is often avoided by collapsing a continuous variable, again say age, into broader categories. People could be classified as young, medium, and old or even just into young and old. Let us say that age were collapsed into two values, young, being 18 to 40, and old, being 41 to 75. If this collapsed version of the variable age were now controlled, two tables would result: one for young people (18 to 40) and one for old people (41 to 75), which is a great improvement over 57. But in these tables the third variable, age, is *not* constant within each table; in one table people could vary in age from 18 to 40, and in the other, people could vary in age from 41 to 75. Thus, the ideal of holding the third variable constant is not achieved by this "collapsing" procedure. Age can vary somewhat in each table, so it could still be having some effect on whatever association appears in the "controlled" tables.

The range over which age can vary in each of the controlled tables, however, has been cut about in half. If age were in some way affecting the association between the two original variables, this should reduce its impact. Thus, in controlling for a continuous third variable (or any variable that has a large number of values), it is necessary to arrive at a compromise between the need to reduce the number of "controlled" tables that will be generated (so as to keep the size of the number of observations in each table large enough to be meaningful) and the ideal of holding the third variable constant.

Before we begin the analysis of the tables, it should be noted that Tables 13–2 and 13–3 "add up" to Table 13–1. If the same cell in each of the three tables is considered, it will be seen that the values for that cell in Tables 13–2 and 13–3 add up to the value in the corresponding cell in Table 13–1. For the upper-left cells in the tables, for example, $3 + 57 = 60$. For the lower-right cells, $54 + 6 = 60$. This is also true of the marginals and the totals of the tables. For the first row marginals in the tables, $92 + 8 = 100$. For the totals, $108 + 92 = 200$. Finally, the totals of Tables 13–2 and 13–3 tell us how many males and females there are in the original table. Of the 200 people in Table 13–1, 108 are males (the total of Table 13–2) and 92 are females (the total of Table 13–3).

In Table 13–2, 85.1% of the males who are high on political information have an interventionist foreign policy attitude, compared to 85.4% of the males who are low on political information. In Table 13–3, 9.1% of the females who are high on political information have an interventionist foreign policy attitude, compared to 8.5% of the females who are low on political information. In both tables, then, level of political information makes little difference in the holding of interventionist or isolationist foreign policy attitudes. The percentages indicate that there is no association in either table. This is reinforced by the fact that if chi-square were calculated, it would not be statistically significant for either table.

The original relationship between level of political information and foreign policy attitude in Table 13–1 has disappeared. This is not a magician's trick but a consequence of the way the three variables are interrelated. The interrelationship becomes clear if we look at two more tables. Tables 13–4 and 13–5 examine the relationship between sex and the two original variables. They show that the variable sex is strongly related to level of political information (females having less political information than males) and to foreign policy attitude (females being much more isolationist in attitude than males).

Table 13–4

		Sex		
		Male	Female	
Level of Political Information	High	67 62%	33 35.9%	100
	Low	41 38%	59 64.1%	100
		108	92	200

Table 13–1 shows the association between level of political information and foreign policy attitude uncontrolled for any third variable. Tables 13–2 and 13–3 examine the same association but at the same time control for the effects of a third variable, sex. Such tables are called *partials* because they each present the part of the original association that remains, if any, when controlled for one of the values of a third variable. Tables 13–4 and 13–5 are called *marginals*. They present the association between the third variable and one of the original variables. These three types of tables, the original, the partials, and the marginals, are all interrelated. Although it will not be demonstrated here, it can be shown that the *associations* in the partial and marginal tables can be combined to equal the association in the original table. It is this fact which provides the mathematical justification for making analyses through the controlling process. We have already seen that the partial tables must "add up" to equal the original table. The partial tables are also related to the marginal tables. The marginal tables can be constructed from the row and column marginals of the partial tables. Thus, the column marginals of Tables 13–2 (67 and 41) and 13–3 (33 and 59) are the cell values of Tables 13–4; and the row marginals of Tables 13–2

Table 13–5

		Sex		
		Male	Female	
Foreign Policy Attitude	Interventionist	92 85.2%	8 8.7%	100
	Isolationist	16 14.8%	84 91.3%	100
		108	92	200

(92 and 16) and 13–3 (8 and 84) are the cell values of Table 13–5. These interrelationships among the tables point up the fact that the five tables as a group attempt to untangle the associations that may exist among the three variables.

The analysis of Tables 13–1 through 13–5 leads to the conclusion that level of political information is not *causally* related to foreign policy attitude (Tables 13–2 and 13–3) but that the third variable, sex, is strongly related to each of these variables (Tables 13–4 and 13–5); so that when sex is uncontrolled, a *spurious* association between level of political information and foreign policy attitude appears (Table 13–1). Further interpretation depends upon whatever theories and hypotheses might be in mind.

It is unlikely that the biological determination of people's sex affects their level of political information or the nature of their attitudes on foreign policy. But it may be hypothesized that in our society there are different types of roles that are associated with a person's sex and that in living out the roles associated with being a male, a person tends to acquire a "high" level of political information and an interventionist foreign policy attitude, while in living out the roles associated with being a female, a person tends to acquire a "low" level of political information and an isolationist foreign policy attitude. Thus the variable of sex provides an "index" that taps the factors which cause differences in both level of political information and type of foreign policy attitude. Consequently, the association between level of political information and foreign policy attitude found in Table 13–1 is really the result of other factors which can be indexed through the variable of sex. The situation may be diagrammed as follows:

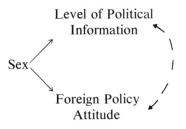

In this figure the "causal" relations are represented by solid arrows and the "spurious" association by a dashed arrow.

The idea of a spurious association may be made clear by the following illustration. It has been shown that the more firemen that come to a fire, the more damage that is done. This is an undeniable association. It is a spurious association, however, because it is not the firemen who are doing the damage — it is the size of the fire. The larger the fire, the greater the number of firemen who come to the scene; and the larger the fire, the more damage that is caused. The association between the number of firemen and amount of damage is an artifact of these two relations and is *explained away* by taking the size of the fire into account (or controlling for it).

Obviously we do not want to formulate generalizations on the basis of spurious associations. What we want are causal associations, relationships in which there is something more than just an association. Philosophically, the notion of cause and effect is quite complicated. But the important point here is that it is not possible to demonstrate causation using statistics, since statistics measure only the extent to which one variable is *associated with* or *correlated with* another. The idea that one variable is causally related to another is an inference that the researcher must make based upon his substantive understanding of the problem. This is why it was said that the interpretation of Tables 13 – 1 through 13 – 5 depended upon the theories and hypotheses that we had in mind.

In identifying causal relationships it is most important that we have a theory or hypothesis from which it can be deduced that there should be a logical connection between the variables. If this can be done, we can have some confidence in asserting a causal relationship between variables that are statistically associated; at least until someone demonstrates our theory to be wrong. Generally, causal associations cannot be *explained away* as was the association between the number of firemen and damage. (They may, however, be explained.) From this discussion we should become wary of accepting any association as a causal relationship even if the association is statistically significant. A common maxim in statistics is that association or correlation does not demonstrate causation.

Returning to the problem of level of political information and foreign policy attitude, the statistically significant association in Table 13 – 1 is explained away when controlled for a third variable, sex. As a result of this further analysis, we will *not* accept Table 13 – 1 as evidence in support of the generalization that increasing levels of political information lead to a more interventionist foreign policy attitude. A simple bivariate analysis might have led us to accept this conclusion. We were only able to discover the spuriousness of the association by untangling the web of interrelationships among the variables through the multivariate technique of controlling.

Question 13–1. A survey of 26 people was conducted in which information on party identification, attitude toward integration, and level of education was obtained. The data for each person on these variables are as follows:

Observation Number	Party Identification	Attitude toward Integration	Level of Education
1	R	Favor	High
2	D	Oppose	High
3	R	Favor	High
4	D	Favor	High
5	D	Oppose	Low
6	R	Oppose	Low
7	D	Oppose	Low
8	R	Favor	High
9	R	Oppose	High
10	R	Oppose	High
11	D	Oppose	Low
12	D	Oppose	Low
13	R	Favor	High
14	D	Oppose	Low
15	D	Favor	High
16	D	Oppose	Low
17	R	Favor	High
18	D	Oppose	Low
19	R	Favor	High
20	D	Favor	High
21	R	Oppose	Low
22	D	Oppose	Low
23	D	Favor	High
24	R	Favor	High
25	D	Oppose	Low
26	R	Favor	High

Use these data to fill in the cell values of the following tables:

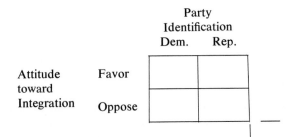

Less Education

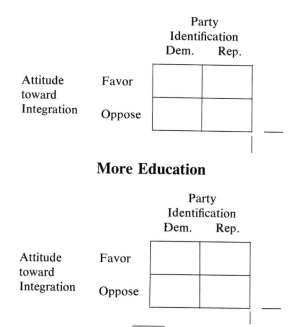

Party
Identification
Dem.   Rep.

Attitude
toward
Integration

Favor

Oppose

More Education

Party
Identification
Dem.   Rep.

Attitude
toward
Integration

Favor

Oppose

STATISTICAL ELABORATION

The process of controlling an association between two variables for a third variable is part of a more general technique called *statistical elaboration*. In statistical elaboration the variable that is thought to be the independent variable is called X; the dependent variable, Y; and the third variable, T. In Tables 13–1 through 13–5, X is level of political information, Y is foreign policy attitude, and T is sex. The third variable is often called the *test variable* since we test to see what happens to the relationship between X and Y when controlling for T.

The first problem in statistical elaboration is the choice of a test variable. This is a substantive question which statistics cannot answer. Basically, a variable is chosen as a test variable because the researcher has a guess or a hypothesis that it is somehow related in a relevant way to the association between X and Y. Theoretically, any variable for which the researcher has data may be used as a test variable. And if the data were available, it might be possible to elaborate a particular association with hundreds of different test variables. But to do this would be dangerous methodologically and conceptually sloppy. It will be recalled that if a large number of significance tests are done, a few of them will turn out to be statistically significant just by chance. This danger would also arise if we were to elaborate one association with regard to many test variables. Therefore, a researcher should only elaborate a relationship with regard to test variables which he has good reason to think are relevant. What makes certain variables relevant are the researcher's ideas about the nature of the problem.

Once a test variable is selected, it is important to identify the time order of the test variable to X and Y. In the case of the variables level of political information, foreign policy attitude, and sex, it can be argued fairly straightforwardly that the effects of sex roles are present prior in time to both the development of an adult's current level of political information and his attitudes on foreign policy. The identification of the time order is important because a variable cannot affect anything that occurs prior in time to itself. Thus variables that are known to occur later in time than both X and Y should not be considered as test variables since they can have no effect on any association between X and Y.

Two other time orderings of T to X and Y are possible. The test variable could occur prior in time to both X and Y, as was the case with sex, level of political information, and foreign policy attitude. In such a situation T is said to be *antecedent* to X and Y. The second possibility is that the test variable occurs later in time than X but prior in time to Y. In this situation T is said to be *intervening* between X and Y. In both cases the test variable is in a position to affect the relationship between X and Y. Whether it actually does affect the association is, of course, what the researcher is attempting to discover. But if an effect is observed, its interpretation will depend upon the researcher's identification of the time ordering involved.

THE FOUR TYPES
OF STATISTICAL ELABORATION

Depending upon the time ordering among X, Y, and T and what happens to the association between X and Y when T is controlled, there are four possible types of statistical elaboration.

1. *Explanation.* In explanation there originally is an association between X and Y which disappears or is substantially reduced in magnitude when T is controlled. The time order among the variables is that T is antecedent to X and Y. Explanation was the type of statistical elaboration that occurred in Tables $13-1$ through $13-5$. This type of elaboration is diagrammed as:

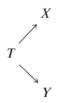

2. *Interpretation.* Suppose that a researcher did a survey of 500 adult women to examine the effects of marriage upon level of political participation. The results, as shown in Table 13-6, revealed that unmarried women are more likely to participate in politics than are married women. Examining these results, it might be concluded that getting married leads women to participate less in politics. The researcher speculated, however, that it is not marriage per se which causes less participation, but that

Table 13-6

		Marriage		
		Unmarried Women	Married Women	
Political Participation	High	120 60%	126 42%	246
	Low	80 40%	174 58%	254
		200	300	500

Table 13-7
No Children

		Marriage		
		Unmarried Women	Married Women	
Political Participation	High	114 63.7%	75 63.6%	189
	Low	65 36.3%	43 36.4%	108
		179	118	297

Table 13–8
Children

		Marriage		
		Unmarried Women	Married Women	
Political Participation	High	6 28.6%	51 28.0%	57
	Low	15 71.4%	131 72.0%	146
		21	182	203

married women are more likely to have children and that it is the care that children require which causes less political participation. To test this hypothesis, the researcher elaborated the relationship in Table 13 – 6 with respect to the test variable of having children by reexamining the relationship between marriage and political participation for women with children and for women without children. The results are presented in Tables 13 – 7 and 13 – 8. In each of these tables, being married has no effect upon level of political participation. Once again the relationship that existed in the original table has disappeared when controlled for a test variable. As can be expected, there are associations between the test variable (T) and the two original variables, marriage (X) and political participation (Y). These associations are shown in Tables 13 – 9 and 13 – 10. Tables 13 – 6 through 13 – 10 present a pattern similar to that in Tables 13 – 1 through 13 – 5. The two situations are not the same, however, because there is a

Table 13–9

		Marriage		
		Unmarried Women	Married Women	
Children	No	179 89.5%	118 39.3%	297
	Yes	21 10.5%	182 60.7%	203
		200	300	500

Table 13–10

		Children		
		Yes	No	
Political Participation	High	57 28.1%	189 63.6%	246
	Low	146 71.7%	108 36.4%	254
		203	297	500

different time order among the second set of variables than there was in the first.

In the first set of tables, T (sex) was antecedent to X (level of political information) and Y (foreign policy attitude). In the present situation, T, having children, usually occurs after marriage, X. This, of course, is not true 100% of the time, but, as can be seen in Table 13–8, is the case for the vast majority of women in the survey. Therefore, we can say that the time order of the variables in this situation is X first, then T, then Y, or that T is intervening between X and Y. An examination of Tables 13–6 through 13–10 shows that it is not marriage which itself causes lower levels of political participation among women but the fact that getting married leads to having children and having children leads to less participation, perhaps because children deprive mothers of time for political participation. This chain of events can be represented in symbols as

$$X \rightarrow T \rightarrow Y$$

What we have done is to "interpret" the original relationship between X and Y through an intervening test variable. The "causal" sequence, then, is not directly from X to Y, as Table 13–6 would make it seem, but from X to T and from T to Y. This type of statistical elaboration is called *interpretation*.

In both explanation and interpretation as types of statistical elaboration, an association between X and Y is present when no third variable is being controlled. When a particular test variable is controlled by the researcher, the association between X and Y disappears or is reduced in magnitude. The time ordering among X, Y, and T determines whether the elaboration is one of explanation or interpretation. It is the responsibility of the researcher, based upon his substantive knowledge of the problem, to account for the results.

Table 13–11

		Exposure to News Media		
		High	Low	
Knowledge of Political Issues	High	140 25.5%	65 26.0%	205
	Low	410 74.5%	185 74.0%	595
		550	250	800

In some situations it is possible that there will be no association between X and Y originally, but an association will appear when the table is elaborated with regard to a particular test variable. There are again two different types of statistical elaboration that are possible in these situations, depending upon the time ordering among the variables. We now turn to an examination of such situations.

3. *Specification.* A researcher is interested in whether exposure to the news media leads to increasing knowledge of political issues among college students. He surveyed a random sample of 800 college students, the results of which are presented in Table 13 – 11. This table shows that there is just about no difference between students who have a high or low exposure to the news media in their level of knowledge about political issues. Considering this finding, the researcher recalls some elementary psychology that people learn most effectively about things in which they are interested. He therefore hypothesizes that if people have a general interest in politics, increasing exposure to the news media will lead them to learn more about political issues. But that if people are generally uninterested in politics, increasing exposure to the news media will not result in their learning more about political issues.

To test these hypotheses, the researcher reexamines the association between news media exposure and knowledge of political issues controlling for level of interest in politics. The results are presented in Tables 13 – 12 and 13 – 13. The tables support the researcher's new hypotheses. For people who have a low general interest in politics, amount of exposure to the news media makes no difference in their level of knowledge about political issues. But among people who have a high general interest in politics, those who have a high level of exposure to the news media are much more likely to have a high knowledge of political issues than are those with a low level of exposure to the news

Table 13–12
Low Interest in Politics

		Exposure to News Media		
		High	Low	
Knowledge of Political Issues	High	50 12.5%	15 12.5%	65
	Low	350 87.5%	105 87.5%	455
		400	120	520

media. Thus, where originally there was no association between amount of exposure to the news media (X) and knowledge of political issues (Y), an association between these variables appeared for one condition of a test variable, general level of interest in politics. This example shows that we may miss important relationships among variables if we stop the research process at the point of examining the association between two variables without exploring the effects of test variables.

One of the important points about statistical elaboration is the increase in refinement and sophistication which it causes in our ideas. To suppose that exposure to the news media affects a person's knowledge of political issues is a nice, but simple, idea. To say that the effect is present for those with a high general interest in politics but not for those with a low level of interest in politics is a much more sophisticated idea and one that does more justice to the complexities of the world than does the original more simple idea.

Table 13–13
High Interest in Politics

		Exposure to News Media		
		High	Low	
Knowledge of Political Issues	High	90 60.0%	50 38.5%	140
	Low	60 40.0%	80 61.5%	140
		150	130	280

In situations where there is no association between X and Y when uncontrolled for any test variable but where an association appears when a particular test variable (T) is controlled, the type of statistical elaboration involved again depends upon the time order among X, Y, and T. In the present problem, the researcher has hypothesized that a person's general level of interest in politics, whether high or low, is determined before the occurrence of either the variable of exposure to the news media or the variable of knowledge about political issues. Thus the time order in this situation is that T is antecedent to X and Y. This type of statistical elaboration is called *specification* because it specifies a condition that first must exist for a relationship to appear between X and Y. It is symbolized as:

$$T: X \rightarrow Y$$

In our problem the elaboration has told us that the prior condition of a high general interest in politics (a specific value of T) must hold if there is to be an association between level of exposure to the news media (X) and amount of knowledge about political issues (Y).

4. *Contingency.* Suppose that a researcher was interested in seeing if the amount of concern people felt about which candidate won an election was related to whether or not they voted on election day. He surveyed a sample of 400 people and asked them if they felt a great deal of concern over who won the election. The results are presented in Table 13–14.

Table 13–14 shows that in the sample those people who were greatly concerned over who wins were somewhat more likely to vote than those who did not express a great deal of concern. The difference of 6.5%, however, is slight, and it is likely that it is a result of sampling error. The value of chi-square for this table is 1.49 and is not statistically significant. More impor-

Table 13–14

		Great Concern Over Who Wins		
		Yes	No	
Vote	Voted	170 60.7%	65 54.2%	235
	Did Not Vote	110 39.3%	55 45.8%	165
		280	120	400

tantly, such a slight difference is not of much substantive interest. Consequently, it must be concluded that the amount of concern people have about who wins has no effect upon whether or not people vote.

In analyzing this result, the researcher concluded that people probably feel that voting is a duty which they learn through years of "citizenship training" in the schools and the media. And that this training would tend to override the effects of differences caused by concern over who won. Further reflection led him to speculate that among people who were in situations where for some reason it was difficult for them to get to the polls, the extra motivation of being very concerned over who won would lead such people to make an extra effort to get to the polls. The researcher hypothesized, then, that among people who found themselves in situations where it was difficult to get to the polls on election day, concern over who won would be related to whether or not a person voted. One factor that makes a difference in ease of voting is urban–rural residence. People in rural areas generally have more difficulty in getting to the polls than do urban residents.

The researcher reexamined the association between concern over who wins and voting, controlling for urban or rural residence. The results are presented in Tables 13–15 and 13–16. Table 13–15 shows that in urban areas there is no relationship between degree of concern over who wins and voting or not voting. Table 13–16, however, shows that in rural areas there is a strong relationship between these variables. In rural areas concern over who wins makes a 23.3% difference in voting. In this elaboration, X is concern over who wins, Y is voting or not voting, and T is urban or rural residence. Although there was no association between X and Y originally, an association appeared when the relationship was controlled for T.

Table 13–15

Urban

		Great Concern Over Who Wins		
		Yes	No	
Vote	Voted	115 56.1%	50 55.6%	165
	Did Not Vote	90 43.9%	40 44.4%	130
		205	90	295

Table 13–16

Rural

| | | Great Concern Over Who Wins | | |
		Yes	No	
Vote	Voted	55 73.3%	15 50.0%	70
	Did Not Vote	20 26.7%	15 50.0%	35
		75	30	105

The time order among these variables is tricky and is dependent upon the interpretation of the problem. For most people, residence, urban or rural, occurs prior in time to their developing a concern over who wins a particular election and to voting in that election. But what we want to get at by urban–rural residence is difficulty in getting to the polls on election day, not just where a person lives. If a farmer happened to be in town for business reasons on election day he has less difficulty in getting to the polls than a farmer who has to plow his fields that day. Difficulty in getting to the polls is something which occurs on election day and probably after a person has developed a high or low level of concern about who wins. The variable of urban–rural residence is being used in this situation as an index of probable difficulty in getting to the polls. Therefore, we will interpret the time ordering among the variables as X first, then T, and then Y.

What the analysis has uncovered, then, is that an association appears between X and Y when a contingent condition is fulfilled: There is a relationship between degree of concern over who wins and voting contingent upon the presence of some difficulty in getting to the polls. According to the hypotheses, this happens because concern provides some extra motive power that makes a difference in whether or not people vote when conditions (rural residence) make it difficult for them to get to the polls. Thus the causal sequence is not from X to T to Y as in interpretation, but from X to Y contingent upon the presence of a certain value of T which in this case is difficulty in getting to the polls *as measured* by rural residence. This type of elaboration, then, is called *contingency* and is symbolized as

$$X \xrightarrow{T} Y$$

IDENTIFYING TYPES
OF STATISTICAL ELABORATION

The examples that have been considered in this discussion have presented rather clear-cut instances of each type of elaboration. It is likely that in real problems the patterns will not be so sharp. The thing to look for is what happens to the strength of the association between X and Y when the test variable is uncontrolled and when it is controlled. It is the presence of a definite change in the strength of association in these situations which indicates the relevance of the test variable to the relationship between X and Y and signals the possibility of an interpretation according to one of the types of statistical elaboration.

In the examples used to illustrate explanation and interpretation, an association between X and Y disappeared when T was controlled. It is not necessary that the association completely disappear to have one of these types of statistical elaboration. If controlling for T substantially reduces the size of the original association between X and Y, the elaboration is one of explanation or interpretation. Similarly, the examples used to illustrate specification and contingency showed an association appearing when T was controlled when there was no association between X and Y originally. But it is also possible to have some degree of association between X and Y originally which substantially increases in size when T is controlled. Such situations will also be elaborations of either specification or contingency. What constitutes a substantial change in the magnitude of an association depends upon what size change the researcher thinks is important in terms of the substance of his problem.

What type of statistical elaboration is present in a particular situation is determined by two factors. The first is whether or not there is an association present between X and Y when uncontrolled for T and what change takes place in that association when T is controlled. If controlling for T has no effect upon the association, then T, at least by itself, is not relevant to the problem. If there is an association between X and Y and controlling for T causes that association to disappear or be substantially reduced in magnitude, the elaboration is either explanation or interpretation. If there is no association between X and Y and one appears when T is controlled or if there is an association between X and Y which substantially increases in size when T is controlled, the elaboration is either specification or contingency. The second factor is the time order as interpreted by the researcher among X, Y, and T. If T is antecedent to both X and Y, the elaboration is either explanation or contingency. If T is intervening between X and Y, the elaboration is either interpretation

Table 13–17
Types of Statistical Elaboration

Is There an Association between X and Y when Uncontrolled for T?	What Happens to Original Association between X and Y when T Is Controlled?	Is T Intervening or Antecedent to X and Y?	Type of Elaboration	Symbolic Representation
1. Association	Association disappears	Antecedent ⟶ Explanation Intervening ⟶ Interpretation		$T \nearrow^{X}_{\searrow Y}$
2. Association	Association is substantially reduced in size	Antecedent ⟶ Interpretation Intervening ⟶		$X \rightarrow T \rightarrow Y$
3. No association	Association appears	Antecedent ⟶ Specification Intervening ⟶ Contingency		$T{:}X \rightarrow Y$
4. Assocation	Association substantially increases size	Antecedent ⟶ Intervening ⟶		T $X \longrightarrow Y$

or contingency. Combining these two factors determines which one of the four types of elaboration is present in a particular situation. Table 13 – 17 summarizes this process.

Question 13–2. Percentage the three tables constructed in Question 13 – 1 in such a way as to examine the effect of the independent variable on the dependent variable.

a. Fill in these percentages in the following tables:

		Party Identification	
		Dem.	Rep.
Attitude toward Integration	Favor		
	Oppose		

Less Education

		Party Identification	
		Dem.	Rep.
Attitude toward Integration	Favor		
	Oppose		

More Education

		Party Identification	
		Dem.	Rep.
Attitude toward Integration	Favor		
	Oppose		

b. Name the independent variable, the dependent variable, and the test variable in these tables.

c. Use the percentages calculated above to explain verbally what is happening in the uncontrolled table.
d. What is happening in the control tables?
e. Explain what type of statistical elaboration is involved in these tables.
f. Construct a verbal explanation that accounts for the results in all three tables. _____

PROBLEMS IN
STATISTICAL ELABORATION

In the examples, percentages have been used to measure the strength of association in the tables. Differences in percentages are fairly easy to interpret in the case of 2×2 tables. In tables larger than 2×2, percentages may become quite difficult to interpret. This problem may be avoided by using a measure of association such as one of those discussed in Chapter 10. The measure of association is used to measure the strength of association between X and Y when the test variable is uncontrolled and when it is controlled. Instead of comparing changes in the difference of percentages, changes in the magnitude of the measure of association are compared. The same patterns of changes determine the type of elaboration.

It is possible to expand the elaboration process to control for two or more variables at the same time. If it were desired to control for, say, sex and religion, where religion had the values of Protestant, Catholic, Jew, other, and none, 10 control tables would be generated. The first would be for male Protestants, the second for female Protestants, the third for male Catholics, the fourth for female Catholics, and so on. It can be seen that to control for three or more variables at the same time increases the number of control tables geometrically. The use of computers makes it quite easy to generate tables using lots of controls. It is still necessary, though, that a researcher judiciously construct and select control variables or he may be overwhelmed by an unmanageable number of tables.

The use of tables imposes a serious limitation upon the process of statistical elaboration. This limitation arises from the fact that when a test variable is controlled, a different table is constructed for each value of the control variable. As was discussed before, variables can be employed which have more than two values. Consequently, it is possible to generate three, four, or more tables when controlling for a single test variable. When the test variable is controlled, the sample is physically divided into subgroups so that observations in each subgroup have the same value on the test variable. Although there is no theoretical limit on the number of subgroups into which the sample may be divided, we are limited by the practical need to keep each subgroup large enough so that reliable information may be derived from it. Thus, if an association in a sample of 100 people were being considered, to control for sex would generate two tables with about 50 people in each. But if religion were the test variable, the controlled tables might consist of 60 Protestants, 20 Catholics, 5 Jews, 10 others, and 5 none. Except for the Protestant subgroup, none of the other control tables contain enough observations to yield reliable information. If, in addition, the X and Y variables have more than two values, the problem becomes even more serious because the observations within each table have to be distributed among a greater number of cells. This same difficulty limits the number of variables that may be controlled for simultaneously. When controlling for sex and religion simultaneously in the illustration discussed above, the sample would have to be divided into 10 subgroups. In the next section we consider a controlling process for interval and ratio level variables which does not require the sample to be physically divided into subgroups and thus avoids these problems.

PARTIAL CORRELATION

In Chapter 11 the Pearson product-moment correlation coefficient was introduced as a measure of the association between two interval or ratio level variables. It is possible to extend the Pearson correlation coefficient to measure the association between two variables while simultaneously controlling for the effects of a third variable. This process is called *partial correlation*.

In Chapter 11 the problem was posed of determining the extent to which education accounted for income. In order to simplify our illustrations, let us assume that in another set of data, the correlation between education and income is found to be .50. If we call income variable 1 and years of education variable 2, this correlation will be written as

$$r_{12} = .50$$

This is to be read as "the correlation between variables 1 and 2 is .50." As we found, this coefficient can be interpreted to mean that 25% ($r_{12}^2 = .25$) of the variation in people's incomes in this set of data is associated with or accounted for by the variable years of education.

Suppose, however, that a researcher feels that age may be influencing the value of this correlation. Poorly educated older people may have a job that originally was low-paying but is now relatively high-paying because of their long service. If this is true, some relatively high paying jobs are determined by age and not because of education. The researcher checks this possibility by calculating the correlation between income and age and finds it to be .40. If age is called variable 3, this correlation is written as

$$r_{13} = .40$$

Having established that age as well as education is related to income, the researcher wishes to find the extent to which education affects income independent of any effects of age. This can be done by calculating what is called the "partial" correlation coefficient. The *partial correlation coefficient* measures the amount of association between two variables after the effects of a third variable on the first two have been removed. The calculation of a partial correlation coefficient is another way of controlling for a test variable. In statistical elaboration using tables, the observations are physically divided into subgroups to hold a third variable constant. In partial correlation the observations are not split into subgroups. Rather, the amount of variation in the independent (education) and dependent (income) variables that is associated with the third variable (age) is removed mathematically and the correlation between the independent and dependent variables is computed on the variation that remains. Consequently, it is not necessary to divide the observations into subgroups.

There is an important difference in the interpretation of controlling through partial correlation and controlling through tables. When control tables are constructed, ideally the value of the third (test) variable in any control table is constant. Thus, within any one control table the third variable can have no effect, since it is not allowed to vary. Controlling through tables is accomplished, then, by holding the third variable constant. In partial correlation, however, the third variable is not held constant but is allowed to vary. What is done is to let the third variable account for as much of the variation in the original two variables as it can. Then its effects are removed from consideration and the correlation between the original two variables is calculated on the variation that remains.

The formula for the partial correlation coefficient is

$$r_{ij \cdot k} = \frac{r_{ij} - (r_{ik})(r_{jk})}{\sqrt{1 - r_{ik}^2} \sqrt{1 - r_{jk}^2}}$$

where $r_{ij \cdot k}$ is the partial correlation between variables i and j controlling for variable k. It is conventional to use a dot in writing the partial correlation coefficient to indicate that the variable to the right of the dot is the one that is being controlled.

Examination of the formula for the partial correlation coefficient shows that its use requires the values of the correlations among all three variables. In our present problem two of these coefficients have already been determined. The last one is the correlation between age and education. Suppose that the researcher calculates this to be

$$r_{23} = -.30$$

This negative correlation indicates that younger people tend to have higher levels of education than older people.

We can now rewrite the formula for the partial correlation in terms of numbers that have been used to designate each variable:

$$r_{12 \cdot 3} = \frac{r_{12} - (r_{13})(r_{23})}{\sqrt{1 - r_{13}^2}\ \sqrt{1 - r_{23}^2}}$$

Substituting the values given above for the various coefficients yields

$$r_{12 \cdot 3} = \frac{.50 - (.40)(-.30)}{\sqrt{1 - (.40)^2}\ \sqrt{1 - (-.30)^2}}$$

$$= \frac{.50 + .12}{\sqrt{1 - .16}\ \sqrt{1 - .09}}$$

$$= \frac{.62}{\sqrt{.84}\ \sqrt{.91}}$$

$$= \frac{.62}{(.92)(.95)}$$

$$= \frac{.62}{.87}$$

$$= .71$$

The partial correlation, then, between education and income controlling for the effects of age is .71. The situation in this problem is a "contingency" type of elaboration. The original association between education and income uncontrolled for any test variable was .50, as measured by the correlation coefficient. When the test variable, age, was controlled, the association between education and income increased to .71, as measured by the partial correlation coefficient. Age is an intervening variable to education and income. When its effects are controlled, education is more strongly related to income.

The partial correlation coefficient is interpreted in a way similar to the ordinary correlation coefficient. When the partial correlation coefficient is squared, it indicates the percentage of the variance remaining in the dependent variable that can be accounted for by the independent variable *after* the variance in these two variables that is associated with the control variable has been removed from consideration. Thus, after the effects of age have been removed, education accounts for 50% ($r^2_{12 \cdot 3} = .50$) of the remaining variance in income.

It is possible to extend partial correlation to control for more than one variable at the same time. For two control variables the symbol for the partial correlation is

$$r_{ij \cdot kl}$$

indicating the association between variables i and j after removing the effects of variables k and l. Such a partial correlation is called a *second-order partial correlation* because there are two control variables. The partial correlation between education and income controlling for age is called a *first-order partial correlation* because there is only one control variable. The original correlation between education and income is called a *zero-order correlation* because no control variables were involved. It is possible to compute higher-order partial correlations (third-order, three control variables; and so on), although the arithmetic becomes overbearing. In any case, the interpretation of a partial correlation is the same, the extent to which two variables are associated after removing the effects of the control variables.

Question 13–3. In a random sample of male high school students, the correlation between height and time in the mile run is $r_{12} = -.75$. This correlation shows that the taller boys are (variable 1), the faster they run the mile (variable 2). The correlation between height and age is $r_{13} = .89$, and the correlation between time in running the mile and age is $r_{23} = -.83$.

a. Compute the partial correlation between height and time in running the mile controlling for age. Use the formula

$$r_{12 \cdot 3} = \frac{r_{12} - (r_{13})(r_{23})}{\sqrt{1 - r^2_{13}} \, \sqrt{1 - r^2_{23}}}$$

b. What pattern of statistical elaboration is this? Explain.

c. Construct an explanation that can account for the zero-order correlations and the partial correlation. _____

MULTIPLE
CORRELATION AND REGRESSION

In Chapter 11 linear regression was considered as a means to predict the scores of an interval or ratio level dependent variable from one interval or ratio level independent variable. In linear regression, the prediction is based upon the formula for a straight line that "best fits" the data. The general formula for this regression is

$$Y = a + bX$$

The basic regression model can be extended in a number of ways. One extension is to situations where two or more interval or ratio level independent variables are used to predict the scores of one interval or ratio level dependent variable. This technique is called *multiple regression*. If there are two independent variables, X_1 and X_2, the multiple regression model will be:

$$Y = a + b_1X_1 + b_2X_2$$

This formula does not represent a straight line, as was the case in linear regression with a single variable, but a plane. Thus, instead of speaking about a regression line as we did before, we speak of a regression plane when two independent variables are involved. But, as in the case of one independent variable, the regression plane is the plane that best fits the data in the sense of least squares.

If two independent variables are involved, Figure 13 – 1 shows the regression plane and the data points in three dimensions. In Figure 13 – 1, the two independent variables are represented on the axes labeled X_1 and X_2 and the dependent variable, Y, on the vertical axis. The data points are scattered through three dimensions. The regression plane is that plane which passes through the data points in such a way as to minimize the sum of the squared Y deviations from the plane. This reasoning is exactly analogous to the selection of a regression line in the case of a single independent variable. In cases where there are more than two independent variables, a geometric representation is not possible. In such cases, the data points may be thought of as scattered through a space of more than three dimensions, the number of dimensions being equal to 1 plus the number of independent variables. But in all cases, the regression formula defines a *plane* which best fits the data points in the least-squares sense. (With more than two independent variables, the plane involved is not a two-dimensional one and is properly called a *hyperplane*.) By calculating the coefficients of this best-fit plane, a formula can be specified that best predicts the scores of the dependent variable from a knowledge of scores on the independent variable.

In multiple regression, the a coefficient of the formula is the point where the regression plane intersects the Y axis. Thus, a is the Y intercept just as it was in the case of linear regression with one independent variable. The b coefficients are a little more complicated. If a plane is drawn perpendicular to one of the X axes, its intersection with the re-

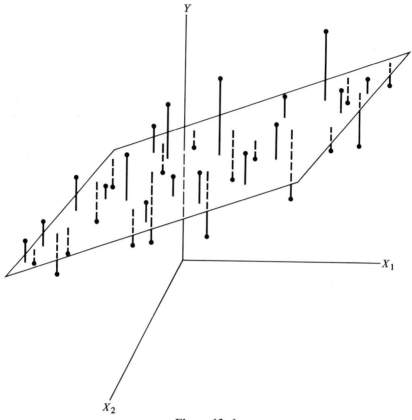

Figure 13–1
Representation of the Least Squares Multiple Regression Plane in a Scatter of Data Points with Two Independent Variables

gression plane will be a straight line. Figure 13–2 illustrates this for a plane drawn perpendicular to the X_2 axis. Since the vertical plane in Figure 13–2 is drawn perpendicular to the X_2 axis, variable X_2 is, in effect, being held constant, since all points in this plane have the same X_2 value. The intersection of this plane with the regression plane forms a straight line. The points on this line also all have the same X_2 value, since they are part of the plane that was drawn perpendicular to the X_2 axis. The points on the line, then, can vary only in their X_1 values and their Y values. This line is the regression line of Y on X_1 with X_2 being held constant. The slope of this line is b_1. In other words, if variable X_2 is held constant, b_1 is the slope of the regression line of Y on X_1. It does not matter at what point we draw a plane perpendicular to the X_2 axis. The intersection of all planes perpendicular to the X_2 axis will be straight

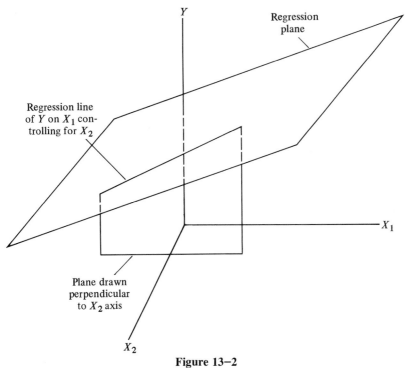

Figure 13–2
Representation of the Regression Line of Y on X_1 in a Multiple Regression with Two Independent Variables

lines with the same slopes, b_1. Similarly, a plane may be drawn perpendicular to the X_1 axis. Now, variable X_1 is held constant and the intersection of such a plane with the regression plane yields a regression line of Y on X_2 whose slope is b_2. The slopes b_1 and b_2 are called *partial slopes*. They give the slope of the regression line of Y on one of the independent variables controlling for all the other independent variables in the equation.

Up to this point, the dependent variable has been referred to as Y, the first independent variable as X_1, the second independent variable as X_2, and so on. It will now be useful to change the terminology somewhat. We will refer to the variables simply as 1, 2, 3, . . ., n, where 1 is the *dependent* variable, 2 is the *first independent* variable, 3 is the *second independent* variable, and n is the last independent variable. All variables will now be represented by X. In partial correlation, $r_{12 \cdot 34}$ represents the correlation between variables 1 and 2 after controlling the effects of variables 3 and 4. Since a partial slope in multiple regression refers to the regression line of the dependent variable on one of the independent variables controlling for all other independent variables, the new notation can be used to reflect this in the same way as is done with partial correlation coefficients. Thus, the slope for the first independent variable in a multiple regression where there are three independent variables may be written as $b_{12 \cdot 34}$. Since variable 1 is the dependent variable and variable 2 is the first independent variable, this symbol represents the slope of the regression line of the dependent variable on the first independent variable controlling for the second and third independent variables. The regression equation for three independent variables would be written as

$$X_1 = a_{1 \cdot 234} + b_{12 \cdot 34}X_2 + b_{13 \cdot 24}X_3 + b_{14 \cdot 23}X_4$$

where $a_{1 \cdot 234}$ is used to indicate that we are predicting scores on variable 1 from a knowledge of variables 2, 3, and 4. The partial slopes in a multiple regression equation tell us how much of a change in the dependent variable will be caused by a given change in one independent variable, holding all the other independent variables constant.

Suppose that we were trying to predict people's salaries in a company from a knowledge of their years of service and their years of education. Symbolically, the regression equation would be

$$X_1 = a_{1 \cdot 23} + b_{12 \cdot 3}X_2 + b_{13 \cdot 2}X_3$$

Suppose that the coefficients of the regression equation were calculated and found to be

$$X_1 = -5000 + 200X_2 + 1000X_3$$

For a person who worked for the company 15 years and had 12 years of education, the equation would predict his salary as

$$X_1 = -5000 + (200)(15) + (1000)(12)$$
$$= -5000 + 3000 + 12,000$$
$$= 10,000$$

For a person who worked for the company 5 years and had 16 years of education, the prediction would be

$$X_1 = -5000 + (200)(5) + (1000)(16)$$
$$= -5000 + 1000 + 16,000$$
$$= 12,000$$

The value of a in this regression equation is not very meaningful in this problem. It says that if a person had no years of education and no years of service with the company, he would have a salary of -5000 dollars, which is nonsensical, of course. The problem is that everyone working for the company has a substantial number of years of education, probably at least eight, and most have some number of years of service. Thus, there really is no one who is close to zero on both these variables. The value of a then is obtained by extending the regression plane which best fits the data points until it hits the Y axis. And it happens that the regression plane is inclined in such a way that when it is extended, it hits the Y axis at the point -5000.

The partial slope for variable 2 (years of service) indicates that for an increase in 1 year of service, the predicted change in salary will be an increase of \$200, when education is held constant. The partial slope for variable 3 (years of education) indicates that for an increase of 1 year of education, the predicted change in salary will be an increase of \$1000 when years of service is held constant.

At this point we might inquire as to which of the independent variables in a regression equation is the most important: that is, which independent variable is the most important determinant of scores on the dependent variable. In most cases this information cannot be obtained from the partial slopes. The reasons are that (1) ordinarily, the units of measurement of the various independent variables are not the same, and that (2) the ranges of the values of the independent variables are usually not the same. In the present example, even though "years" are used in both independent variables, the units are not the same: A year of service is not the same thing as a year of education. In addition, the two variables are spread over different ranges. It is likely that employees have at least a grade school education. Thus, the variable "years of education" is probably restricted to a range from 8 to about 16 (college degree) or at most 20 (a doctorate). The variable of years of service, however, may range from 0 to 40 or even 45. Thus, years of education has a much more restricted range than does years of service. Another way of saying this is that the standard deviation for the variable "years of education" is smaller than the standard deviation for the variable "years of service." Consequently, a 1-unit (year) change is a bigger change for the variable years of education than it is for the variable years of service, and absolute amounts of change in the two variables are not directly comparable.

Let us assume that the standard deviations of the three variables are

$$s_1(\text{salary}) = 6000$$

$$s_2(\text{years of service}) = 8$$

$$s_3(\text{years of education}) = 3$$

To determine the relative importance of the independent variables, it is necessary to obtain a common unit of measurement for the two variables. This is done by multiplying the partial slopes by the standard deviation of the independent variable, divided by the standard deviation of the dependent variable. The resulting coefficients are called *beta weights*, symbolized as β. Thus

$$\beta_{12 \cdot 3} = b_{12 \cdot 3}\left(\frac{s_2}{s_1}\right)$$

$$\beta_{13 \cdot 2} = b_{13 \cdot 2}\left(\frac{s_3}{s_1}\right)$$

For the present example, the beta weights would be

$$\beta_{12 \cdot 3} = (200)\left(\frac{8}{6000}\right)$$

$$= .27$$

$$\beta_{13 \cdot 2} = (1000)\left(\frac{8}{6000}\right)$$

$$= .50$$

Beta weights are best interpreted in terms of standard deviations. For variable X_2, the beta weight tells us that if years of service increases by 1 standard deviation (8 years), salary will increase by .27 standard deviation, or $1620 [(.27)(6000) = 1620]$. If years of education increases by 1 standard deviation (3 years), salary will increase by .50 standard deviation, or $3000 [(.50)(6000) = 3000]$. The beta weights tell us what will happen to the dependent variable if an independent variable is increased 1 standard deviation in value and all other independent variables are held constant. Beta weights are directly comparable since they indicate what happens when a standardized change – 1 standard deviation – takes place in an independent variable. When 1 standard deviation of change takes place in years of service, salary increases by 1620. But when 1 standard deviation of change takes place in years of education, salary increases by $3000. Thus, years of education is 1.85 times as important as years of service in determining salary ($\frac{3000}{1620} = 1.85$). This information can be obtained from a direct comparison of the beta weights: $.50/.27 = 1.85$.

Finally, partial correlation coefficients could be used to assess the impact of the individual independent variables on the dependent variable. The partial correlation of salary with years of service controlling for years of education is

$$r_{12 \cdot 3} = .316$$

And the partial correlation of salary with years of education controlling for years of service is

$$r_{13 \cdot 2} = .525$$

Squaring these partial correlations will tell us how much of the variance in the dependent variable is accounted for by one independent variable after removing the effects of the other independent variable:

$$r_{12 \cdot 3}^2 = (.316)^2 = .100$$

$$r_{13 \cdot 2}^2 = (.525)^2 = .276$$

According to the squared partials, years of service accounts for 10% of the variation in salary after controlling for education and years of education accounts for 27.6% of the variation in salary after controlling for service. The partial correlations, then, also indicate that years of education is more important than years of service in determining salary. As can be seen, partial correlations and beta weights do not give precisely the same evaluation of the relative importance of the independent variables, and this is true generally. Partial correlations and beta weights, though, will usually rank variables in the same order of importance.

In multiple regression, then, there are three different sets of coefficients that yield information about the relation of the independent variables to the dependent variable:

1. *Partial slopes.* Partial slopes are the unstandardized regression coefficients. A partial slope is the slope of the regression line of the dependent variable on one of the independent variables, controlling for all other independent variables. A partial slope tells us how much a 1-unit (unstandardized) change in this independent variable will affect the dependent variable when the other independent variables are controlled.

2. *Beta weights.* A beta weight is a standardized partial slope. A partial slope for a particular independent variable is standardized by dividing the standard deviation of that independent variable by the standard deviation of the dependent variable and multiplying the partial slope by the result. The beta weight tells us how much change in the dependent variable will take place as a result of a standardized change in one of the independent variables controlling for all other independent variables. Because it is the effects of a standardized amount of change that are measured (1 standard deviation) by the beta weights, they can be used to assess the relative importance of independent variables on the dependent variable.

3. *Partial correlations.* Partial correlations are not part of the multiple regression equation. The coefficients in a multiple regression

equation, whether partial slopes or beta weights, measure how much change in the dependent variable is produced by changes in the independent variables. A partial correlation coefficient measures how much of the variation in the dependent variable is associated with an independent variable, after the effects of the other independent variables have been removed.

Question 13–4. A researcher is interested in predicting how often congressmen vote Democratic (as defined by how a majority of Democratic congressmen vote). Multiple regression is used and the following regression equation is obtained:

$$Y = 15 + 1.3X_1 - .002X_2 - .05X_3$$

where Y = percentage of the time voting Democratic
X_1 = percentage Democratic vote in congressional districts in the last election
X_2 = median income in congressional districts
X_3 = average age of adults in congressional districts

a. Explain what each of the coefficients in the equation means.
b. The standard deviations of the variables in the regression equation are as follows:

$$s_Y = 18 \qquad s_{X_2} = 5000$$
$$s_{X_1} = 8 \qquad s_{X_3} = 7$$

Compute the beta weights for the independent variables. Which independent variable is most important in determining whether congressmen vote Democratic? Explain what the beta weight for this variable means.

———

THE MULTIPLE
CORRELATION COEFFICIENT

While partial slopes, beta weights, and partial correlation coefficients can be used to yield information about how the independent variables in the multiple regression equation individually affect the dependent variable, we would like to have a measure of how all the independent variables, taken together, affect the dependent variable. This is done through the multiple correlation coefficient. The multiple correlation coefficient is a measure of the extent to which two or more independent variables taken together can account for the scores in a dependent variable.

One use of a multiple regression equation is to attempt to predict the scores of the dependent variable from a knowledge of the independent variables. In the example from the preceding section, the multiple regression equation was

$$Y' = -5000 + 200X_1 + 1000X_2$$

where Y' is now used to indicate what the multiple regression equation predicts an employee's salary to be. Suppose that a particular employee has worked for the company for 7 years and has 15 years of education and his actual salary is $11,000. The regression equation predicts his salary to be

$$Y' = -5000 + (200)(7) + (1000)(15)$$
$$= -5000 + 1400 + 15,000$$
$$= 11,400$$

The prediction from the regression equation is in error by $400 since the employee's actual salary (Y) is $11,000. If we were to obtain the actual salary (Y) and the predicted salary (Y') for every employee and correlate them, this correlation is the multiple correlation coefficient, symbolized as R. It is a measure of how well the independent variables in a regression equation predict the actual scores on the dependent variable. If the actual scores of the dependent variable lay precisely in the least-squares regression plane, the regression equation would predict the actual scores exactly and the correlation between predicted and actual scores would be 1.0. The more the actual scores are scattered about the least-squares regression plane, the poorer will the regression equation predict the scores and the lower the correlation will be between predicted scores and actual scores.

Since the predicted scores are obtained from the regression plane that best fits the data, the multiple correlation coefficient will always be positive in value. And since it is possible that some of the variables in a regression equation are positively related to the dependent variable and others negatively related, the multiple correlation coefficient is not in itself particularly useful. More meaningful is the square of the multiple correlation coefficient, R^2. It specifies how much of the variation in the dependent variable can be accounted for by all the independent variables taken together. In the present example, $R = .63$ and $R^2 = .40$. This says that years of service and years of education taken together can account for 40% of the variation in salary. Consequently, 60% of the variation in salary still remains unaccounted for and is obviously due to other variables, not present in the regression equation. If other variables thought to affect salary could be identified, they could be added to the regression equation and we could see if the power to predict scores on salary was increased by seeing if the value of R^2 increased.

R^2 is *not* equal to the sum of the separate squared correlations of the independent variables with the dependent variable, *nor* to the sum of the separate squared partial correlations of the independent variables with the dependent variable. In the example these are

$$r_{12}^2 = .176 \qquad\qquad r_{12\cdot3}^2 = .100$$

$$\underline{r_{13}^2 = .336} \qquad\qquad \underline{r_{13\cdot2}^2 = .276}$$

$$r_{12}^2 + r_{13}^2 = .512 \qquad r_{12\cdot3}^2 + r_{13\cdot2}^2 = .376$$

As can be seen, neither of these sums equals .40, which is what R^2 was stated to be for this problem.

R^2, though, can be obtained from these figures. R^2 is the percentage of the variation in the dependent variable that can be "explained" by all the independent variables. This percentage of "explained" variation can be obtained through the following steps. The first independent variable is allowed to account for as much of the variation in the dependent variable as it can. This will be r_{12}^2. The percentage of variation that remains to be accounted for must be $1 - r_{12}^2$. We now let the second independent variable account for as much of this remaining variation as it can. But in doing so, we must control for the first independent variable to avoid duplication. Thus, we use the squared partial correlation $r_{13\cdot2}^2$ times the remaining variation $(1 - r_{12}^2)$. If more than two independent variables were involved, we would simply continue adding in the partial correlation of each independent variable, controlling for the previous ones already entered times the amount of variation remaining after the previous step. For our example with two independent variables, this would be

$$R_{1\cdot23}^2 = r_{12}^2 + r_{13\cdot2}^2(1 - r_{12}^2)$$

$$= .176 + (.276)(1 - .176)$$

$$= .176 + (.276)(.824)$$

$$= .176 + .227$$

$$= .403$$

When rounded, this is equal to the value of R^2 given earlier. (Rounding was performed in the earlier calculations to keep the figures simple.)

The computations for calculating the coefficients in multiple regression quickly become quite onerous as the number of independent variables increases. With the development of sophisticated computer programs, researchers are likely to compute multiple regressions on a computer. It is probably most useful, therefore, to have a conceptual understanding of the technique and of the coefficients that a multiple regression produces. Consequently, the actual computing formulas will not be presented here.

ASSUMPTIONS
IN MULTIPLE REGRESSION

The technique of multiple regression is not applicable to all situations in which a dependent variable is to be predicted from two or more independent variables. The types of multiple regression discussed here assume that the independent variables are linearly (straight-line) related to the dependent variable. But there is no reason why curvilinear relationships could not exist. There are curvilinear regression techniques to handle such situations. The important point is that linear multiple regression imposes the assumption of linearity on the data and we should be aware of this assumption when we use it.

A second assumption of linear multiple regression is that the independent variables are not highly intercorrelated with each other. Unfortunately, it often happens in the social sciences that the independent variables in a problem display high intercorrelations. For example, if we use "social-economic status of parent's family" and "prestige of the college from which a person graduated" to predict income earned, the two independent variables are probably highly correlated. Such a situation is called *multicollinearity*. When multicollinearity exists, multiple regression is likely to produce unstable and misleading results. Such situations present difficult problems, and more sophisticated techniques must be used.

Finally, multiple regression assumes that while the independent variables affect the dependent variable, there is no (or very little) feedback from changes in the dependent variable that affect the independent variables. If a substantial amount of feedback does exist, a different technique must be used.

EXERCISES FOR CHAPTER 13

1. Suppose that a researcher obtained the following tables from a survey of 500 people:

		Age Young	Old	
Political Attitude	Liberal	150	150	300
	Conservative	100	100	200
		250	250	500

Females

		Age Young	Old	
Political Attitude	Liberal	50	100	150
	Conservative	50	50	100
		100	150	250

Males

		Age Young	Old	
Political Attitude	Liberal	100	50	150
	Conservative	50	50	100
		150	100	250

Specify the independent, dependent, and test variables in this situation. Percentage the three tables to examine the effect of the independent variable on the dependent variable. What type of statistical elaboration is present in the tables? Why? Construct an explanation that can account for the results in all three tables.

2. In a survey of 300 people a researcher determines whether or not

people voted a straight ticket in the last election. In attempting to discover the causes of straight-ticket voting, the following tables are obtained:

		Education		
		High	Low	
Straight-Ticket Voting	Yes	70	115	185
	No	65	50	115
		135	165	300

Low Level of Information about Candidates

		Education		
		High	Low	
Straight-Ticket Voting	Yes	50	80	130
	No	15	25	40
		65	105	170

High Level of Information about Candidates

		Education		
		High	Low	
Straight-Ticket Voting	Yes	20	35	55
	No	50	25	75
		70	60	130

Specify the independent, dependent, and test variables in this situation. Percentage the three tables to examine the effect of the independent variable on the dependent variable. What type of statistical elaboration is present in this situation? Why? Construct an explanation that can account for the results in all three tables.

3. In a survey of college students, variables were: 1, cumulative grade-point average; 2, hours of study per week; and 3, years of education of parents. The following correlations were found among the three variables:

$$r_{12} = .50$$

$$r_{13} = .60$$

$$r_{23} = .70$$

Compute $r_{12.3}$. Interpret the value of this partial correlation in terms of the context of the situation. What type of statistical elaboration is present in this situation? Construct an explanation that can account for the zero-order and first-order correlations.

4. Data were obtained from the counties in a state on the following variables:

Y = educational expenditure per pupil

X_1 = percentage of adults registered as Democrats in counties

X_2 = median income of counties

A multiple regression yields the equation

$$Y = -200 + 5X_1 + .02X_2$$

What are the values of a, $b_{12.3}$, and $b_{13.2}$? Explain what these coefficients mean. If $s_Y = 70$, $s_{X_1} = 8$, and $s_{X_2} = 3000$, compute the beta weights for the regression equation. Which independent variable is most important in determining education expenditures in the counties? Why?

ANSWERS TO
QUESTIONS IN CHAPTER 13

13–1:

		Party Identification		
		Dem.	Rep.	
Attitude toward Integration	Favor	4	8	12
	Oppose	10	4	14
		14	12	26

Less Education

Party
Identification

		Dem.	Rep.	
Attitude toward Integration	Favor	0	0	0
	Oppose	9	2	11
		9	2	11

More Education

Party
Identification

		Dem.	Rep.	
Attitude toward Integration	Favor	4	8	12
	Oppose	1	2	3
		5	10	15

13–2: a.

Party
Identification

		Dem.	Rep.
Attitude toward Integration	Favor	28.6%	66.7%
	Oppose	71.4%	33.3%
		100%	100%

Less Education

Party
Identification

		Dem.	Rep.
Attitude toward Integration	Favor	0%	0%
	Oppose	100%	100%
		100%	100%

More Education

		Party Identification	
		Dem.	Rep.
Attitude toward Integration	Favor	80%	80%
	Oppose	20%	20%
		100%	100%

b. The independent variable is "party identification," the dependent variable is "attitude toward integration," and the test variable is "education."

c. The table shows that there is an association between party identification and attitude toward integration, with 66.7% of the Republicans in favor of integration compared to 28.6% of the Democrats. Thus, Republicans tend to be more in favor of integration than Democrats, and, conversely, Democrats tend to be more opposed to integration than Republicans.

d. In both control tables there is no association between party identification and attitude toward integration. Within each control table there is no difference in the percentages of Republicans and Democrats in favoring or opposing integration.

e. "Explanation."

f. One interpretation of these tables is that people with less education tend also to be Democratic party identifiers and people with more education tend to be Republican party identifiers, and that educational level affects people's attitude toward integration. Thus, it is educational level that really affects attitude toward integration, and it is the difference in educational level between Democrats and Republicans that makes a spurious association appear between party identification and attitude toward integration.

13–3: a. $r_{12 \cdot 3} = .043$.

b. "Explanation" or "interpretation," depending upon the interpretation of the problem and the time ordering among the variables.

c. If "interpretation," we might say that age leads to more height, which in turn leads to faster times. If "explanation," we might say that age leads to greater stamina, strength, and coordination, which in turn yields faster times; and it is these or similar things, but not height, which brings about faster times. Thus, the association between height and time in running the mile is explained away.

13–4: a. 15; this is the Y intercept, the point at which the regression plane intercepts the Y axis.

1.3; this is the partial slope of percentage time voting Democratic for the Democratic vote in congressional district in the last election. It indicates that for every 1-percentage-point increase in the Democratic vote in their district, congressmen will increase their Democratic voting in the House by 1.3% given that the other independent variables are held constant.

$-.002$; this is the partial slope of percentage time voting Democratic for median income in the congressional district. It indicates that for every $1 increase in median income in their congressional district, congressmen will decrease their voting Democratic in the House by .002%.

$-.05$; this is the partial slope of percentage time voting Democratic on average age of adults in congressional districts. It indicates that for every 1-year increase in average age of their congressional district, congressmen will decrease their voting Democratic in the House by .05%.

b. $\beta_{12 \cdot 34} = .52$, $\beta_{13 \cdot 24} = .55$, $\beta_{14 \cdot 23} = .02$. The median income in the congressional district is the most important determinant of the percentage of time congressmen vote Democratic in the House among the three independent variables. The beta weight for this variable indicates that for a 1-standard-deviation increase in median income ($5000), congressmen decrease their voting Democratic by .55 standard deviation (.55 of 18).

Appendix

Table A-1
Random Numbers*

10097	32533	76520	13586	34673	54876	80959	09117	39292	74945
37542	04805	64894	74296	24805	24037	20636	10402	00822	91665
08422	68953	19645	09303	23209	02560	15953	34764	35080	33606
99019	02529	09376	70715	38311	31165	88676	74397	04436	27659
12807	99970	80157	36147	64032	36653	98951	16877	12171	76833
66065	74717	34072	76850	36697	36170	65813	39885	11199	29170
31060	10805	45571	82406	35303	42614	86799	07439	23403	09732
85269	77602	02051	65692	68665	74818	73053	85247	18623	88579
63573	32135	05325	47048	90553	57548	28468	28709	83491	25624
73796	45753	03529	64778	35808	34282	60935	20344	35273	88435
98520	17767	14905	68607	22109	40558	60970	93433	50500	73998
11805	05431	39808	27732	50725	68248	29405	24201	52775	67851
83452	99634	06288	98083	13746	70078	18475	40610	68711	77817
88685	40200	86507	58401	36766	67951	90364	76493	29609	11062
99594	67348	87517	64969	91826	08928	93785	61368	23478	34113
65481	17674	17468	50950	58047	76974	73039	57186	40218	16544
80124	35635	17727	08015	45318	22374	21115	78253	14385	53763
74350	99817	77402	77214	43236	00210	45521	64237	96286	02655
69916	26803	66252	29148	36936	87203	76621	13990	94400	56418
09893	20505	14225	68514	46427	56788	96297	78822	54382	14598
91499	14523	68479	27686	46162	83554	94750	89923	37089	20048
80336	94598	26940	36858	70297	34135	53140	33340	42050	82341
44104	81949	85157	47954	32979	26575	57600	40881	22222	06413
12550	73742	11100	02040	12860	74697	96644	89439	28707	25815
63606	49329	16505	34484	40219	52563	43651	77082	07207	31790
61196	90446	26457	47774	51924	33729	65394	59593	42582	60527
15474	45266	95270	79953	59367	83848	82396	10118	33211	59466

*This table is reproduced with permission from the RAND Corporation, *A Million Random Digits.* Santa Monica, Calif., 1955.

Table A–1. *Continued*

94557	28573	67897	54387	54622	44431	91190	42592	92927	45973
42481	16213	97344	08721	16868	48767	03071	12059	25701	46670
23523	78317	73208	89837	68935	91416	26252	29663	05522	82562
04493	52494	75246	33824	45862	51025	61962	79335	65337	12472
00549	97654	64051	88159	96119	63896	54692	82391	23287	29529
35963	15307	26898	09354	33351	35462	77974	50024	90103	39333
59808	08391	45427	26842	83609	49700	13021	24892	78565	20106
46058	85236	01390	92286	77281	44077	93910	83647	70617	42941
32179	00597	87379	25241	05567	07007	86743	17157	85394	11838
69234	61406	20117	45204	15956	60000	18743	92423	97118	96338
19565	41430	01758	75379	40419	21585	66674	36806	84962	85207
45155	14938	19476	07246	43667	94543	59047	90033	20826	69541
94864	31994	36168	10851	34888	81553	01540	35456	05014	51176
98086	24826	45240	28404	44999	08896	39094	73407	35441	31880
33185	16232	41941	50949	89435	48581	88695	41994	37548	73043
80951	00406	96382	70774	20151	23387	25016	25298	94624	61171
79752	49140	71961	28296	69861	02591	74852	20539	00387	59579
18633	32537	98145	06571	31010	24674	05455	61427	77938	91936
74029	43902	77557	32270	97790	17119	52527	58021	80814	51748
54178	45611	80993	37143	05335	12969	56127	19255	36040	90324
11664	49883	52079	84827	59381	71539	09973	33440	88461	23356
48324	77928	31249	64710	02295	36870	32307	57546	15020	09994
69074	94138	87637	91976	35584	04401	10518	21615	01848	76938
03991	10461	93716	16894	66083	24653	84609	58232	88618	19161
38555	95554	32886	59780	08355	60860	29735	47762	71299	23853
17546	73704	92052	46215	55121	29281	59076	07936	27954	58909
32643	52861	95819	06831	00911	98936	76355	93779	80863	00514
69572	68777	39510	35905	14060	40619	29549	69616	33564	60780
24122	66591	27699	06494	14845	46672	61958	77100	90899	75754
61196	30231	92962	61773	41839	55382	17267	70943	78038	70267
30532	21704	10274	12202	39685	23309	10061	68829	55986	66485
03788	97599	75867	20717	74416	53166	35208	33374	87539	08823
48228	63379	85783	47619	53152	67433	35663	52972	16818	60311
60365	94653	35075	33949	42614	29297	01918	28316	98953	73231
83799	42402	56623	34442	34994	41374	70071	14736	09958	18065
32960	07405	36409	83232	99385	41600	11133	07586	15917	06253
19322	53845	57620	52606	66497	68646	78138	66559	19640	99413
11220	94747	07399	37408	48509	23929	27482	45476	85244	35159

Table A-1. *Continued*

31751	57260	68980	05339	15470	48355	88651	22596	03152	19121
88492	99382	14454	04504	20094	98977	74843	93413	22109	78508
30934	47744	07481	83828	73788	06533	28597	20405	94205	20380
22888	48893	27499	98748	60530	45128	74022	84617	82037	10268
78212	16993	35902	91386	44372	15486	65741	14014	87481	37220
41849	84547	46850	52326	34677	58300	74910	64345	19325	81549
46352	33049	69248	93460	45305	07521	61318	31855	14413	70951
11087	96294	14013	31792	59747	67277	76503	34513	39663	77544
52701	08337	56303	87315	16520	69676	11654	99893	02181	68161
57275	36898	81304	48585	68652	27376	92852	55866	88448	03584
20857	73156	70284	24326	79375	95220	01159	63267	10622	48391
15633	84924	90415	93614	33521	26665	55823	47641	86225	31704
92694	48297	39904	02115	59589	49067	66821	41575	49767	04037
77613	19019	88152	00080	20554	91409	96277	48257	50816	97616
38688	32486	45134	63545	59404	72059	43947	51680	43852	59693
25163	01889	70014	15021	41290	67312	71857	15957	68971	11403
65251	07629	37239	33295	05870	01119	92784	26340	18477	65622
36815	43625	18637	37509	82444	99005	04921	73701	14707	93997
64397	11692	05327	82162	20247	81759	45197	25332	83745	22567
04515	25624	95096	67946	48460	85558	15191	18782	16930	33361
83761	60873	43253	84145	60833	25983	01291	41349	20368	07126
14387	06345	80854	09279	43529	06318	38384	74761	41196	37480
51321	92246	80088	77074	88722	56736	66164	49431	66919	31678
72472	00008	80890	18002	94813	31900	54155	83436	35352	54131
05466	55306	93128	18464	74457	90561	72848	11834	79982	68416
39528	72484	82474	25593	48545	35247	18619	13674	18611	19241
81616	18711	53342	44276	75122	11724	74627	73707	58319	15997
07586	16120	82641	22820	92904	13141	32392	19763	61199	67940
90767	04235	13574	17200	69902	63742	78464	22501	18627	90872
40188	28193	29593	88627	94972	11598	62095	36787	00441	58997
34414	82157	86887	55087	19152	00023	12302	80783	32624	68691
63439	75363	44989	16822	36024	00867	76378	41605	65961	73488
67049	09070	93399	45547	94458	74284	05041	49807	20288	34060
79495	04146	52162	90286	54158	34243	46978	35482	59362	95938
91704	30552	04737	21031	75051	93029	47665	64382	99782	93478
19612	78430	11661	94770	77603	65669	86868	12665	30012	75989
39141	77400	28000	64238	73258	71794	31340	26256	66453	37016
64756	80457	08747	12836	03469	50678	03274	43423	66677	82556

Table A–1. *Continued*

92901	51878	56441	22998	29718	38447	06453	25311	07565	53771
03551	90070	09483	94050	45938	18135	36908	43321	11073	51803
98884	66209	06830	53656	14663	56346	71430	04909	19818	05707
27369	86882	53473	07541	53633	70863	03748	12822	19360	49088
59066	75974	63335	20483	43514	37481	58278	26967	49325	43951
91647	93783	64169	49022	98588	09495	49829	59068	38831	04838
83605	92419	39542	07772	71568	75673	35185	89759	44901	74291
24895	88530	70774	35439	46758	70472	70207	92675	91623	61275
35720	26556	95596	20094	73750	85788	34264	01703	46833	65248
14141	53410	38649	06343	57256	61342	72709	75318	90379	37562
27416	75670	92176	72535	93119	56077	06886	18244	92344	31374
82071	07429	81007	47749	40744	56974	23336	88821	53841	10536
21445	82793	24831	93241	14199	76268	70883	68002	03829	17443
72513	76400	52225	92348	62308	98481	29744	33165	33141	61020
71479	45027	76160	57411	13780	13632	52308	77762	88874	33697
83210	51466	09088	50395	26743	05306	21706	70001	99439	80767
68749	95148	94897	78636	96750	09024	94538	91143	96693	61886
05184	75763	47075	88158	05313	53439	14908	08830	60096	21551
13651	62546	96892	25240	47511	58483	87342	78818	07855	39269
00566	21220	00292	24069	25072	29519	52548	54091	21282	21296
50958	17695	58072	68990	60329	95955	71586	63417	35947	67807
57621	64547	46850	37981	38527	09037	64756	03324	04986	83666
09282	25844	79139	78435	35428	43561	69799	63314	12991	93516
23394	94206	93432	37836	94919	26846	02555	74410	94915	48199
05280	37470	93622	04345	15092	19510	18094	16613	78234	50001
95491	97976	38306	32192	82639	54624	72434	92606	23191	74693
78521	00104	18248	75583	90326	50785	54034	66251	35774	14692
96345	44579	85932	44053	75704	20840	86583	83944	52456	73766
77963	31151	32364	91691	47357	40338	23435	24065	08458	95366
07520	11294	23238	01748	41690	67328	54814	37777	10057	42332
38423	02309	70703	85736	46148	14258	29236	12152	05088	65825
02463	65533	21199	60555	33928	01817	07396	89215	30722	22102
15880	92261	17292	88190	61781	48898	92525	21283	88581	60098
71926	00819	59144	00224	30570	90194	18329	06999	26857	19238
64425	28108	16554	16016	00042	83229	10333	36168	65617	94834
79782	23924	49440	30432	81077	31543	95216	64865	13658	51081
35337	74538	44553	64672	90960	41849	93865	44608	93176	34851

Table A–1. *Continued*

05249	29329	19715	94082	14738	86667	43708	66354	93692	25527
56463	99380	38793	85774	19056	13939	46062	27647	66146	63210
96296	33121	54196	34108	75814	85986	71171	15102	28992	63165
98380	36269	60014	07201	62448	46385	42175	88350	46182	49126
52567	64350	16315	53969	80395	81114	54358	64578	47269	15747
78498	90830	25955	99236	43286	91064	99969	95144	64424	77377
49553	24241	08150	89535	08703	91041	77323	81079	45127	93686
32151	07075	83155	10252	73100	88618	23891	87418	45417	20268
11314	50363	26860	27799	49416	83534	19187	08059	76677	02110
12364	71210	87052	50241	90785	97889	81399	58130	64439	05614
94015	46874	32444	48277	59820	96163	64654	25843	41145	42820
74108	88222	88570	74015	25704	91035	01755	14750	48968	38603
62880	87873	95160	59221	22304	90314	72877	17334	39283	04149
11748	12102	80580	41867	17710	59621	06554	07850	73950	79552
17944	05600	60478	03343	25852	58905	57216	39618	49856	99326
66067	42792	95043	52680	46780	56487	09971	59481	37006	22186
54244	91030	45547	70818	59849	96169	61459	21647	87417	17198
30945	57589	31732	57260	47670	07654	46376	25366	94746	49580
69170	37403	86995	90307	94304	71803	26825	05511	12459	91314
08345	88975	35841	85771	08105	59987	87112	21476	14713	71181
27767	43584	85301	88977	29490	69714	73035	41207	74699	09310
13025	14338	54066	15243	47724	66733	47431	43905	31048	56699
80217	36292	98525	24335	24432	24896	43277	58874	11466	16082
10875	62004	90391	61105	57411	06368	53856	30743	08670	84741
54127	57326	26629	19087	24472	88779	30540	27886	61732	75454
60311	42824	37301	42678	45990	43242	17374	52003	70707	70214
49739	71484	92003	98086	76668	73209	59202	11973	02902	33250
78626	51594	16453	94614	39014	97066	83012	09832	25571	77628
66692	13986	99837	00582	81232	44987	09504	96412	90193	79568
44071	28091	07362	97703	76447	42537	98524	97831	65704	09514
41468	85149	49554	17994	14924	39650	95294	00556	70481	06905
94559	37559	49678	53119	70312	05682	66986	34099	74474	20740
41615	70360	64114	58660	90850	64618	80620	51790	11436	38072
50273	93113	41794	86861	24781	89683	55411	85667	77535	99892
41396	80504	90670	08289	40902	05069	95083	06783	28102	57816
25807	24260	71529	78920	72682	07385	90726	57166	98884	08583
06170	97965	88302	98041	21443	41808	68984	83620	89747	98882

Table A–1. *Continued*

60808	54444	74412	81105	01176	28838	36421	16489	18059	51061
80940	44893	10408	36222	80582	71944	92638	40333	67054	16067
19516	90120	46759	71643	13177	55292	21036	82808	77501	97427
49386	54480	23604	23554	21785	41101	91178	10174	29420	90438
06312	88940	15995	69321	47458	64809	98189	81851	29651	84215
60942	00307	11897	92674	40405	68032	96717	54244	10701	41393
92329	98932	78284	46347	71209	92061	39448	93136	25722	08564
77936	63574	31384	51924	85561	29671	58137	17820	22751	36518
38101	77756	11657	13897	95889	57067	47648	13885	70669	93406
39641	69457	91339	22502	92613	89719	11947	56203	19324	20504
84054	40455	99396	63680	67667	60631	69181	96845	38525	11600
47468	03577	57649	63266	24700	71594	14004	23153	69249	05747
43321	31370	28977	23896	76479	68562	62342	07589	08899	05985
64281	61826	18555	64937	13173	33365	78851	16499	87064	13075
66847	70495	32350	02985	86716	38746	26313	77463	55387	72681
72461	33230	21529	53424	92581	02262	78438	66276	18396	73538
21032	91050	13058	16218	12470	56500	15292	76139	59526	52113
95362	67011	06651	16136	01016	00857	55018	56374	35824	71708
49712	97380	10404	55452	34030	60726	75211	10271	36633	68424
58275	61764	97586	54716	50259	46345	87195	46092	26787	60939
89514	11788	68224	23417	73959	76145	30342	40277	11049	72049
15472	50669	48139	36732	46874	37088	73465	09819	58869	35220
12120	86124	51247	44302	60883	52109	21437	36786	49226	77837
09188	20097	32825	39527	04220	86304	83389	87374	64278	58044
90045	85497	51981	50654	94938	81997	91870	76150	68476	64659
73189	50207	47677	26269	62290	64464	27124	67018	41361	82760
75768	76490	20971	87749	90429	12272	95375	05871	93823	43178
54016	44056	66281	31003	00682	27398	20714	53295	07706	17813
08358	69910	78542	42785	13661	58873	04618	97553	31223	08420
28306	03264	81333	10591	40510	07893	32604	60475	94119	01840
53840	86233	81594	13628	51215	90290	28466	68795	77762	20791
91757	53741	61613	62269	50263	90212	55781	76514	83483	47055
89415	92694	00397	58391	12607	17646	48949	72306	94541	37408
77513	03820	86864	29901	68414	82774	51908	13980	72893	55507
19502	37174	69979	20288	55210	29773	74287	75251	65344	67415
21818	59313	93278	81757	05686	73156	07082	85046	31853	38452
51474	66499	68107	23621	94049	91345	42836	09191	08007	45449
99559	68331	62535	24170	69777	12830	74819	78142	43860	72834

Table A-1. *Continued*

33713	48007	93584	72869	51926	64721	58303	29822	93174	93972
85274	86893	11303	22970	28834	34137	73515	90400	71148	43643
84133	89640	44035	52166	73852	70091	61222	60561	62327	18423
56732	16234	17395	96131	10123	91622	85496	57560	81604	18880
65138	56806	87648	85261	34313	65861	45875	21069	85644	47277
38001	02176	81719	11711	71602	92937	74219	64049	65584	49698
37402	96397	01304	77586	56271	10086	47324	62605	40030	37438
97125	40348	87083	31417	21815	39250	75237	62047	15501	29578
21826	41134	47143	34072	64638	85902	49139	06441	03856	54552
73135	42742	95719	09035	85794	74296	08789	88156	64691	19202
07638	77929	03061	18072	96207	44156	23821	99538	04713	66994
60528	83441	07954	19814	59175	20695	05533	52139	61212	06455
83596	35655	06958	92983	05128	09719	77433	53783	92301	50498
10850	62746	99599	10507	13499	06319	53075	71839	06410	19362
39820	98952	43622	63147	64421	80814	43800	09351	31024	73167
59580	06478	75569	78800	88835	54486	23768	06156	04111	08408
38508	07341	23793	48763	90822	97022	17719	04207	95954	49953
30692	70668	94688	16127	56196	80091	82067	63400	05462	69200
65443	95659	18288	27437	49632	24041	08337	65676	96299	90836
27267	50264	13192	72294	07477	44606	17985	48911	97341	30358
91307	06991	19072	24210	36699	53728	28825	35793	28976	66252
68434	94688	84473	13622	62126	98408	12843	82590	09815	93146
48908	15877	54745	24591	35700	04754	83824	52692	54130	55160
06913	45197	42672	78601	11883	09528	63011	98901	14974	40344
10455	16019	14210	33712	91342	37821	88325	80851	43667	70883
12883	97343	65027	61184	04285	01392	17974	15077	90712	26769
21778	30976	38807	36961	31649	42096	63281	02023	08816	47449
19523	59515	65122	59659	86283	68258	69572	13798	16435	91529
67245	52670	35583	16563	79246	86686	76463	34222	26655	90802
60584	47377	07500	37992	45134	26529	26760	83637	41326	44344
53853	41377	36066	94850	58838	73859	49364	73331	96240	43642
24637	38736	74384	89342	52623	07992	12369	18601	03742	83873
83080	12451	38992	22815	07759	51777	97377	27585	51972	37867
16444	24334	36151	99073	27493	70939	85130	32552	54846	54759
60790	18157	57178	65762	11161	78576	45819	52979	65130	04860

Table A-2
Squares and Square Roots

Number	Square	Square Root	Number	Square	Square Root
1	1	1.0000	31	9 61	5.5678
2	4	1.4142	32	10 24	5.6569
3	9	1.7321	33	10 89	5.7446
4	16	2.0000	34	11 56	5.8310
5	25	2.2361	35	12 25	5.9161
6	36	2.4495	36	12 96	6.0000
7	49	2.6458	37	13 69	6.0828
8	64	2.8284	38	14 44	6.1644
9	81	3.0000	39	15 21	6.2450
10	1 00	3.1623	40	16 00	6.3246
11	1 21	3.3166	41	16 81	6.4031
12	1 44	3.4641	42	17 64	6.4807
13	1 69	3.6056	43	18 49	6.5574
14	1 96	3.7417	44	19 36	6.6332
15	2 25	3.8730	45	20 25	6.7082
16	2 56	4.0000	46	21 16	6.7823
17	2 89	4.1231	47	22 09	6.8557
18	3 24	4.2426	48	23 04	6.9282
19	3 61	4.3589	49	24 01	7.0000
20	4 00	4.4721	50	25 00	7.0711
21	4 41	4.5826	51	26 01	7.1414
22	4 84	4.6904	52	27 04	7.2111
23	5 29	4.7958	53	28 09	7.2801
24	5 76	4.8990	54	29 16	7.3485
25	6 25	5.0000	55	30 25	7.4162
26	6 76	5.0990	56	31 36	7.4833
27	7 29	5.1962	57	32 49	7.5498
28	7 84	5.2915	58	33 64	7.6158
29	8 41	5.3852	59	34 81	7.6811
30	9 00	5.4772	60	36 00	7.7460

Table A—2. *Continued*

Number	Square	Square Root	Number	Square	Square Root
61	37 21	7.8102	101	1 02 01	10.0499
62	38 44	7.8740	102	1 04 04	10.0995
63	39 69	7.9373	103	1 06 09	10.1489
64	40 96	8.0000	104	1 08 16	10.1980
65	42 25	8.0623	105	1 10 25	10.2470
66	43 56	8.1240	106	1 12 36	10.2956
67	44 89	8.1854	107	1 14 49	10.3441
68	46 24	8.2462	108	1 16 64	10.3923
69	47 61	8.3066	109	1 18 81	10.4403
70	49 00	8.3666	110	1 21 00	10.4881
71	50 41	8.4261	111	1 23 21	10.5357
72	51 84	8.4853	112	1 25 44	10.5830
73	53 29	8.5440	113	1 27 69	10.6301
74	54 76	8.6023	114	1 29 96	10.6771
75	56 25	8.6603	115	1 32 25	10.7238
76	57 76	8.7178	116	1 34 56	10.7703
77	59 29	8.7750	117	1 36 89	10.8167
78	60 84	8.8318	118	1 39 24	10.8628
79	62 41	8.8882	119	1 41 61	10.9087
80	64 00	8.9443	120	1 44 00	10.9545
81	65 61	9.0000	121	1 46 41	11.0000
82	67 24	9.0554	122	1 48 84	11.0454
83	68 89	9.1104	123	1 51 29	11.0905
84	70 56	9.1652	124	1 53 76	11.1355
85	72 25	9.2195	125	1 56 25	11.1803
86	73 96	9.2736	126	1 58 76	11.2250
87	75 69	9.3274	127	1 61 29	11.2694
88	77 44	9.3808	128	1 63 84	11.3137
89	79 21	9.4340	129	1 66 41	11.3578
90	81 00	9.4868	130	1 69 00	11.4018
91	82 81	9.5394	131	1 71 61	11.4455
92	84 64	9.5917	132	1 74 24	11.4891
93	86 49	9.6437	133	1 76 89	11.5326
94	88 36	9.6954	134	1 79 56	11.5758
95	90 25	9.7468	135	1 82 25	11.6190
96	92 16	9.7980	136	1 84 96	11.6619
97	94 09	9.8489	137	1 87 69	11.7047
98	96 04	9.8995	138	1 90 44	11.7473
99	98 01	9.9499	139	1 93 21	11.7898
100	1 00 00	10.0000	140	1 96 00	11.8322

Table A−2. *Continued*

Number	Square	Square Root	Number	Square	Square Root
141	1 98 81	11.8743	181	3 27 61	13.4536
142	2 01 64	11.9164	182	3 31 24	13.4907
143	2 04 49	11.9583	183	3 34 89	13.5277
144	2 07 36	12.0000	184	3 38 56	13.5647
145	2 10 25	12.0416	185	3 42 25	13.6015
146	2 13 16	12.0830	186	3 45 96	13.6382
147	2 16 09	12.1244	187	3 49 69	13.6748
148	2 19 04	12.1655	188	3 53 44	13.7113
149	2 22 01	12.2066	189	3 57 21	13.7477
150	2 25 00	12.2474	190	3 61 00	13.7840
151	2 28 01	12.2882	191	3 64 81	13.8203
152	2 31 04	12.3288	192	3 68 64	13.8564
153	2 34 09	12.3693	193	3 72 49	13.8924
154	2 37 16	12.4097	194	3 76 36	13.9284
155	2 40 25	12.4499	195	3 80 25	13.9642
156	2 43 36	12.4900	196	3 84 16	14.0000
157	2 46 49	12.5300	197	3 88 09	14.0357
158	2 49 64	12.5698	198	3 92 04	14.0712
159	2 52 81	12.6095	199	3 96 01	14.1067
160	2 56 00	12.6491	200	4 00 00	14.1421
161	2 59 21	12.6886	201	4 04 01	14.1774
162	2 62 44	12.7279	202	4 08 04	14.2127
163	2 65 69	12.7671	203	4 12 09	14.2478
164	2 68 96	12.8062	204	4 16 16	14.2829
165	2 72 25	12.8452	205	4 20 25	14.3178
166	2 75 56	12.8841	206	4 24 36	14.3527
167	2 78 89	12.9228	207	4 28 49	14.3875
168	2 82 24	12.9615	208	4 32 64	14.4222
169	2 85 61	13.0000	209	4 36 81	14.4568
170	2 89 00	13.0384	210	4 41 00	14.4914
171	2 92 41	13.0767	211	4 45 21	14.5258
172	2 95 84	13.1149	212	4 49 44	14.5602
173	2 99 29	13.1529	213	4 53 69	14.5945
174	3 02 76	13.1909	214	4 57 96	14.6287
175	3 06 25	13.2288	215	4 62 25	14.6629
176	3 09 76	13.2665	216	4 66 56	14.6969
177	3 13 29	13.3041	217	4 70 89	14.7309
178	3 16 84	13.3417	218	4 75 24	14.7648
179	3 20 41	13.3791	219	4 79 61	14.7986
180	3 24 00	13.4164	220	4 84 00	14.8324

Table A−2. *Continued*

Number	Square	Square Root	Number	Square	Square Root
221	4 88 41	14.8661	261	6 81 21	16.1555
222	4 92 84	14.8997	262	6 86 44	16.1864
223	4 97 29	14.9332	263	6 91 69	16.2173
224	5 01 76	14.9666	264	6 96 96	16.2481
225	5 06 25	15.0000	265	7 02 25	16.2788
226	5 10 76	15.0333	266	7 07 56	16.3095
227	5 15 29	15.0665	267	7 12 89	16.3401
228	5 19 84	15.0997	268	7 18 24	16.3707
229	5 24 41	15.1327	269	7 23 61	16.4012
230	5 29 00	15.1658	270	7 29 00	16.4317
231	5 33 61	15.1987	271	7 34 41	16.4621
232	5 38 24	15.2315	272	7 39 84	16.4924
233	5 42 89	15.2643	273	7 45 29	16.5227
234	5 47 56	15.2971	274	7 50 76	16.5529
235	5 52 25	15.3297	275	7 56 25	16.5831
236	5 56 96	15.3623	276	7 61 76	16.6132
237	5 61 69	15.3948	277	7 67 29	16.6433
238	5 66 44	15.4272	278	7 72 84	16.6733
239	5 71 21	15.4596	279	7 78 41	16.7033
240	5 76 00	15.4919	280	7 84 00	16.7332
241	5 80 81	15.5242	281	7 89 61	16.7631
242	5 85 64	15.5563	282	7 95 24	16.7929
243	5 90 49	15.5885	283	8 00 89	16.8226
244	5 95 36	15.6205	284	8 06 56	16.8523
245	6 00 25	15.6525	285	8 12 25	16.8819
246	6 05 16	15.6844	286	8 17 96	16.9115
247	6 10 09	15.7162	287	8 23 69	16.9411
248	6 15 04	15.7480	288	8 29 44	16.9706
249	6 20 01	15.7797	289	8 35 21	17.0000
250	6 25 00	15.8114	290	8 41 00	17.0294
251	6 30 01	15.8430	291	8 46 81	17.0587
252	6 35 04	15.8745	292	8 52 64	17.0880
253	6 40 09	15.9060	293	8 58 49	17.1172
254	6 45 16	15.9374	294	8 64 36	17.1464
255	6 50 25	15.9687	295	8 70 25	17.1756
256	6 55 36	16.0000	296	8 76 16	17.2047
257	6 60 49	16.0312	297	8 82 09	17.2337
258	6 65 64	16.0624	298	8 88 04	17.2627
259	6 70 81	16.0935	299	8 94 01	17.2916
260	6 76 00	16.1245	300	9 00 00	17.3205

Table A−2. *Continued*

Number	Square	Square Root	Number	Square	Square Root
301	9 06 01	17.3494	341	11 62 81	18.4662
302	9 12 04	17.3781	342	11 69 64	18.4932
303	9 18 09	17.4069	343	11 76 49	18.5203
304	9 24 16	17.4356	344	11 83 36	18.5472
305	9 30 25	17.4642	345	11 90 25	18.5742
306	9 36 36	17.4929	346	11 97 16	18.6011
307	9 42 49	17.5214	347	12 04 09	18.6279
308	9 48 64	17.5499	348	12 11 04	18.6548
309	9 54 81	17.5784	349	12 18 01	18.6815
310	9 61 00	17.6068	350	12 25 00	18.7083
311	9 67 21	17.6352	351	12 32 01	18.7350
312	9 73 44	17.6635	352	12 39 04	18.7617
313	9 79 69	17.6918	353	12 46 09	18.7883
314	9 85 96	17.7200	354	12 53 16	18.8149
315	9 92 25	17.7482	355	12 60 25	18.8414
316	9 98 56	17.7764	356	12 67 36	18.8680
317	10 04 89	17.8045	357	12 74 49	18.8944
318	10 11 24	17.8326	358	12 81 64	18.9209
319	10 17 61	17.8606	359	12 88 81	18.9473
320	10 24 00	17.8885	360	12 96 00	18.9737
321	10 30 41	17.9165	361	13 03 21	19.0000
322	10 36 84	17.9444	362	13 10 44	19.0263
323	10 43 29	17.9722	363	13 17 69	19.0526
324	10 49 76	18.0000	364	13 24 96	19.0788
325	10 56 25	18.0278	365	13 32 25	19.1050
326	10 62 76	18.0555	366	13 39 56	19.1311
327	10 69 29	18.0831	367	13 46 89	19.1572
328	10 75 84	18.1108	368	13 54 24	19.1833
329	10 82 41	18.1384	369	13 61 61	19.2094
330	10 89 00	18.1659	370	13 69 00	19.2354
331	10 95 61	18.1934	371	13 76 41	19.2614
332	11 02 24	18.2209	372	13 83 84	19.2873
333	11 08 89	18.2483	373	13 91 29	19.3132
334	11 15 56	18.2757	374	13 98 76	19.3391
335	11 22 25	18.3030	375	14 06 25	19.3649
336	11 28 96	18.3303	376	14 13 76	19.3907
337	11 35 69	18.3576	377	14 21 29	19.4165
338	11 42 44	18.3848	378	14 28 84	19.4422
339	11 49 21	18.4120	379	14 36 41	19.4679
340	11 56 00	18.4391	380	14 44 00	19.4936

Table A−2. *Continued*

Number	Square	Square Root	Number	Square	Square Root
381	14 51 61	19.5192	421	17 72 41	20.5183
382	14 59 24	19.5448	422	17 80 84	20.5426
383	14 66 89	19.5704	423	17 89 29	20.5670
384	14 74 56	19.5959	424	17 97 76	20.5913
385	14 82 25	19.6214	425	18 06 25	20.6155
386	14 89 96	19.6469	426	18 14 76	20.6398
387	14 97 69	19.6723	427	18 23 29	20.6640
388	15 05 44	19.6977	428	18 31 84	20.6882
389	15 13 21	19.7231	429	18 40 41	20.7123
390	15 21 00	19.7484	430	18 49 00	20.7364
391	15 28 81	19.7737	431	18 57 61	20.7605
392	15 36 64	19.7990	432	18 66 24	20.7846
393	15 44 49	19.8242	433	18 74 89	20.8087
394	15 52 36	19.8494	434	18 83 56	20.8327
395	15 60 25	19.8746	435	18 92 25	20.8567
396	15 68 16	19.8997	436	19 00 96	20.8806
397	15 76 09	19.9249	437	19 09 69	20.9045
398	15 84 04	19.9499	438	19 18 44	20.9284
399	15 92 01	19.9750	439	19 27 21	20.9523
400	16 00 00	20.0000	440	19 36 00	20.9762
401	16 08 01	20.0250	441	19 44 81	21.0000
402	16 16 04	20.0499	442	19 53 64	21.0238
403	16 24 09	20.0749	443	19 62 49	21.0476
404	16 32 16	20.0998	444	19 71 36	21.0713
405	16 40 25	20.1246	445	19 80 25	21.0950
406	16 48 36	20.1494	446	19 89 16	21.1187
407	16 56 49	20.1742	447	19 98 09	21.1424
408	16 64 64	20.1990	448	20 07 04	21.1660
409	16 72 81	20.2237	449	20 16 01	21.1896
410	16 81 00	20.2485	450	20 25 00	21.2132
411	16 89 21	20.2731	451	20 34 01	21.2368
412	16 97 44	20.2978	452	20 43 04	21.2603
413	17 05 69	20.3224	453	20 52 09	21.2838
414	17 13 96	20.3470	454	20 61 16	21.3073
415	17 22 25	20.3715	455	20 70 25	21.3307
416	17 30 56	20.3961	456	20 79 36	21.3542
417	17 38 89	20.4206	457	20 88 49	21.3776
418	17 47 24	20.4450	458	20 97 64	21.4009
419	17 55 61	20.4695	459	21 06 81	21.4243
420	17 64 00	20.4939	460	21 16 00	21.4476

Table A-2. *Continued*

Number	Square	Square Root	Number	Square	Square Root
461	21 25 21	21.4709	501	25 10 01	22.3830
462	21 34 44	21.4942	502	25 20 04	22.4054
463	21 43 69	21.5174	503	25 30 09	22.4277
464	21 52 96	21.5407	504	25 40 16	22.4499
465	21 62 25	21.5639	505	25 50 25	22.4722
466	21 71 56	21.5870	506	25 60 36	22.4944
467	21 80 89	21.6102	507	25 70 49	22.5167
468	21 90 24	21.6333	508	25 80 64	22.5389
469	21 99 61	21.6564	509	25 90 81	22.5610
470	22 09 00	21.6795	510	26 01 00	22.5832
471	22 18 41	21.7025	511	26 11 21	22.6053
472	22 27 84	21.7256	512	26 21 44	22.6274
473	22 37 29	21.7486	513	26 31 69	22.6495
474	22 46 76	21.7715	514	26 41 96	22.6716
475	22 56 25	21.7945	515	26 52 25	22.6936
476	22 65 76	21.8174	516	26 62 56	22.7156
477	22 75 29	21.8403	517	26 72 89	22.7376
478	22 84 84	21.8632	518	26 83 24	22.7596
479	22 94 41	21.8861	519	26 93 61	22.7816
480	23 04 00	21.9089	520	27 04 00	22.8035
481	23 13 61	21.9317	521	27 14 41	22.8254
482	23 23 24	21.9545	522	27 24 84	22.8473
483	23 32 89	21.9773	523	27 35 29	22.8692
484	23 42 56	22.0000	524	27 45 76	22.8910
485	23 52 25	22.0227	525	27 56 25	22.9129
486	23 61 96	22.0454	526	27 66 76	22.9347
487	23 71 69	22.0681	527	27 77 29	22.9565
488	23 81 44	22.0907	528	27 87 84	22.9783
489	23 91 21	22.1133	529	27 98 41	23.0000
490	24 01 00	22.1359	530	28 09 00	23.0217
491	24 10 81	22.1585	531	28 19 61	23.0434
492	24 20 64	22.1811	532	28 30 24	23.0651
493	24 30 49	22.2036	533	28 40 89	23.0868
494	24 40 36	22.2261	534	28 51 56	23.1084
495	24 50 25	22.2486	535	28 62 25	23.1301
496	24 60 16	22.2711	536	28 72 96	23.1517
497	24 70 09	22.2935	537	28 83 69	23.1733
498	24 80 04	22.3159	538	28 94 44	23.1948
499	24 90 01	22.3383	539	29 05 21	23.2164
500	25 00 00	22.3607	540	29 16 00	23.2379

Table A−2. *Continued*

Number	Square	Square Root	Number	Square	Square Root
541	29 26 81	23.2594	581	33 75 61	24.1039
542	29 37 64	23.2809	582	33 87 24	24.1247
543	29 48 49	23.3024	583	33 98 89	24.1454
544	29 59 36	23.3238	584	34 10 56	24.1661
545	29 70 25	23.3452	585	34 22 25	24.1868
546	29 81 16	23.3666	586	34 33 96	24.2074
547	29 92 09	23.3880	587	34 45 69	24.2281
548	30 03 04	23.4094	588	34 57 44	24.2487
549	30 14 01	23.4307	589	34 69 21	24.2693
550	30 25 00	23.4521	590	34 81 00	24.2899
551	30 36 01	23.4734	591	34 92 81	24.3105
552	30 47 04	23.4947	592	35 04 64	24.3311
553	30 58 09	23.5160	593	35 16 49	24.3516
554	30 69 16	23.5372	594	35 28 36	24.3721
555	30 80 25	23.5584	595	35 40 25	24.3926
556	30 91 36	23.5797	596	35 52 16	24.4131
557	31 02 49	23.6008	597	35 64 09	24.4336
558	31 13 64	23.6220	598	35 76 04	24.4540
559	31 24 81	23.6432	599	35 88 01	24.4745
560	31 36 00	23.6643	600	36 00 00	24.4949
561	31 47 21	23.6854	601	36 12 01	24.5153
562	31 58 44	23.7065	602	36 24 04	24.5357
563	31 69 69	23.7276	603	36 36 09	24.5561
564	31 80 96	23.7487	604	36 48 16	24.5764
565	31 92 25	23.7697	605	36 60 25	24.5967
566	32 03 56	23.7908	606	36 72 36	24.6171
567	32 14 89	23.8118	607	36 84 49	24.6374
568	32 26 24	23.8328	608	36 96 64	24.6577
569	32 37 61	23.8537	609	37 08 81	24.6779
570	32 49 00	23.8747	610	37 21 00	24.6982
571	32 60 41	23.8956	611	37 33 21	24.7184
572	32 71 84	23.9165	612	37 45 44	24.7385
573	32 83 29	23.9374	613	37 57 69	24.7588
574	32 94 76	23.9583	614	37 69 96	24.7790
575	33 06 25	23.9792	615	37 82 25	24.7992
576	33 17 76	24.0000	616	37 94 56	24.8193
577	33 29 29	24.0208	617	38 06 89	24.8395
578	33 40 84	24.0416	618	38 19 24	24.8596
579	33 52 41	24.0624	619	38 31 61	24.8797
580	33 64 00	24.0832	620	38 44 00	24.8998

Table A-2. *Continued*

Number	Square	Square Root	Number	Square	Square Root
621	38 56 41	24.9199	661	43 69 21	25.7099
622	38 68 84	24.9399	662	43 82 44	25.7294
623	38 81 29	24.9600	663	43 95 69	25.7488
624	38 93 76	24.9800	664	44 08 96	25.7682
625	39 06 25	25.0000	665	44 22 25	25.7876
626	39 18 76	25.0200	666	44 35 56	25.8070
627	39 31 29	25.0400	667	44 48 89	25.8263
628	39 43 84	25.0599	668	44 62 24	25.8457
629	39 56 41	25.0799	669	44 75 61	25.8650
630	39 69 00	25.0998	670	44 89 00	25.8844
631	39 81 61	25.1197	671	45 02 41	25.9037
632	39 94 24	25.1396	672	45 15 84	25.9230
633	40 06 89	25.1595	673	45 29 29	25.9422
634	40 19 56	25.1794	674	45 42 76	25.9615
635	40 32 25	25.1992	675	45 56 25	25.9808
636	40 44 96	25.2190	676	45 69 76	26.0000
637	40 57 69	25.2389	677	45 83 29	26.0192
638	40 70 44	25.2587	678	45 96 84	26.0384
639	40 83 21	25.2784	679	46 10 41	26.0576
640	40 96 00	25.2982	680	46 24 00	26.0768
641	41 08 81	25.3180	681	46 37 61	26.0960
642	41 21 64	25.3377	682	46 51 24	26.1151
643	41 34 49	25.3574	683	46 64 89	26.1343
644	41 47 36	25.3772	684	46 78 56	26.1534
645	41 60 25	25.3969	685	46 92 25	26.1725
646	41 73 16	25.4165	686	47 05 96	26.1916
647	41 86 09	25.4362	687	47 19 69	26.2107
648	41 99 04	25.4558	688	47 33 44	26.2298
649	42 12 01	25.4755	689	47 47 21	26.2488
650	42 25 00	25.4951	690	47 61 00	26.2679
651	42 38 01	25.5147	691	47 74 81	26.2869
652	42 51 04	25.5343	692	47 88 64	26.3059
653	42 64 09	25.5539	693	48 02 49	26.3249
654	42 77 16	25.5734	694	48 16 36	26.3439
655	42 90 25	25.5930	695	48 30 25	26.3629
656	43 03 36	25.6125	696	48 44 16	26.3818
657	43 16 49	25.6320	697	48 58 09	26.4008
658	43 29 64	25.6515	698	48 72 04	26.4197
659	43 42 81	25.6710	699	48 86 01	26.4386
660	43 56 00	25.6905	700	49 00 00	26.4575

Table A–2. *Continued*

Number	Square	Square Root	Number	Square	Square Root
701	49 14 01	26.4764	741	54 90 81	27.2213
702	49 28 04	26.4953	742	55 05 64	27.2397
703	49 42 09	26.5141	743	55 20 49	27.2580
704	49 56 16	26.5330	744	55 35 36	27.2764
705	49 70 25	26.5518	745	55 50 25	27.2947
706	49 84 36	26.5707	746	55 65 16	27.3130
707	49 98 49	26.5895	747	55 80 09	27.3313
708	50 12 64	26.6083	748	55 95 04	27.3496
709	50 26 81	26.6271	749	56 10 01	27.3679
710	50 41 00	26.6458	750	56 25 00	27.3861
711	50 55 21	26.6646	751	56 40 01	27.4044
712	50 69 44	26.6833	752	56 55 04	27.4226
713	50 83 69	26.7021	753	56 70 09	27.4408
714	50 97 96	26.7208	754	56 85 16	27.4591
715	51 12 25	26.7395	755	57 00 25	27.4773
716	51 26 56	26.7582	756	57 15 36	27.4955
717	51 40 89	26.7769	757	57 30 49	27.5136
718	51 55 24	26.7955	758	57 45 64	27.5318
719	51 69 61	26.8142	759	57 60 81	27.5500
720	51 84 00	26.8328	760	57 76 00	27.5681
721	51 98 41	26.8514	761	57 91 21	27.5862
722	52 12 84	26.8701	762	58 06 44	27.6043
723	52 27 29	26.8887	763	58 21 69	27.6225
724	52 41 76	26.9072	764	58 36 96	27.6405
725	52 56 25	26.9258	765	58 52 25	27.6586
726	52 70 76	26.9444	766	58 67 56	27.6767
727	52 85 29	26.9629	767	58 82 89	27.6948
728	52 99 84	26.9815	768	58 98 24	27.7128
729	53 14 41	27.0000	769	59 13 61	27.7308
730	53 29 00	27.0185	770	59 29 00	27.7489
731	53 43 61	27.0370	771	59 44 41	27.7669
732	53 58 24	27.0555	772	59 59 84	27.7849
733	53 72 89	27.0740	773	59 75 29	27.8029
734	53 87 56	27.0924	774	59 90 76	27.8209
735	54 02 25	27.1109	775	60 06 25	27.8388
736	54 16 96	27.1293	776	60 21 76	27.8568
737	54 31 69	27.1477	777	60 37 29	27.8747
738	54 46 44	27.1662	778	60 52 84	27.8927
739	54 61 27	27.1846	779	60 68 41	27.9106
740	54 76 00	27.2029	780	60 84 00	27.9285

Table A–2. *Continued*

Number	Square	Square Root	Number	Square	Square Root
781	60 99 61	27.9464	821	67 40 41	28.6531
782	61 15 24	27.9643	822	67 56 84	28.6705
783	61 30 89	27.9821	823	67 73 29	28.6880
784	61 46 56	28.0000	824	67 89 76	28.7054
785	61 62 25	28.0179	825	68 06 25	28.7228
786	61 77 96	28.0357	826	68 22 76	28.7402
787	61 93 69	28.0535	827	68 39 29	28.7576
788	62 09 44	28.0713	828	68 55 84	28.7750
789	62 25 21	28.0891	829	68 72 41	28.7924
790	62 41 00	28.1069	830	68 89 00	28.8097
791	62 56 81	28.1247	831	69 05 61	28.8271
792	62 72 64	28.1425	832	69 22 24	28.8444
793	62 88 49	28.1603	833	69 38 89	28.8617
794	63 04 36	28.1780	834	69 55 56	28.8791
795	63 20 25	28.1957	835	69 72 25	28.8964
796	63 36 16	28.2135	836	69 88 96	28.9137
797	63 52 09	28.2312	837	70 05 69	28.9310
798	63 68 04	28.2489	838	70 22 44	28.9482
799	63 84 01	28.2666	839	70 39 21	28.9655
800	64 00 00	28.2843	840	70 56 00	28.9828
801	64 16 01	28.3019	841	70 72 81	29.0000
802	64 32 04	28.3196	842	70 89 64	29.0172
803	64 48 09	28.3373	843	71 06 49	29.0345
804	64 64 16	28.3549	844	71 23 36	29.0517
805	64 80 25	28.3725	845	71 40 25	29.0689
806	64 96 36	28.3901	846	71 57 16	29.0861
807	65 12 49	28.4077	847	71 74 09	29.1033
808	65 28 64	28.4253	848	71 91 04	29.1204
809	65 44 81	28.4429	849	72 08 01	29.1376
810	65 61 00	28.4605	850	72 25 00	29.1548
811	65 77 21	28.4781	851	72 42 01	29.1719
812	65 93 44	28.4956	852	72 59 04	29.1890
813	66 09 69	28.5132	853	72 76 09	29.2062
814	66 25 96	28.5307	854	72 93 16	29.2233
815	66 42 25	28.5482	855	73 10 25	29.2404
816	66 58 56	28.5657	856	73 27 36	29.2575
817	66 74 89	28.5832	857	73 44 49	29.2746
818	66 91 24	28.6007	858	73 61 64	29.2916
819	67 07 61	28.6082	859	73 78 81	29.3087
820	67 24 00	28.6356	860	73 96 00	29.3258

Table A–2. *Continued*

Number	Square	Square Root	Number	Square	Square Root
861	74 13 21	29.3428	901	81 18 01	30.0167
862	74 30 44	29.3598	902	81 36 04	30.0333
863	74 47 69	29.3769	903	81 54 09	30.0500
864	74 64 96	29.3939	904	81 72 16	30.0666
865	74 82 25	29.4109	905	81 90 25	30.0832
866	74 99 56	29.4279	906	82 08 36	30.0998
867	75 16 89	29.4449	907	82 26 49	30.1164
868	75 34 24	29.4618	908	82 44 64	30.1330
869	75 51 61	29.4788	909	82 62 81	30.1496
870	75 69 00	29.4958	910	82 81 00	30.1662
871	75 86 41	29.5127	911	82 99 21	30.1828
872	76 03 84	29.5296	912	83 17 44	30.1993
873	76 21 29	29.5466	913	83 35 69	30.2159
874	76 38 76	29.5635	914	83 53 96	30.2324
875	76 56 25	29.5804	915	83 72 25	30.2490
876	76 73 76	29.5973	916	83 90 56	30.2655
877	76 91 29	29.6142	917	84 08 89	30.2820
878	77 08 84	29.6311	918	84 27 24	30.2985
879	77 26 41	29.6479	919	84 45 61	30.3150
880	77 44 00	29.6648	920	84 64 00	30.3315
881	77 61 61	29.6816	921	84 82 41	30.3480
882	77 79 24	29.6985	922	85 00 84	30.3645
883	77 96 89	29.7153	923	85 19 29	30.3809
884	78 14 56	29.7321	924	85 37 76	30.3974
885	78 32 25	29.7489	925	85 56 25	30.4138
886	78 49 96	29.7658	926	85 74 76	30.4302
887	78 67 69	29.7825	927	85 93 29	30.4467
888	78 85 44	29.7993	928	86 11 84	30.4631
889	79 03 21	29.8161	929	86 30 41	30.4795
890	79 21 00	29.8329	930	86 49 00	30.4959
891	79 38 81	29.8496	931	86 67 61	30.5123
892	79 56 64	29.8664	932	86 86 24	30.5287
893	79 74 49	29.8831	933	87 04 89	30.5450
894	79 92 36	29.8998	934	87 23 56	30.5614
895	80 10 25	29.9166	935	87 42 25	30.5778
896	80 28 16	29.9333	936	87 60 96	30.5941
897	80 46 09	29.9500	937	87 79 69	30.6105
898	80 64 04	29.9666	938	87 98 44	30.6268
899	80 82 01	29.9833	939	88 17 21	30.6431
900	81 00 00	30.0000	940	88 36 00	30.6594

Table A−2. *Continued*

Number	Square	Square Root	Number	Square	Square Root
941	88 54 81	30.6757	971	94 28 41	31.1609
942	88 73 64	30.6920	972	94 47 84	31.1769
943	88 92 49	30.7083	973	94 67 29	31.1929
944	89 11 36	30.7246	974	94 86 76	31.2090
945	89 30 25	30.7409	975	95 06 25	31.2250
946	89 49 16	30.7571	976	95 25 76	31.2410
947	89 68 09	30.7734	977	95 45 29	31.2570
948	89 87 04	30.7896	978	95 64 84	31.2730
949	90 06 01	30.8058	979	95 84 41	31.2890
950	90 25 00	30.8221	980	96 04 00	31.3050
951	90 44 01	30.8383	981	96 23 61	31.3209
952	90 63 04	30.8545	982	96 43 24	31.3369
953	90 82 09	30.8707	983	96 62 89	31.3528
954	91 01 16	30.8869	984	96 82 56	31.3688
955	91 20 25	30.9031	985	97 02 25	31.3847
956	91 39 36	30.9192	986	97 21 96	31.4006
957	91 58 49	30.9354	987	97 41 69	31.4166
958	91 77 64	30.9516	988	97 61 44	31.4325
959	91 96 81	30.9677	989	97 81 21	31.4484
960	92 16 00	30.9839	990	98 01 00	31.4643
961	92 35 21	31.0000	991	98 20 81	31.4802
962	92 54 44	31.0161	992	98 40 64	31.4960
963	92 73 69	31.0322	993	98 60 49	31.5119
964	92 92 96	31.0483	994	98 80 36	31.5278
965	93 12 25	31.0644	995	99 00 25	31.5436
966	93 31 56	31.0805	996	99 20 16	31.5595
967	93 50 89	31.0966	997	99 40 09	31.5753
968	93 70 24	31.1127	998	99 60 04	31.5911
969	93 89 61	31.1288	999	99 80 01	31.6070
970	94 09 00	31.1448	1000	100 00 00	31.6228

Table A–3
Areas of a Standard Normal Distribution*

An entry in the table is the proportion under the entire curve which is between $z = 0$ and a positive value of z. Areas for negative values of z are obtained by symmetry.

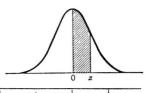

z	.00	.01	.02	.03	.04	.05	.06	.07	.08	.09
0.0	.0000	.0040	.0080	.0120	.0160	.0199	.0239	.0279	.0319	.0359
0.1	.0398	.0438	.0478	.0517	.0557	.0596	.0636	.0675	.0714	.0753
0.2	.0793	.0832	.0871	.0910	.0948	.0987	.1026	.1064	.1103	.1141
0.3	.1179	.1217	.1255	.1293	.1331	.1368	.1406	.1443	.1480	.1517
0.4	.1554	.1591	.1628	.1664	.1700	.1736	.1772	.1808	.1844	.1879
0.5	.1915	.1950	.1985	.2019	.2054	.2088	.2123	.2157	.2190	.2224
0.6	.2257	.2291	.2324	.2357	.2389	.2422	.2454	.2486	.2517	.2549
0.7	.2580	.2611	.2642	.2673	.2703	.2734	.2764	.2794	.2823	.2852
0.8	.2881	.2910	.2939	.2967	.2995	.3023	.3051	.3078	.3106	.3133
0.9	.3159	.3186	.3212	.3238	.3264	.3289	.3315	.3340	.3365	.3389
1.0	.3413	.3438	.3461	.3485	.3508	.3531	.3554	.3577	.3599	.3621
1.1	.3643	.3665	.3686	.3708	.3729	.3749	.3770	.3790	.3810	.3830
1.2	.3849	.3869	.3888	.3907	.3925	.3944	.3962	.3980	.3997	.4015
1.3	.4032	.4049	.4066	.4082	.4099	.4115	.4131	.4147	.4162	.4177
1.4	.4192	.4207	.4222	.4236	.4251	.4265	.4279	.4292	.4306	.4319
1.5	.4332	.4345	.4357	.4370	.4382	.4394	.4406	.4418	.4429	.4441
1.6	.4452	.4463	.4474	.4484	.4495	.4505	.4515	.4525	.4535	.4545
1.7	.4554	.4564	.4573	.4582	.4591	.4599	.4608	.4616	.4625	.4633
1.8	.4641	.4649	.4656	.4664	.4671	.4678	.4686	.4693	.4699	.4706
1.9	.4713	.4719	.4726	.4732	.4738	.4744	.4750	.4756	.4761	.4767
2.0	.4772	.4778	.4783	.4788	.4793	.4798	.4803	.4808	.4812	.4817
2.1	.4821	.4826	.4830	.4834	.4838	.4842	.4846	.4850	.4854	.4857
2.2	.4861	.4864	.4868	.4871	.4875	.4878	.4881	.4884	.4887	.4890
2.3	.4893	.4896	.4898	.4901	.4904	.4906	.4909	.4911	.4913	.4916
2.4	.4918	.4920	.4922	.4925	.4927	.4929	.4931	.4932	.4934	.4936
2.5	.4938	.4940	.4941	.4943	.4945	.4946	.4948	.4949	.4951	.4952
2.6	.4953	.4955	.4956	.4957	.4959	.4960	.4961	.4962	.4963	.4964
2.7	.4965	.4966	.4967	.4968	.4969	.4970	.4971	.4972	.4973	.4974
2.8	.4974	.4975	.4976	.4977	.4977	.4978	.4979	.4979	.4980	.4981
2.9	.4981	.4982	.4982	.4983	.4984	.4984	.4985	.4985	.4986	.4986
3.0	.4987	.4987	.4987	.4988	.4988	.4989	.4989	.4989	.4990	.4990

*Reproduced with permission from Hoel, P. G.: *Elementary Statistics*, 2nd Ed. New York: John Wiley & Sons, Inc., 1966.

Table A–4
Student's t Distribution*

The first column lists the number of degrees of freedom (df). The headings of the other columns give probabilities (P) for t to exceed numerically the entry value.

df \ P	0.50	0.25	0.10	0.05	0.025	0.01	0.005
1	1.00000	2.4142	6.3138	12.706	25.452	63.657	127.32
2	0.81650	1.6036	2.9200	4.3027	6.2053	9.9248	14.089
3	0.76489	1.4226	2.3534	3.1825	4.1765	5.8409	7.4533
4	0.74070	1.3444	2.1318	2.7764	3.4954	4.6041	5.5976
5	0.72669	1.3009	2.0150	2.5706	3.1634	4.0321	4.7733
6	0.71756	1.2733	1.9432	2.4469	2.9687	3.7074	4.3168
7	0.71114	1.2543	1.8946	2.3646	2.8412	3.4995	4.0293
8	0.70639	1.2403	1.8595	2.3060	2.7515	3.3554	3.8325
9	0.70272	1.2297	1.8331	2.2622	2.6850	3.2498	3.6897
10	0.69981	1.2213	1.8125	2.2281	2.6338	3.1693	3.5814
11	0.69745	1.2145	1.7959	2.2010	2.5931	3.1058	3.4966
12	0.69548	1.2089	1.7823	2.1788	2.5600	3.0545	3.4284
13	0.69384	1.2041	1.7709	2.1604	2.5326	3.0123	3.3725
14	0.69242	1.2001	1.7613	2.1448	2.5096	2.9768	3.3257
15	0.69120	1.1967	1.7530	2.1315	2.4899	2.9467	3.2860
16	0.69013	1.1937	1.7459	2.1199	2.4729	2.9208	3.2520
17	0.68919	1.1910	1.7396	2.1098	2.4581	2.8982	3.2225
18	0.68837	1.1887	1.7341	2.1009	2.4450	2.8784	3.1966
19	0.68763	1.1866	1.7291	2.0930	2.4334	2.8609	3.1737
20	0.68696	1.1848	1.7247	2.0860	2.4231	2.8453	3.1534
21	0.68635	1.1831	1.7207	2.0796	2.4138	2.8314	3.1352
22	0.68580	1.1816	1.7171	2.0739	2.4055	2.8188	3.1188
23	0.68531	1.1802	1.7139	2.0687	2.3979	2.8073	3.1040
24	0.68485	1.1789	1.7109	2.0639	2.3910	2.7969	3.0905
25	0.68443	1.1777	1.7081	2.0595	2.3846	2.7874	3.0782
26	0.68405	1.1766	1.7056	2.0555	2.3788	2.7787	3.0669
27	0.68370	1.1757	1.7033	2.0518	2.3734	2.7707	3.0565
28	0.68335	1.1748	1.7011	2.0484	2.3685	2.7633	3.0469
29	0.68304	1.1739	1.6991	2.0452	2.3638	2.7564	3.0380
30	0.68276	1.1731	1.6973	2.0423	2.3596	2.7500	3.0298
40	0.68066	1.1673	1.6839	2.0211	2.3289	2.7045	2.9712
60	0.67862	1.1616	1.6707	2.0003	2.2991	2.6603	2.9146
120	0.67656	1.1559	1.6577	1.9799	2.2699	2.6174	2.8599
∞	0.67449	1.1503	1.6449	1.9600	2.2414	2.5758	2.8070

* Reproduced with permission from Hoel, P. G.: *Elementary Statistics*, 2nd Ed. John Wiley and Sons, Inc., 1966.

Table A-5
Distribution of x^2*

The first column contains the number of degrees of freedom. The values in the body of the table are the values of x^2 required for the listed probability levels.

df	P = .99	.98	.95	.90	.80	.70	.50
1	.00016	.00063	.0039	.016	.064	.15	.46
2	.02	.04	.10	.21	.45	.71	1.39
3	.12	.18	.35	.58	1.00	1.42	2.37
4	.30	.43	.71	1.06	1.65	2.20	3.36
5	.55	.75	1.14	1.61	2.34	3.00	4.35
6	.87	1.13	1.64	2.20	3.07	3.83	5.35
7	1.24	1.56	2.17	2.83	3.82	4.67	6.35
8	1.65	2.03	2.73	3.49	4.59	5.53	7.34
9	2.09	2.53	3.32	4.17	5.38	6.39	8.34
10	2.56	3.06	3.94	4.86	6.18	7.27	9.34
11	3.05	3.61	4.58	5.58	6.99	8.15	10.34
12	3.57	4.18	5.23	6.30	7.81	9.03	11.34
13	4.11	4.76	5.89	7.04	8.63	9.93	12.34
14	4.66	5.37	6.57	7.79	9.47	10.82	13.34
15	5.23	5.98	7.26	8.55	10.31	11.72	14.34
16	5.81	6.61	7.96	9.31	11.15	12.62	15.34
17	6.41	7.26	8.67	10.08	12.00	13.53	16.34
18	7.02	7.91	9.39	10.86	12.86	14.44	17.34
19	7.63	8.57	10.12	11.65	13.72	15.35	18.34
20	8.26	9.24	10.85	12.44	14.58	16.27	19.34
21	8.90	9.92	11.59	13.24	15.44	17.18	20.34
22	9.54	10.60	12.34	14.04	16.31	18.10	21.34
23	10.20	11.29	13.09	14.85	17.19	19.02	22.34
24	10.86	11.99	13.85	15.66	18.06	19.94	23.34
25	11.52	12.70	14.61	16.47	18.94	20.87	24.34
26	12.20	13.41	15.38	17.29	19.82	21.79	25.34
27	12.88	14.12	16.15	18.11	20.70	22.72	26.34
28	13.56	14.85	16.93	18.94	21.59	23.65	27.34
29	14.26	15.57	17.71	19.77	22.48	24.58	28.34
30	14.95	16.31	18.49	20.60	23.36	25.51	29.34

*Reproduced with permission from McNemar, Q.: *Psychological Statistics*, 3rd Ed. New York: John Wiley and Sons, Inc. 1962. This table is abridged from Table IV of Fisher and Yates: *Statistical Tables for Biological, Agricultural and Medical Research*. Edinburgh: Oliver and Boyd, Ltd.

Table A–5. *Continued*

df	.30	.20	.10	.05	.02	.01	.001
1	1.07	1.64	2.71	3.84	5.41	6.64	10.83
2	2.41	3.22	4.60	5.99	7.82	9.21	13.82
3	3.66	4.64	6.25	7.82	9.84	11.34	16.27
4	4.88	5.99	7.78	9.49	11.67	13.28	18.46
5	6.06	7.29	9.24	11.07	13.39	15.09	20.52
6	7.23	8.56	10.64	12.59	15.03	16.81	22.46
7	8.38	9.80	12.02	14.07	16.62	18.48	24.32
8	9.52	11.03	13.36	15.51	18.17	20.09	26.12
9	10.66	12.24	14.68	16.92	19.68	21.67	27.88
10	11.78	13.44	15.99	18.31	21.16	23.21	29.59
11	12.90	14.63	17.28	19.68	22.62	24.72	31.26
12	14.01	15.81	18.55	21.03	24.05	26.22	32.91
13	15.12	16.98	19.81	22.36	25.47	27.69	34.53
14	16.22	18.15	21.06	23.68	26.87	29.14	36.12
15	17.32	19.31	22.31	25.00	28.26	30.58	37.70
16	18.42	20.46	23.54	26.30	29.63	32.00	39.25
17	19.51	21.62	24.77	27.59	31.00	33.41	40.79
18	20.60	22.76	25.99	28.87	32.35	34.80	42.31
19	21.69	23.90	27.20	30.14	33.69	36.19	43.82
20	22.78	25.04	28.41	31.41	35.02	37.57	45.32
21	23.86	26.17	29.62	32.67	36.34	38.93	46.80
22	24.94	27.30	30.81	33.92	37.66	40.29	48.27
23	26.02	28.43	32.01	35.17	38.97	41.64	49.73
24	27.10	29.55	33.20	36.42	40.27	42.98	51.18
25	28.17	30.68	34.38	37.65	41.57	44.31	52.62
26	29.25	31.80	35.56	38.88	42.86	45.64	54.05
27	30.32	32.91	36.74	40.11	44.14	46.96	55.48
28	31.39	34.03	37.92	41.34	45.42	48.28	56.89
29	32.46	35.14	39.09	42.56	46.69	49.59	58.30
30	33.53	36.25	40.26	43.77	47.96	50.89	59.70

Table A–6
F Distribution*

5% (Roman Type) and 1% (Bold Face Type) Points for the Distribution of F

f_1 Degrees of Freedom (for greater mean square)

Each cell shows 5% (Roman) over 1% (Bold Face).

f_2	1	2	3	4	5	6	7	8	9	10	11	12	14	16	20	24	30	40	50	75	100	200	500	∞
1	161 / **4,052**	200 / **4,999**	216 / **5,403**	225 / **5,625**	230 / **5,764**	234 / **5,859**	237 / **5,928**	239 / **5,981**	241 / **6,022**	242 / **6,056**	243 / **6,082**	244 / **6,106**	245 / **6,142**	246 / **6,169**	248 / **6,208**	249 / **6,234**	250 / **6,261**	251 / **6,286**	252 / **6,302**	253 / **6,323**	253 / **6,334**	254 / **6,352**	254 / **6,361**	254 / **6,366**
2	18.51 / **98.49**	19.00 / **99.00**	19.16 / **99.17**	19.25 / **99.25**	19.30 / **99.30**	19.33 / **99.33**	19.36 / **99.36**	19.37 / **99.37**	19.38 / **99.39**	19.39 / **99.40**	19.40 / **99.41**	19.41 / **99.42**	19.42 / **99.43**	19.43 / **99.44**	19.44 / **99.45**	19.45 / **99.46**	19.46 / **99.47**	19.47 / **99.48**	19.47 / **99.48**	19.48 / **99.49**	19.49 / **99.49**	19.49 / **99.49**	19.50 / **99.50**	19.50 / **99.50**
3	10.13 / **34.12**	9.55 / **30.82**	9.28 / **29.46**	9.12 / **28.71**	9.01 / **28.24**	8.94 / **27.91**	8.88 / **27.67**	8.84 / **27.49**	8.81 / **27.34**	8.78 / **27.23**	8.76 / **27.13**	8.74 / **27.05**	8.71 / **26.92**	8.69 / **26.83**	8.66 / **26.69**	8.64 / **26.60**	8.62 / **26.50**	8.60 / **26.41**	8.58 / **26.35**	8.57 / **26.27**	8.56 / **26.23**	8.54 / **26.18**	8.54 / **26.14**	8.53 / **26.12**
4	7.71 / **21.20**	6.94 / **18.00**	6.59 / **16.69**	6.39 / **15.98**	6.26 / **15.52**	6.16 / **15.21**	6.09 / **14.98**	6.04 / **14.80**	6.00 / **14.66**	5.96 / **14.54**	5.93 / **14.45**	5.91 / **14.37**	5.87 / **14.24**	5.84 / **14.15**	5.80 / **14.02**	5.77 / **13.93**	5.74 / **13.83**	5.71 / **13.74**	5.70 / **13.69**	5.68 / **13.61**	5.66 / **13.57**	5.65 / **13.52**	5.64 / **13.48**	5.63 / **13.46**
5	6.61 / **16.26**	5.79 / **13.27**	5.41 / **12.06**	5.19 / **11.39**	5.05 / **10.97**	4.95 / **10.67**	4.88 / **10.45**	4.82 / **10.29**	4.78 / **10.15**	4.74 / **10.05**	4.70 / **9.96**	4.68 / **9.89**	4.64 / **9.77**	4.60 / **9.68**	4.56 / **9.55**	4.53 / **9.47**	4.50 / **9.38**	4.46 / **9.29**	4.44 / **9.24**	4.42 / **9.17**	4.40 / **9.13**	4.38 / **9.07**	4.37 / **9.04**	4.36 / **9.02**
6	5.99 / **13.74**	5.14 / **10.92**	4.76 / **9.78**	4.53 / **9.15**	4.39 / **8.75**	4.28 / **8.47**	4.21 / **8.26**	4.15 / **8.10**	4.10 / **7.98**	4.06 / **7.87**	4.03 / **7.79**	4.00 / **7.72**	3.96 / **7.60**	3.92 / **7.52**	3.87 / **7.39**	3.84 / **7.31**	3.81 / **7.23**	3.77 / **7.14**	3.75 / **7.09**	3.72 / **7.02**	3.71 / **6.99**	3.69 / **6.94**	3.68 / **6.90**	3.67 / **6.88**
7	5.59 / **12.25**	4.74 / **9.55**	4.35 / **8.45**	4.12 / **7.85**	3.97 / **7.46**	3.87 / **7.19**	3.79 / **7.00**	3.73 / **6.84**	3.68 / **6.71**	3.63 / **6.62**	3.60 / **6.54**	3.57 / **6.47**	3.52 / **6.35**	3.49 / **6.27**	3.44 / **6.15**	3.41 / **6.07**	3.38 / **5.98**	3.34 / **5.90**	3.32 / **5.85**	3.29 / **5.78**	3.28 / **5.75**	3.25 / **5.70**	3.24 / **5.67**	3.23 / **5.65**
8	5.32 / **11.26**	4.46 / **8.65**	4.07 / **7.59**	3.84 / **7.01**	3.69 / **6.63**	3.58 / **6.37**	3.50 / **6.19**	3.44 / **6.03**	3.39 / **5.91**	3.34 / **5.82**	3.31 / **5.74**	3.28 / **5.67**	3.23 / **5.56**	3.20 / **5.48**	3.15 / **5.36**	3.12 / **5.28**	3.08 / **5.20**	3.05 / **5.11**	3.03 / **5.06**	3.00 / **5.00**	2.98 / **4.96**	2.96 / **4.91**	2.94 / **4.88**	2.93 / **4.86**
9	5.12 / **10.56**	4.26 / **8.02**	3.86 / **6.99**	3.63 / **6.42**	3.48 / **6.06**	3.37 / **5.80**	3.29 / **5.62**	3.23 / **5.47**	3.18 / **5.35**	3.13 / **5.26**	3.10 / **5.18**	3.07 / **5.11**	3.02 / **5.00**	2.98 / **4.92**	2.93 / **4.80**	2.90 / **4.73**	2.86 / **4.64**	2.82 / **4.56**	2.80 / **4.51**	2.77 / **4.45**	2.76 / **4.41**	2.73 / **4.36**	2.72 / **4.33**	2.71 / **4.31**
10	4.96 / **10.04**	4.10 / **7.56**	3.71 / **6.55**	3.48 / **5.99**	3.33 / **5.64**	3.22 / **5.39**	3.14 / **5.21**	3.07 / **5.06**	3.02 / **4.95**	2.97 / **4.85**	2.94 / **4.78**	2.91 / **4.71**	2.86 / **4.60**	2.82 / **4.52**	2.77 / **4.41**	2.74 / **4.33**	2.70 / **4.25**	2.67 / **4.17**	2.64 / **4.12**	2.61 / **4.05**	2.59 / **4.01**	2.56 / **3.96**	2.55 / **3.93**	2.54 / **3.91**

*Reproduced with permission from Hoel, P. G.: Elementary Statistics, 2nd Ed. New York: John Wiley and Sons, Inc., 1966.

Table A–6. Continued

f_1 Degrees of Freedom (for greater mean square)

f_2	1	2	3	4	5	6	7	8	9	10	11	12	14	16	20	24	30	40	50	75	100	200	500	∞	f_2
11	4.84 / 9.65	3.98 / 7.20	3.59 / 6.22	3.36 / 5.67	3.20 / 5.32	3.09 / 5.07	3.01 / 4.88	2.95 / 4.74	2.90 / 4.63	2.86 / 4.54	2.82 / 4.46	2.79 / 4.40	2.74 / 4.29	2.70 / 4.21	2.65 / 4.10	2.61 / 4.02	2.57 / 3.94	2.53 / 3.86	2.50 / 3.80	2.47 / 3.74	2.45 / 3.70	2.42 / 3.66	2.41 / 3.62	2.40 / 3.60	11
12	4.75 / 9.33	3.88 / 6.93	3.49 / 5.95	3.26 / 5.41	3.11 / 5.06	3.00 / 4.82	2.92 / 4.65	2.85 / 4.50	2.80 / 4.39	2.76 / 4.30	2.72 / 4.22	2.69 / 4.16	2.64 / 4.05	2.60 / 3.98	2.54 / 3.86	2.50 / 3.78	2.46 / 3.70	2.42 / 3.61	2.40 / 3.56	2.36 / 3.49	2.35 / 3.46	2.32 / 3.41	2.31 / 3.38	2.30 / 3.36	12
13	4.67 / 9.07	3.80 / 6.70	3.41 / 5.74	3.18 / 5.20	3.02 / 4.86	2.92 / 4.62	2.84 / 4.44	2.77 / 4.30	2.72 / 4.19	2.67 / 4.10	2.63 / 4.02	2.60 / 3.96	2.55 / 3.85	2.51 / 3.78	2.46 / 3.67	2.42 / 3.59	2.38 / 3.51	2.34 / 3.42	2.32 / 3.37	2.28 / 3.30	2.26 / 3.27	2.24 / 3.21	2.22 / 3.18	2.21 / 3.16	13
14	4.60 / 8.86	3.74 / 6.51	3.34 / 5.56	3.11 / 5.03	2.96 / 4.69	2.85 / 4.46	2.77 / 4.28	2.70 / 4.14	2.65 / 4.03	2.60 / 3.94	2.56 / 3.86	2.53 / 3.80	2.48 / 3.70	2.44 / 3.62	2.39 / 3.51	2.35 / 3.43	2.31 / 3.34	2.27 / 3.26	2.24 / 3.21	2.21 / 3.14	2.19 / 3.11	2.16 / 3.06	2.14 / 3.02	2.13 / 3.00	14
15	4.54 / 8.68	3.68 / 6.36	3.29 / 5.42	3.06 / 4.89	2.90 / 4.56	2.79 / 4.32	2.70 / 4.14	2.64 / 4.00	2.59 / 3.89	2.55 / 3.80	2.51 / 3.73	2.48 / 3.67	2.43 / 3.56	2.39 / 3.48	2.33 / 3.36	2.29 / 3.29	2.25 / 3.20	2.21 / 3.12	2.18 / 3.07	2.15 / 3.00	2.12 / 2.97	2.10 / 2.92	2.08 / 2.89	2.07 / 2.87	15
16	4.49 / 8.53	3.63 / 6.23	3.24 / 5.29	3.01 / 4.77	2.85 / 4.44	2.74 / 4.20	2.66 / 4.03	2.59 / 3.89	2.54 / 3.78	2.49 / 3.69	2.45 / 3.61	2.42 / 3.55	2.37 / 3.45	2.33 / 3.37	2.28 / 3.25	2.24 / 3.18	2.20 / 3.10	2.16 / 3.01	2.13 / 2.96	2.09 / 2.98	2.07 / 2.86	2.04 / 2.80	2.02 / 2.77	2.01 / 2.75	16
17	4.45 / 8.40	3.59 / 6.11	3.20 / 5.18	2.96 / 4.67	2.81 / 4.34	2.70 / 4.10	2.62 / 3.93	2.55 / 3.79	2.50 / 3.68	2.45 / 3.59	2.41 / 3.52	2.38 / 3.45	2.33 / 3.35	2.29 / 3.27	2.23 / 3.16	2.19 / 3.08	2.15 / 3.00	2.11 / 2.92	2.08 / 2.86	2.04 / 2.79	2.02 / 2.76	1.99 / 2.70	1.97 / 2.67	1.96 / 2.65	17
18	4.41 / 8.28	3.55 / 6.01	3.16 / 5.09	2.93 / 4.58	2.77 / 4.25	2.66 / 4.01	2.58 / 3.85	2.51 / 3.71	2.46 / 3.60	2.41 / 3.51	2.37 / 3.44	2.34 / 3.37	2.29 / 3.27	2.25 / 3.19	2.19 / 3.07	2.15 / 3.00	2.11 / 2.91	2.07 / 2.83	2.04 / 2.78	2.00 / 2.71	1.98 / 2.68	1.95 / 2.62	1.93 / 2.59	1.92 / 2.57	18
19	4.38 / 8.18	3.52 / 5.93	3.13 / 5.01	2.90 / 4.50	2.74 / 4.17	2.63 / 3.94	2.55 / 3.77	2.48 / 3.63	2.43 / 3.52	2.38 / 3.43	2.34 / 3.36	2.31 / 3.30	2.26 / 3.19	2.21 / 3.12	2.15 / 3.00	2.11 / 2.92	2.07 / 2.84	2.02 / 2.76	2.00 / 2.70	1.96 / 2.63	1.94 / 2.60	1.91 / 2.54	1.90 / 2.51	1.88 / 2.49	19
20	4.35 / 8.10	3.49 / 5.85	3.10 / 4.94	2.87 / 4.43	2.71 / 4.10	2.60 / 3.87	2.52 / 3.71	2.45 / 3.56	2.40 / 3.45	2.35 / 3.37	2.31 / 3.30	2.28 / 3.23	2.23 / 3.13	2.18 / 3.05	2.12 / 2.94	2.08 / 2.86	2.04 / 2.77	1.99 / 2.69	1.96 / 2.63	1.92 / 2.56	1.90 / 2.53	1.87 / 2.47	1.85 / 2.44	1.84 / 2.42	20
21	4.32 / 8.02	3.47 / 5.78	3.07 / 4.87	2.84 / 4.37	2.68 / 4.04	2.57 / 3.81	2.49 / 3.65	2.42 / 3.51	2.37 / 3.40	2.32 / 3.31	2.28 / 3.24	2.25 / 3.17	2.20 / 3.07	2.15 / 2.99	2.09 / 2.88	2.05 / 2.80	2.00 / 2.72	1.96 / 2.63	1.93 / 2.58	1.89 / 2.51	1.87 / 2.47	1.84 / 2.42	1.82 / 2.38	1.81 / 2.36	21
22	4.30 / 7.94	3.44 / 5.72	3.05 / 4.82	2.82 / 4.31	2.66 / 3.99	2.55 / 3.76	2.47 / 3.59	2.40 / 3.45	2.35 / 3.35	2.30 / 3.26	2.26 / 3.18	2.23 / 3.12	2.18 / 3.02	2.13 / 2.94	2.07 / 2.83	2.03 / 2.75	1.98 / 2.67	1.93 / 2.58	1.91 / 2.53	1.87 / 2.46	1.84 / 2.42	1.81 / 2.37	1.80 / 2.33	1.78 / 2.31	22
23	4.28 / 7.88	3.42 / 5.66	3.03 / 4.76	2.80 / 4.26	2.64 / 3.94	2.53 / 3.71	2.45 / 3.54	2.38 / 3.41	2.32 / 3.30	2.28 / 3.21	2.24 / 3.14	2.20 / 3.07	2.14 / 2.97	2.10 / 2.89	2.04 / 2.78	2.00 / 2.70	1.96 / 2.62	1.91 / 2.53	1.88 / 2.48	1.84 / 2.41	1.82 / 2.37	1.79 / 2.32	1.77 / 2.28	1.76 / 2.26	23
24	4.26 / 7.82	3.40 / 5.61	3.01 / 4.72	2.78 / 4.22	2.62 / 3.90	2.51 / 3.67	2.43 / 3.50	2.36 / 3.36	2.30 / 3.25	2.26 / 3.17	2.22 / 3.09	2.18 / 3.03	2.13 / 2.93	2.09 / 2.85	2.02 / 2.74	1.98 / 2.66	1.94 / 2.58	1.89 / 2.49	1.86 / 2.44	1.82 / 2.36	1.80 / 2.33	1.76 / 2.27	1.74 / 2.23	1.73 / 2.21	24
25	4.24 / 7.77	3.38 / 5.57	2.99 / 4.68	2.76 / 4.18	2.60 / 3.86	2.49 / 3.63	2.41 / 3.46	2.34 / 3.32	2.28 / 3.21	2.24 / 3.13	2.20 / 3.05	2.16 / 2.99	2.11 / 2.89	2.06 / 2.81	2.00 / 2.70	1.96 / 2.62	1.92 / 2.54	1.87 / 2.45	1.84 / 2.40	1.80 / 2.32	1.77 / 2.29	1.74 / 2.23	1.72 / 2.19	1.71 / 2.17	25

476

Table A-6. Continued

f_1 Degrees of Freedom (for greater mean square)

f_2	1	2	3	4	5	6	7	8	9	10	11	12	14	16	20	24	30	40	50	75	100	200	500	∞	f_2
26	4.22 / 7.72	3.37 / 5.53	2.98 / 4.64	2.74 / 4.14	2.59 / 3.82	2.47 / 3.59	2.39 / 3.42	2.32 / 3.29	2.27 / 3.17	2.22 / 3.09	2.18 / 3.02	2.15 / 2.96	2.10 / 2.86	2.05 / 2.77	1.99 / 2.66	1.95 / 2.58	1.90 / 2.50	1.85 / 2.41	1.82 / 2.36	1.78 / 2.28	1.76 / 2.25	1.72 / 2.19	1.70 / 2.15	1.69 / 2.13	26
27	4.21 / 7.68	3.35 / 5.49	2.96 / 4.60	2.73 / 4.11	2.57 / 3.79	2.46 / 3.56	2.37 / 3.39	2.30 / 3.26	2.25 / 3.14	2.20 / 3.06	2.16 / 2.98	2.13 / 2.93	2.08 / 2.83	2.03 / 2.74	1.97 / 2.63	1.93 / 2.55	1.88 / 2.47	1.84 / 2.38	1.80 / 2.33	1.76 / 2.25	1.74 / 2.21	1.71 / 2.16	1.68 / 2.12	1.67 / 2.10	27
28	4.20 / 7.64	3.34 / 5.45	2.95 / 4.57	2.71 / 4.07	2.56 / 3.76	2.44 / 3.53	2.36 / 3.36	2.29 / 3.23	2.24 / 3.11	2.19 / 3.03	2.15 / 2.95	2.12 / 2.90	2.06 / 2.80	2.02 / 2.71	1.96 / 2.60	1.91 / 2.52	1.87 / 2.44	1.81 / 2.35	1.78 / 2.30	1.75 / 2.22	1.72 / 2.18	1.69 / 2.13	1.67 / 2.09	1.65 / 2.06	28
29	4.18 / 7.60	3.33 / 5.42	2.93 / 4.54	2.70 / 4.04	2.54 / 3.73	2.43 / 3.50	2.35 / 3.33	2.28 / 3.20	2.22 / 3.08	2.18 / 3.00	2.14 / 2.92	2.10 / 2.87	2.05 / 2.77	2.00 / 2.68	1.94 / 2.57	1.90 / 2.49	1.85 / 2.41	1.80 / 2.32	1.77 / 2.27	1.73 / 2.19	1.71 / 2.15	1.68 / 2.10	1.65 / 2.06	1.64 / 2.03	29
30	4.17 / 7.56	3.32 / 5.39	2.92 / 4.51	2.69 / 4.02	2.53 / 3.70	2.42 / 3.47	2.34 / 3.30	2.27 / 3.17	2.21 / 3.06	2.16 / 2.98	2.12 / 2.90	2.09 / 2.84	2.04 / 2.74	1.99 / 2.66	1.93 / 2.55	1.89 / 2.47	1.84 / 2.38	1.79 / 2.29	1.76 / 2.24	1.72 / 2.16	1.69 / 2.13	1.66 / 2.07	1.64 / 2.03	1.62 / 2.01	30
32	4.15 / 7.50	3.30 / 5.34	2.90 / 4.46	2.67 / 3.97	2.51 / 3.66	2.40 / 3.42	2.32 / 3.25	2.25 / 3.12	2.19 / 3.01	2.14 / 2.94	2.10 / 2.86	2.07 / 2.80	2.02 / 2.70	1.97 / 2.62	1.91 / 2.51	1.86 / 2.42	1.82 / 2.34	1.76 / 2.25	1.74 / 2.20	1.69 / 2.12	1.67 / 2.08	1.64 / 2.02	1.61 / 1.98	1.59 / 1.96	32
34	4.13 / 7.44	3.28 / 5.29	2.88 / 4.42	2.65 / 3.93	2.49 / 3.61	2.38 / 3.38	2.30 / 3.21	2.23 / 3.08	2.17 / 2.97	2.12 / 2.89	2.08 / 2.82	2.05 / 2.76	2.00 / 2.66	1.95 / 2.58	1.89 / 2.47	1.84 / 2.38	1.80 / 2.30	1.74 / 2.21	1.71 / 2.15	1.67 / 2.08	1.64 / 2.04	1.61 / 1.98	1.59 / 1.94	1.57 / 1.91	34
36	4.11 / 7.39	3.26 / 5.25	2.86 / 4.38	2.63 / 3.89	2.48 / 3.58	2.36 / 3.35	2.28 / 3.18	2.21 / 3.04	2.15 / 2.94	2.10 / 2.86	2.06 / 2.78	2.03 / 2.72	1.98 / 2.62	1.93 / 2.54	1.87 / 2.43	1.82 / 2.35	1.78 / 2.26	1.72 / 2.17	1.69 / 2.12	1.65 / 2.04	1.62 / 2.00	1.59 / 1.94	1.56 / 1.90	1.55 / 1.87	36
38	4.10 / 7.35	3.25 / 5.21	2.85 / 4.34	2.62 / 3.86	2.46 / 3.54	2.35 / 3.32	2.26 / 3.15	2.19 / 3.02	2.14 / 2.91	2.09 / 2.82	2.05 / 2.75	2.02 / 2.69	1.96 / 2.59	1.92 / 2.51	1.85 / 2.40	1.80 / 2.32	1.76 / 2.22	1.71 / 2.14	1.67 / 2.08	1.63 / 2.00	1.60 / 1.97	1.57 / 1.90	1.54 / 1.86	1.53 / 1.84	38
40	4.08 / 7.31	3.23 / 5.18	2.84 / 4.31	2.61 / 3.83	2.45 / 3.51	2.34 / 3.29	2.25 / 3.12	2.18 / 2.99	2.12 / 2.88	2.07 / 2.80	2.04 / 2.73	2.00 / 2.66	1.95 / 2.56	1.90 / 2.49	1.84 / 2.37	1.79 / 2.29	1.74 / 2.20	1.69 / 2.11	1.66 / 2.05	1.61 / 1.97	1.59 / 1.94	1.55 / 1.88	1.53 / 1.84	1.51 / 1.81	40
42	4.07 / 7.27	3.22 / 5.15	2.83 / 4.29	2.59 / 3.80	2.44 / 3.49	2.32 / 3.26	2.24 / 3.10	2.17 / 2.96	2.11 / 2.86	2.06 / 2.77	2.02 / 2.70	1.99 / 2.64	1.94 / 2.54	1.89 / 2.46	1.82 / 2.35	1.78 / 2.26	1.73 / 2.17	1.68 / 2.08	1.64 / 2.02	1.60 / 1.94	1.57 / 1.91	1.54 / 1.85	1.51 / 1.80	1.49 / 1.78	42
44	4.06 / 7.24	3.21 / 5.12	2.82 / 4.26	2.58 / 3.78	2.43 / 3.46	2.31 / 3.24	2.23 / 3.07	2.16 / 2.94	2.10 / 2.84	2.05 / 2.75	2.01 / 2.68	1.98 / 2.62	1.92 / 2.52	1.88 / 2.44	1.81 / 2.32	1.76 / 2.24	1.72 / 2.15	1.66 / 2.06	1.63 / 2.00	1.58 / 1.92	1.56 / 1.88	1.52 / 1.82	1.50 / 1.78	1.48 / 1.75	44
46	4.05 / 7.21	3.20 / 5.10	2.81 / 4.24	2.57 / 3.76	2.42 / 3.44	2.30 / 3.22	2.22 / 3.05	2.14 / 2.92	2.09 / 2.82	2.04 / 2.73	2.00 / 2.66	1.97 / 2.60	1.91 / 2.50	1.87 / 2.42	1.80 / 2.30	1.75 / 2.22	1.71 / 2.13	1.65 / 2.04	1.62 / 1.98	1.57 / 1.90	1.54 / 1.86	1.51 / 1.80	1.48 / 1.76	1.46 / 1.72	46
48	4.04 / 7.19	3.19 / 5.08	2.80 / 4.22	2.56 / 3.74	2.41 / 3.42	2.30 / 3.20	2.21 / 3.04	2.14 / 2.90	2.08 / 2.80	2.03 / 2.71	1.99 / 2.64	1.96 / 2.58	1.90 / 2.48	1.86 / 2.40	1.79 / 2.28	1.74 / 2.20	1.70 / 2.11	1.64 / 2.02	1.61 / 1.96	1.56 / 1.88	1.53 / 1.84	1.50 / 1.78	1.47 / 1.73	1.45 / 1.70	48

Table A–6. *Continued*

f_1 Degrees of Freedom (for greater mean square)

f_2	1	2	3	4	5	6	7	8	9	10	11	12	14	16	20	24	30	40	50	75	100	200	500	∞	f_2
50	4.03 / 7.17	3.18 / 5.06	2.79 / 4.20	2.56 / 3.72	2.40 / 3.41	2.29 / 3.18	2.20 / 3.02	2.13 / 2.88	2.07 / 2.78	2.02 / 2.70	1.98 / 2.62	1.95 / 2.56	1.90 / 2.46	1.85 / 2.39	1.78 / 2.26	1.74 / 2.18	1.69 / 2.10	1.63 / 2.00	1.60 / 1.94	1.55 / 1.86	1.52 / 1.82	1.48 / 1.76	1.46 / 1.71	1.44 / 1.68	50
55	4.02 / 7.12	3.17 / 5.01	2.78 / 4.16	2.54 / 3.68	2.38 / 3.37	2.27 / 3.15	2.18 / 2.98	2.11 / 2.85	2.05 / 2.75	2.00 / 2.66	1.97 / 2.59	1.93 / 2.53	1.88 / 2.43	1.83 / 2.35	1.76 / 2.23	1.72 / 2.15	1.67 / 2.06	1.61 / 1.96	1.58 / 1.90	1.52 / 1.82	1.50 / 1.78	1.46 / 1.71	1.43 / 1.66	1.41 / 1.64	55
60	4.00 / 7.08	3.15 / 4.98	2.76 / 4.13	2.52 / 3.65	2.37 / 3.34	2.25 / 3.12	2.17 / 2.95	2.10 / 2.82	2.04 / 2.72	1.99 / 2.63	1.95 / 2.56	1.92 / 2.50	1.86 / 2.40	1.81 / 2.32	1.75 / 2.20	1.70 / 2.12	1.65 / 2.03	1.59 / 1.93	1.56 / 1.87	1.50 / 1.79	1.48 / 1.74	1.44 / 1.68	1.41 / 1.63	1.39 / 1.60	60
65	3.99 / 7.04	3.14 / 4.95	2.75 / 4.10	2.51 / 3.62	2.36 / 3.31	2.24 / 3.09	2.15 / 2.93	2.08 / 2.79	2.02 / 2.70	1.98 / 2.61	1.94 / 2.54	1.90 / 2.47	1.85 / 2.37	1.80 / 2.30	1.73 / 2.18	1.68 / 2.09	1.63 / 2.00	1.57 / 1.90	1.54 / 1.84	1.49 / 1.76	1.46 / 1.71	1.42 / 1.64	1.39 / 1.60	1.37 / 1.56	65
70	3.98 / 7.01	3.13 / 4.92	2.74 / 4.08	2.50 / 3.60	2.35 / 3.29	2.23 / 3.07	2.14 / 2.91	2.07 / 2.77	2.01 / 2.67	1.97 / 2.59	1.93 / 2.51	1.89 / 2.45	1.84 / 2.35	1.79 / 2.28	1.72 / 2.15	1.67 / 2.07	1.62 / 1.98	1.56 / 1.88	1.53 / 1.82	1.47 / 1.74	1.45 / 1.69	1.40 / 1.62	1.37 / 1.56	1.35 / 1.53	70
80	3.96 / 6.96	3.11 / 4.88	2.72 / 4.04	2.48 / 3.56	2.33 / 3.25	2.21 / 3.04	2.12 / 2.87	2.05 / 2.74	1.99 / 2.64	1.95 / 2.55	1.91 / 2.48	1.88 / 2.41	1.82 / 2.32	1.77 / 2.24	1.70 / 2.11	1.65 / 2.03	1.60 / 1.94	1.54 / 1.84	1.51 / 1.78	1.45 / 1.70	1.42 / 1.65	1.38 / 1.57	1.35 / 1.52	1.32 / 1.49	80
100	3.94 / 6.90	3.09 / 4.82	2.70 / 3.98	2.46 / 3.51	2.30 / 3.20	2.19 / 2.99	2.10 / 2.82	2.03 / 2.69	1.97 / 2.59	1.92 / 2.51	1.88 / 2.43	1.85 / 2.36	1.79 / 2.26	1.75 / 2.19	1.68 / 2.06	1.63 / 1.98	1.57 / 1.89	1.51 / 1.79	1.48 / 1.73	1.42 / 1.64	1.39 / 1.59	1.34 / 1.51	1.30 / 1.46	1.28 / 1.43	100
125	3.92 / 6.84	3.07 / 4.78	2.68 / 3.94	2.44 / 3.47	2.29 / 3.17	2.17 / 2.95	2.08 / 2.79	2.01 / 2.65	1.95 / 2.56	1.90 / 2.47	1.86 / 2.40	1.83 / 2.33	1.77 / 2.23	1.72 / 2.15	1.65 / 2.03	1.60 / 1.94	1.55 / 1.85	1.49 / 1.75	1.45 / 1.68	1.39 / 1.59	1.36 / 1.54	1.31 / 1.46	1.27 / 1.40	1.25 / 1.37	125
150	3.91 / 6.81	3.06 / 4.75	2.67 / 3.91	2.43 / 3.44	2.27 / 3.14	2.16 / 2.92	2.07 / 2.76	2.00 / 2.62	1.94 / 2.53	1.89 / 2.44	1.85 / 2.37	1.82 / 2.30	1.76 / 2.20	1.71 / 2.12	1.64 / 2.00	1.59 / 1.91	1.54 / 1.83	1.47 / 1.72	1.44 / 1.66	1.37 / 1.56	1.34 / 1.51	1.29 / 1.43	1.25 / 1.37	1.22 / 1.33	150
200	3.89 / 6.76	3.04 / 4.71	2.65 / 3.88	2.41 / 3.41	2.26 / 3.11	2.14 / 2.90	2.05 / 2.73	1.98 / 2.60	1.92 / 2.50	1.87 / 2.41	1.83 / 2.34	1.80 / 2.28	1.74 / 2.17	1.69 / 2.09	1.62 / 1.97	1.57 / 1.88	1.52 / 1.79	1.45 / 1.69	1.42 / 1.62	1.35 / 1.53	1.32 / 1.48	1.26 / 1.39	1.22 / 1.33	1.19 / 1.28	200
400	3.86 / 6.70	3.02 / 4.66	2.62 / 3.83	2.39 / 3.36	2.23 / 3.06	2.12 / 2.85	2.03 / 2.69	1.96 / 2.55	1.90 / 2.46	1.85 / 2.37	1.81 / 2.29	1.78 / 2.23	1.72 / 2.12	1.67 / 2.04	1.60 / 1.92	1.54 / 1.84	1.49 / 1.74	1.42 / 1.64	1.38 / 1.57	1.32 / 1.47	1.28 / 1.42	1.22 / 1.32	1.16 / 1.24	1.13 / 1.19	400
1000	3.85 / 6.66	3.00 / 4.62	2.61 / 3.80	2.38 / 3.34	2.22 / 3.04	2.10 / 2.82	2.02 / 2.66	1.95 / 2.53	1.89 / 2.43	1.84 / 2.34	1.80 / 2.26	1.76 / 2.20	1.70 / 2.09	1.65 / 2.01	1.58 / 1.89	1.53 / 1.81	1.47 / 1.71	1.41 / 1.61	1.36 / 1.54	1.30 / 1.44	1.26 / 1.38	1.19 / 1.28	1.13 / 1.19	1.08 / 1.11	1000
∞	3.84 / 6.64	2.99 / 4.60	2.60 / 3.78	2.37 / 3.32	2.21 / 3.02	2.09 / 2.80	2.01 / 2.64	1.94 / 2.51	1.88 / 2.41	1.83 / 2.32	1.79 / 2.24	1.75 / 2.18	1.69 / 2.07	1.64 / 1.99	1.57 / 1.87	1.52 / 1.79	1.46 / 1.69	1.40 / 1.59	1.35 / 1.52	1.28 / 1.41	1.24 / 1.36	1.17 / 1.25	1.11 / 1.15	1.00 / 1.00	∞

478

Answers to Even-Numbered Exercises

CHAPTER 2

2. Samples are used for reasons of practicality and efficient use of resources. It is often impractical or too expensive to study all members of a population. If it is possible to make accurate inferences from a sample to a population, it would be wasteful to study the entire population. The study of entire populations would be appropriate when the number of members in a population is small. An example might be studying the voting behavior of U.S. senators.

4. The probability of an event is the relative frequency of the occurrence of the event to the total number of "trials" in the long run. Probability is a statement about what will happen in the long run, not what will happen on a single trial.

6. A variable is a property of objects which a researcher chooses to study. It is a property whose "value" fluctuates across the objects being studied. Numbers are usually associated with variables (the process of quantification) because they allow statistics to treat variables in a more precise way. An independent variable is one which the researcher feels "causes" or "affects" other things in the problem under consideration. A dependent variable is one which the researcher feels is being "caused" or "affected" in the problem. A hypothesis is a guess that a researcher has about one or more variables being related to another one or more variables in a particular situation. It is the researcher's hypothesis which specifies what the independent and dependent variables are in a particular situation.

8. a. Nominal. b. Ordinal; order of finish. c. Interval or ratio; age of trees. d. Nominal. e. Ordinal; quality of meat.

CHAPTER 3

2. 15.

4. $a - b + c$.

6. 4.

8. 450.

10. $\frac{5}{8}$

12. .923.

14. a. 7. b. 95. c. 95. d. 3. e. 30. f. 50. g. 77.

16. 14.

18. 1.

20. 11.

22.

24.

26.

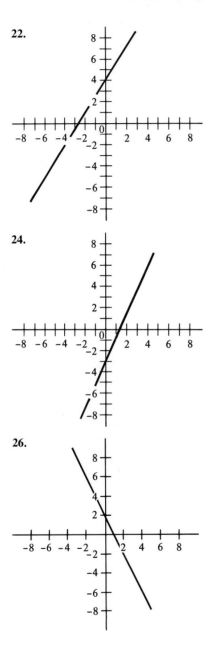

28. 10.6, not 10.56.

CHAPTER 4

2. .55–2.55, 1; 2.55–4.55, 2; 4.55–6.55, 2; 6.55–8.55, 15; 8.55–10.55, 10;

10.55–12.55, 7; 12.55–14.55, 5; 14.55–16.55, 3; 16.55–18.55, 2; 18.55–20.55, 0; 20.55–22.55, 3.

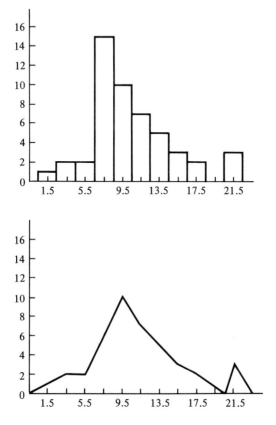

4. Mean = 17, median = 15, mode = 13.

6. 73.33.

8. a. 10. b. 20. c. 14. d. 2. e. 2, −2, 1, 0. f. 12.

CHAPTER 5

2. $s^2 = 9.6$, $s = 3.1$.

4. $s^2 = 209.67$, $s = 14.48$, $[\Sigma X^2 - (\Sigma X)^2/N] = 2516$.

6. Standard deviation = 2, mean = 38. Adding a constant to each score does not change the amount of spread between the scores, but it moves all the scores up the same distance. This movement, then, does not change the value of the standard deviation, but it does change the value of the mean.

8. 140.4 to 179.6, 134.2 to 185.8, 285, 297.

CHAPTER 6

2. 120.

4. The total number of possible outcomes is 128. The number of ways to get all the questions correct is 1. Therefore, the probability that a student would get all seven questions correct by chance alone is $\frac{1}{128} = .0078$. The number of ways to get five of the seven questions correct is 21. Therefore, the probability that a student would get five of the seven questions correct by chance alone is $\frac{21}{128} = .164$. The number of ways to get one of the seven questions correct is 7. Therefore, the probability that a student would get exactly one of the seven questions correct by chance alone is $\frac{7}{128} = .0547$.

6. .04.

8. a. .15. b. .08. c. .65. d. .44.

CHAPTER 7

2. a. 86.65%. b. 1.32%. c. 1.05%. d. 92.51%, 7.49%. e. About 601.

4. a. $UL_{.95} = 1226.3$, $LL_{.95} = 1173.7$.
 b. $UL_{.99} = 1234.6$, $LL_{.99} = 1165.4$.
 c. $UL_{.6826} = 1213.42$, $LL_{.6826} = 1186.58$.

6. $UL_{.95} = 28.78$, $LL_{.95} = 17.22$; $UL_{.90} = 27.75$, $LL_{.90} = 18.25$.

CHAPTER 8

2. Null hypothesis: the mean number of days of church attendance for Protestants and Catholics is the same, H_0: $\mu_P = \mu_C$. Alternative hypothesis: the mean number of days of church attendance differs for Catholics and Protestants, H_1: $\mu_P \neq \mu_C$. A Type I error would occur if the researcher decided that based upon the information in his samples there is a difference in the mean number of days Catholics and Protestants attend church, when in fact there really is no difference. A Type II error would occur if the researcher decided that based upon the information in his samples there is no difference in the mean number of days of church attendance between Catholics and Protestants, when in fact there really is a difference. With an alpha level of .10, the probability of making a Type I error is .10. Since the alpha level is the probability of the test statistic exceeding the critical value by chance alone if the null hypothesis is true, it is also the probability that the researcher would decide that there is a difference between the groups, when in fact there is no difference.

4. Null hypothesis: there is no difference in the mean voter-turnout rates of northern and southern counties, H_0: $\mu_N = \mu_S$. Alternative hypothesis: the mean voter-turnout rate of southern counties is less that of northern counties, H_1: $\mu_N > \mu_S$. Yes the difference is statistically significant by a one-tailed test at an alpha level of .05. $z = 1.66$.

6. Null hypothesis: there is no difference in the mean number of terms served between congressmen from urban and rural districts, H_0: $\mu_U = \mu_R$. Alternative hypothesis: congressmen from rural districts have served a greater mean number of terms than congressmen from urban districts, H_1: $\mu_U < \mu_R$. Yes, the difference is statistically significant by a one-tailed test at an alpha level of .05. $t = 1.92$ using a pooled estimate of the variance.

CHAPTER 9

2. Chi-square $= 7.58$. Yes. The independent variable is party identification and the dependent variable is whether respondents voted because of the researcher's hypothesis.

Party Identification

		Party Identifiers	Independents
Voting Record	Voted	70%	35%
	Did Not Vote	30%	65%
		100%	100%

4. Chi-square $= 6.135$. No, not statistically significant.

CHAPTER 10

2. Since one variable, sex, is a nominal variable, nominal measures of association must be used: $\phi = .16$, $\lambda_B = .16$.
4. Both variables are ordinal. Therefore, gamma is the appropriate measure of association: $\gamma = -.07$.
6. Both variables are nominal: $\phi = .287$, $\lambda_B = .125$.
8. Both variables are ordinal. Yule's $Q = -.50$.

CHAPTER 11

2. $b = .14734$, $a = 8.45$, $r = .694$, $r^2 = .48$.
4. $r = .32$.

CHAPTER 12

2. $\bar{X}_1 = 2.4, \bar{X}_2 = 2.8, \bar{X}_3 = 2.2, \bar{X}_4 = 2.6, \bar{X}_G = 2.5, \sum_i (X_{i1} - \bar{X}_1) = 1.12, \sum_i (X_{i2} - X_2)$

$= 1.06, \sum_i (X_{i3} - \bar{X}_3) = 1.08, \sum_i (X_{i4} - \bar{X}_4) = 1.12, \sum_j \sum_i (X_{ij} - \bar{X}_j) = 4.38,$

$\sum_i (\bar{X}_1 - \bar{X}_G) = .07, \sum_i (\bar{X}_2 - \bar{X}_G) = .63, \sum_i (\bar{X}_3 - \bar{X}_G) = .63, \sum_i (\bar{X}_4 - \bar{X}_G) = .07,$

$\sum_j \sum_i (X_j = X_G) = 1.40.$

Source	Sums of Squares	Degrees of Freedom	Estimate of Variance	F
Total	5.78	$N - 1 = 27$		
Between	1.40	$J - 1 = 3$	.467	
Within	4.38	$N - J = 24$	.182	2.57

The F ratio is not statistically significant.

4.

Source	Sums of Squares	Degrees of Freedom	Estimate of Variance	F
Total	44	$N - 1 = 20$		
Between	14	$J - 1 = 2$	7	
Within	30	$N - J = 18$	1.67	4.19

The F ratio is statistically significant at an alpha level of .05. Therefore, we can conclude that there are differences in the mean number of days workers are absent among the three types of groups. Since the mean number of days absent for the equalitarian group is 2.0, the foreman group 3.0, and the hierarchical group 4.0, we can conclude that different types of organizations do affect absenteeism, with the equalitarian type of organization producing the least amount of absenteeism and the hierarchical type of organization producing the most amount of absenteeism.

CHAPTER 13

2. The independent variable is education, the dependent variable is straight-ticket voting, and the test variable is level of information about candidates. The type of statistical elaboration is contingency; in the presence of more information about candidates, the association between level of education and straight-ticket voting becomes stronger.

4. $a = -200$, $b_{12 \cdot 3} = 5$, $b_{13 \cdot 2} = .02$, $\beta_{12 \cdot 3} = .571$, $\beta_{13 \cdot 2} = .857$. The median income of the counties is the most important variable.

Index